genetic
evolution

Chen
Kang
Chai

genetic
evolution

The University of Chicago Press   chicago and london

CHEN KANG CHAI is a senior staff scientist at
Jackson Laboratory, Bar Harbor, Maine. He is
the author of *Taiwan Aborigines: A Genetic
Study of Tribal Variations*.

THE UNIVERSITY OF CHICAGO PRESS,
CHICAGO 60637
THE UNIVERSITY OF CHICAGO PRESS, LTD.,
LONDON

Library of Congress Cataloging in Publication Data
Chai, Chen Kang, 1916–
    Genetic evolution.
    Bibliography: p.
    Includes index.
    1. Evolution. 2. Genetics. I. Title. [DNLM:
1. Genetics. 2. Evolution. QH430 C434g]
QH371.C47   575   75–27898
ISBN 0–226–10066–9

To Professor Harrison R. Hunt

The WAY gives them life;
Virtue rears them;
Things give them shape,
Circumstances bring them
to maturity.

Lao Tzu, *Tao Te Ching*, 500 B.C.

contents

preface

Life is essentially a self-perpetuating information system. The successive revision of this system—as conditioned by the environment and as manifested in chemical, physiological, and morphological variations in the organism—constitutes evolution. Organic evolution thus represents the history of life. For a scientific inquiry, it is impossible to definitely establish exact events and causal chains from the remote past. On the basis of the available evidence, however, a coherent theory can be formulated and sensible questions asked: How did the genetic information originate? By what mechanisms do organisms transmit and utilize it? What are the processes that lead to species extinction, differentiation, and progressive evolution?

It was approximately forty years ago that Fisher, Haldane, and Wright independently developed mathematical concepts of evolution based on Mendelian inheritance. Their theoretical constructs placed Darwin's theory of natural selection on a genetic foundation. This marriage of Darwinism and Mendelism suggested a modern appellation for the natural selection theory: neo-Darwinism. Natural selection, which deals essentially with random and nonrandom processes in gene frequency change, has remained the dominant theme of evolution, central to studies on evolutionary genetics.

The recent discovery of DNA as the chemical carrier of genetic information has, in its implications, embraced areas far beyond the confines of any single scientific discipline in biology. We now possess much more basic knowledge on the structure, products, and function of genes than before. While enriching the biological sciences as a whole, the DNA discovery has carried genetics in particular into a new era and provided a molecular basis for evolution.

Between species, for example, the number of amino acid differences in a protein coded by a gene provides information on the rate of evolution of this gene since the separation of species. This has been referred to as a protein "clock" and it can be studied in the living species. The virtually unlimited number of species and numerous proteins provide, in fact, "living fossils." Phylogeny, which used to be studied in gross morphological variations, can now also be analyzed on a molecular level. We presently know substantially more than before about chromosome structure and morphology and about the organelles in cells and their evolutionary origins. We also have more information concerning the chemical structures of hormones in higher organisms and, through employment of the theory of gene mutation, can trace their evolutionary origins. This knowledge has produced a methodology capable of elucidating much more about evolution than simple gene frequency dynamics, and it provides a basis for the fundamental theme in neo-Darwinism.

But organic evolution should include not only historical changes in life, but also events leading up to the development of life, to its origin. In most books on evolution, prebiotic synthesis has only been touched on because little was known about it. With the recent advances in geology and laboratory technology, however, primitive earth conditions can be simulated, making much laboratory evidence available toward filling this major gap.

I have attempted to integrate information from genetics and related fields in order to construct a framework for evolution that, although essentially Darwinian, has been modified and updated in the light of current data and more concrete evidence. My discussion is centered around the following areas:

1. *Prebiotic synthesis.* By incorporating contemporary geological and chemical information (such as the synthesis of amino acids and nucleotides under primitive earth conditions) into a concept of the origin of life, it is possible to reproduce the history of that origin in a more accurate and complete form.

2. *Evolution of genes.* Utilizing new techniques, studies of DNA-coded protein molecules in both lower and higher organisms have brought new information to light concerning evolutionary change in genes, gene duplication, gene homology within and between species, mutation, and gene substitution rates.

3. *Symbiosis.* Because of the discovery of the similarity between mitochondria, plastids, and bacteria in DNA and membrane structures, it is now conjectured that symbiosis, a shortcut to acquiring new genetic information, may constitute a prelude to sexual evolution. This discovery has had a great impact in classification and in theories on the earlier evolution of life.

4. *Evolution of chromosomes.* With the considerable body of knowledge on the fine structure and chemistry of chromosomes that has recently become available, it is possible to form a bridge in evolution from prokaryotes to eukaryotes. In light of this, chromosome rearrangements, long believed by cytogeneticists to exercise a specific function in evolution, are reexamined.

5. *Evolution of biological function and structure.* Moving from the genic and chromosomal to the organismic level, gene regulation of growth and development, as well as individual gene effects on characters of interest in terms of fitness and adaptation, are considered. The achievement of major and minor morphological and physiological changes through genetic alterations is discussed.

6. *Evolution in populations.* If a gene should prove advantageous to the organism, how can its transmission through following generations be assured or its frequency increased? The scope of the discussion here broadens from

the organismic to the populational level. Interrelationships between mating individuals, population size and structure, and the selective value of the gene and its allelic/nonallelic interactions all possess considerable significance for the fate of a gene or a population. With respect to these topics, classical work in population genetics retains its validity.

A certain amount of criticism concerning theoretical treatments of population genetics has arisen on the ground that some of these treatments lack practical application for natural populations. I believe that this constitutes an unfortunate gap between theoreticians and field workers, which is at least partly due to the difficulty of some of the mathematical analysis involved. Therefore, I have attempted to clarify this subject in a more elementary and fundamental fashion, following each theoretical discussion with possible examples. Recent work on populations, linkage, polymorphisms, and genetic loads is incorporated in these chapters. Finally, adaptation, speciation, and extinction are discussed, taking into consideration populational genetic theory and the recent molecular genetic evidence.

This text proceeds through an ascending hierarchy of subject matter: chemical compounds, genes, chromosomes, organisms, and, finally, populations; each concerns in some way the evolution of genetic information and the consequences of survival and differentiation of organisms. Since man, as opposed to other species, partly controls his own environment, I have added a final chapter on the character of human evolution, analyzing the role of both genetic and cultural transmission of information in shaping man's evolution, eugenics, and the evolution of consciousness.

This text is not intended as a mere supplement to existing evolutionary theory, but rather as an attempt to integrate modern and classical conceptions into a single coherent interpretation. The synthesis of information proposed requires data from virtually all areas of biology and, in certain cases, actually extends beyond it. Indeed, as one historian points out, there is too much information in the scientific community to ever be fully digested. This is particularly true for biology, both in quantity and in diversity. Nevertheless, the various divisions within this discipline all retain a certain relation to evolution.

I have attempted to use appropriate concepts in the formulation of a unified evolutionary theory, but the impossibility of a complete treatment of this subject is obvious. As a medium to the solution of an interdisciplinary problem, I hope that, beyond numerous errors and gaps, this book may prove to be useful to students of evolution and related fields.

acknowledgments

I would like to express my sincere gratitude to several of my colleagues at the Jackson Laboratory: Dr. D. W. Bailey, for reading the first draft of the manuscript and for the benefit of many valuable comments and informal discussions; Dr. A. Kandutsch, for reading the first three chapters; and Dr. L. C. Stevens, for reading the chapter on development and for his many valuable suggestions. I would also like to thank Dr. James Crow for reading the first draft and for making helpful suggestions on the presentation of the material.

I owe a special debt to Dr. Gabriel Lasker and Dr. Joseph Jensen for their most careful readings of the second draft, as well as for comments and criticisms instrumental in bringing the book into its final form.

Needless to say, I alone am responsible for any errors in the text.

I would like to take this opportunity to thank Mrs. Ruth Soper and Mr. George McKay for their help with the illustrations; Mrs. D. L. Killam for editorial assistance throughout the entire manuscript; Mrs. Isabelle M. Stover and Mrs. Eleanor K. St. Denis for typing the manuscript. I am grateful also to my wife, Ling, and my daughter, Jean, for their cooperation, encouragement, and help; to my son, Leon, for reading many sections and for his comments on the presentation of the material.

Some of the information used in this book came from my research projects. They are supported by grants from the National Institute of Allergy and Infectious Diseases, the Division of Research Resources, the National Cancer Institute, and the National Institutes of Health.

One

prebiotic evolution

Life is an organic unit with form and structure, growing through metabolism, degenerating, self-perpetuating (Urey 1966), and attempting to adapt to its environment through internal changes. How did life as such or with similar properties originate? Obviously the first living organisms would have been different in their essential properties from those of today. But assumptions of certain similarities, suggested by experimental tests, provide valid hypotheses.

The origin of life has long been a matter of speculation. We may never be able to completely ascertain the truth, but our continuous search through geological and experimental evidence provides information by which we can stretch our imagination and render our speculations more logical. Experimental evidence, obtained under simulated geological conditions of prehistoric time, provides a possible connection of evolutionary events, one following another until an organic form comprising the essential properties of life appears. The occurrence of these chemical events prior to the time when the first organisms were developed is referred to as prebiotic evolution.

First, let us review briefly the historical development of thoughts on the origin of life. In biblical accounts, the origin of life was a part of creation. Beginning on the third day, according to illustrations in a sixteenth-century Bible, God made the birds and fish, and then animals and man. But in Genesis, God, instead of creating living things, bade the earth and waters bring them forth. Up to the beginning of the eighteenth century, life was regarded either as having been created supernaturally or as having evolved from nonliving matter. The theological view of a spontaneous generation of life has been accepted more or less without question. Scientists, including Aristotle, Newton, Harvey, Descartes, and van Helmont, accepted it without much doubt.

With the development of biology and chemistry at the beginning of the nineteenth century, the problem of the origin of life came to be regarded from a scientific point of view. In 1809, Lamarck formulated an evolutionary principle related to our present thinking. He proposed that inorganic matter must have contained extremely small, half-liquid bodies of diffuse consistency that developed into cellular forms capable of containing liquid and acquiring the first rudiments of organization. A further materialistic approach began in the second half of the nineteenth century. In the 1860s, in a speech given before the British Association,

1

T. H. Huxley introduced the idea that living protoplasm could develop from non-living matter, such as ammonium carbonates, oxalates, tartrates, and phosphates, without the aid of light. In his lecture at the Sorbonne in 1864, Louis Pasteur raised the question of spontaneous generation of life: Could life originate without parents? He indicated that circumstances unknown to man may have existed and permitted spontaneous generation.

In the latter part of the nineteenth century, the prevailing scientific theory suggested that a complex crystallization of living things occurred in the remote past at a time when the earth's physical and chemical conditions favored such an event. This idea was more precisely developed by E. Haeckel in his theory of archegony. Haeckel believed that primeval organisms must have been entirely homogeneous, formless lumps of protein developed by the interaction between materials in the sea and special external physical forces. He was cautious, however, about offering any detailed hypothesis concerning the origin of life, since there was no information concerning the conditions prevailing on the surface of the earth at that time.

At the turn of the century not only were there a number of hypotheses, but simple experimentation had also begun to deal with the origin of life. Not until the middle of the twentieth century were ideas regarding life's origin tested experimentally by setting up conditions that could be said plausibly to match the cosmological, geological, and chemical conditions of primeval earth. Significant advances have been made as investigators have obtained more precise knowledge about the primeval earth and about genetics and biology in general.

TIME PERIODS OF PREBIOTIC EVOLUTION AND DEVELOPMENT OF PRIMITIVE ORGANISMS

Prebiotic evolution depends basically on geological evolution. Cloud (1974) marked a time scale for the major evolutionary events as shown in figure 1.1.

Earth's formation took place about 4.6 billion years ago (Singer 1966), approximately the time of the origin of the solar system, according to evidence recently adduced from the age of meteorites and the isotopic composition of leads. It was also at this time that the moon began forming, as calculated from the original radioactive materials in lunar rocks brought back by Apollo 12 (Weaver 1973).

Urey (1966) estimated that life was present 3×10^9 to 4.5×10^9 years ago, judging from fossil records (fig. 1.2) and based on his description of life. But Cloud (1968, 1974) postulated that life probably began to develop between 3.3×10^9 and 3.4×10^9 years ago, an assumption based on the time of hematitic banded iron formation (BIF) among sediments deposited between 2 billion and 3.3 billion years ago. The BIF in the lithosphere is explained by the transport of the ferrous iron in solution in the ferrous state, precipitated as ferric or ferroferric iron on combination with biological oxygen. The rhythmic banding could have resulted from a fluctuating balance between oxygen-producing biotas and the

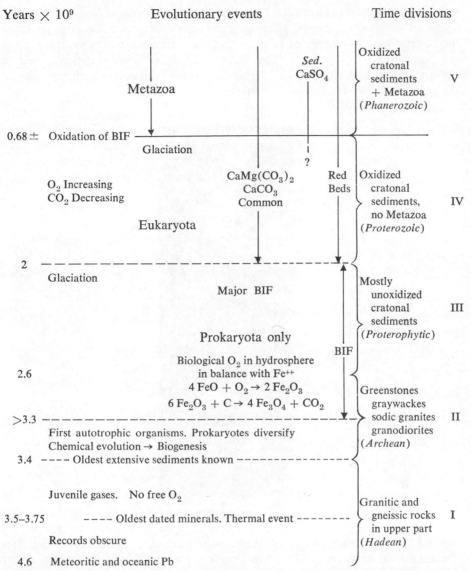

Fig. 1.1. The postulated periods of geological and organic evolution. Autotrophic organisms, such as green plants and some bacteria, require neither organic carbon nor nitrogen for growth. Heterotrophic organisms, such as animals and chlorophyll-free plants, require organic food. Metazoa are multicellular animals including all species of the animal kingdom except the protozoa. Red beds are detrital continental or marginal sediments in which the individual grains are coated with ferric oxides. (Reprinted, by permission, from P. E. Cloud, Jr., *Am. Scientist* 62 (1974): 54–66, journal of Sigma Xi, the Scientific Research Society of North America, Inc.)

supply of ferrous ions. Thus Cloud assumes that some forms of life, which may be called autotrophic organisms, capable of manufacturing their food by photosynthesis, must have been present prior to that time, since oxygen is set free through the process of photosynthesis. The isotopic dating of other sediments, such as sulfides of lead and zinc, which are highly unstable, conforms with the time for the presence of oxygen estimated by BIF.

Fig. 1.2. Electron micrographs of *"Eobacterium isolatum"* about 0.75μ long, from the black chart from rocks of the Fig Tree Series of the Swaziland System, eastern Transvaal, South Africa, three billion years ago. (Reprinted, by permission, from E. S. Barghoorn and J. W. Schopf, *Science* 152(1966):758–63. © 1966 by the American Association for the Advancement of Science.)

Before three billion years ago, the atmosphere was in a reduced state. Thus the first life to develop in this period is generally assumed to have been anaerobic and may be called heterotrophic (that is, dependent on external food sources). Autotrophic organisms were therefore evolved from the heterotrophic organisms.

Paleontological evidence (Cloud 1968) implies that before about 2.0 billion years ago, all organisms were prokaryotes, organisms without nuclear walls and incapable of miotic cell division. This time estimate corresponds with that for BIF. After that came the first accumulation of free oxygen, which, even in small amounts, facilitated the evolution of eukaryotes, organisms that have nuclear walls and well-defined chromosomes capable of cell division and sexual reproduction.

Once this first accumulation of oxygen was begun, oxygen continued to accumulate slowly in the atmosphere; the appearance of the Metazoa, multicelled animal life, being an indicator that the atmospheric oxygen concentration, then about 3% of what exists at present, had reached a level sufficient to support a metazoan oxidative metabolism. The oldest rocks in which eukaryotic fossils have been found are probably more than 0.7 billion years old (Barghoorn and Schopf 1965). Although the minimal physical conditions required for eukaryotes may have appeared much earlier, possibly before the beginning of the Paleozoic age (Cloud 1968), no records of unequivocal Metazoa in rocks of undoubted pre-Paleozoic age have yet been found.

GEOLOGICAL CONDITIONS ON THE PRIMITIVE EARTH

In considering the origin of life on the earth, we need some knowledge of the elements or chemicals present in the primitive lithosphere, hydrosphere (the aqueous envelope of the earth), and atmosphere: From what elements were present, what chemical reaction could have taken place? What products could have been formed? What further reaction could have occurred?

The Lithosphere

The lithosphere may be classified into the core, the mantle, and the crust. The mantle and the core seem to have no direct relation to organic evolution. The geochemical composition of the inorganic lithosphere is as follows (Fox and Dose 1972): crust—silicates of Al, Fe, Ca, Mg, Na, and K; mantle—mainly silicates of Mg and Fe; and core—Fe-Ni alloy. Except for Al, all of the elements in the earth's crust are essential for contemporary organisms. However, there is no direct evidence presently available in the earth's crust regarding the occurrence and origin of carbonaceous materials during the era of prebiotic evolution.

The Atmosphere

During the formation of the earth, various elements, with hydrogen predominant, were distributed in the cosmos. Helium and other inert gases were also present, but in very small amounts in comparison with hydrogen. Oxygen was present as metallic oxides and water, and nitrogen was present as ammonia (Oparin 1957). According to Urey (1952), all free radicals of carbon, nitrogen, and oxygen would have been transformed into stable molecules, such as CH_4, NH_3, and H_2O, due to the catalytic effect of dust and the presence of hydrogen. The composition of the primitive atmosphere of the earth is given below, according to several different authors:

CH_4, NH_3, H_2O, H_2	Oparin (1957)
CH_4, CO_2, NH_3, N_2, H_2O, H_2	Bernal (1951)
CO_2, N_2, H_2S, H_2O	Rubey (1951)
CO_2, CO, N_2, SO_2, HCl, H_2O	Cloud (1968)
and traces of others	

The above composition shows clearly that the primitive atmosphere was in a reduced condition. This reduced state favored the formation of the amino acids necessary for further prebiotic synthesis (Oparin 1957). As will be discussed later, experimental demonstration of the production of amino acids has been carried out under a reduced atmosphere. According to Urey (1952), the proportion of the different gases in the primitive atmosphere was as follows:

CH_4	1 atmosphere
CO_2	3.3×10^{-4} atmosphere (present value)
H_2	1.5×10^{-3} atmosphere
N_2	(in liquid water of unknown concentration)

The Hydrosphere

Since life is dependent on water, some attention must be given to the physical and chemical conditions of the primitive hydrosphere. Concerning the water masses in the primitive oceans, Rubey (1951, 1955) postulated that less than one-tenth of the volume of the present ocean water existed on the surface of the primitive earth. The major body of ocean water was later supplied by outgassing from the interior of the earth. On the basis of the current rate of outgassing of juvenile water (water from the interior of the earth to the surface for the first time) by volcanic activity, hot springs, and fumaroles, Urey (1952) came to a similar conclusion, although there was some disagreement concerning the rate of increase of the oceans during the various periods of earth history (Rubey 1951).

According to Abelson (1966), the primitive ocean was slightly alkaline, with pH 8 to 9, as a result of the interaction of the acid volatiles from outgassing with the alkaline components of the predominantly basaltic earth crust. Rubey's (1951) view was that the initially acid ocean, resulting from acid effluvia of volcanoes, was neutralized by cations leached from the igneous rocks of the land by rain. Fox and Dose (1972) claim that most of the continental rocks on the surface of the crust are sedimentary, and not so alkaline as basaltic rocks. They believe that the pH of the ocean probably was slightly above 8 during most of the earth's history. The generally held view of the alkaline condition of the primitive ocean is consistent with the assumption of a nonoxidizing atmosphere containing substantial amounts of CO_2 (Fox and Dose 1972).

It is generally agreed (Bernal 1951; Rubey 1951) that the initial composition of the primitive hydrosphere was CO_2, NH_3, H_2S, and H_2O. However, a sharp controversy recently arose on the later accumulation of different types of biochemical compounds in the primitive waters, called "thick soup" or "primordial broth." The concept of the primitive soup as one of the most persistent ideas (Oparin 1957) has been criticized on a thermodynamic basis by a number of investigators (Abelson 1966; Hull 1960; Sillen 1965). The reasons, as summarized by Fox and Dose (1972), are as follows: "First, amino acids, aldehydes, cyanides, and other such reactive materials are especially unstable in aqueous solution. Second, most of the resulting products are far from identical with biologically important molecules. Third, the primitive ocean was steadily irradiated with a relatively high dose of solar ultraviolet light. A steady irradiation of a rather homogeneous solution results in degradative rather than synthetic reactions; a photochemical equilibrium stage is finally approached." Thus it is claimed that an accumulation of different types of biochemical compounds was extremely unlikely. Bernal (1951) suggested that the ways of concentrating organic matter in amounts useful for evolution included absorption by minerals on the shores; and a number of other possibilities were discussed by Fox and Dose (1972, p. 38).

Sources of Energy for Prebiotic Syntheses

Energy was required for chemical syntheses. On the surface of primitive earth, were four sources of free energy: solar radiation, electrical discharges, radioactive decay, and volcanic emissions. Fox and Dose (1972) pointed out that the quality of the energy is more important than the quantity for prebiotic synthesis. Although the production of formaldehyde from carbon dioxide as a consequence of the effect of solar energy was shown early in the twentieth century (Oparin 1957), among the different energy sources, electrical discharge is by far the most important. It was believed to be the energy source for the formation of hydrocarbons, the earth's original carbon compound, and for the production of amino acids, the building blocks for peptides and proteinlike macromolecules.

The amount of energy that came from electrical discharges is difficult to estimate, but it would be several orders of magnitude lower than that of ultraviolet light. While the energy of silent electrical discharges (electrons generated by collisions of positive ions with the molecules of gas) may have been considerable, it is believed that lightning was more frequent in primeval times than it is now (Oparin 1957). The energy from the disintegration of atoms by naturally radioactive substances was for the most part in the granitic envelope of the lithosphere, amounting to 2×10^{19} kcal/yr. The heat passed from the center of the earth to its surface at the rate of about 2.5×10^{17} kcal/yr. The total amount of solar radiation energy to reach the outer limits of the atmosphere was 1.2×10^{21} kcal/yr (Oparin 1957), about half of which was absorbed by the atmosphere and the ground. As far as the total amount of energy is concerned, solar energy was the major source of energy on the primitive earth, just as it is today. Fox and Dose (1972) made their estimate of energies available from different sources for the earth of from 4×10^9 to 4.5×10^9 years ago based on those available for the contemporary earth (table 1.1).

TABLE 1.1 Energy Available for Primitive Earth Estimated from Energy Sources for Contemporary Earth

Sources	Total per Year (calories)	Total per Square Centimeter Surface per Year (calories)
Total solar radiation	$850,000 \times 10^{18}$	170,000
Solar radiation below 2,000 Å	150×10^{18}	30
High-energy radiation	240×10^{18}	47
Heat from volcanic emissions	$> 0.75 \times 10^{18}$	> 0.15
Electric discharges	20×10^{18}	4

SOURCE: Reprinted, by permission, from S. W. Fox and K. Dose. *Molecular Evolution and the Origin of Life.* © 1972 by W. H. Freeman & Co.

FORMATION OF SIMPLE ORGANIC COMPOUNDS

In 1918, H. F. Osborn hypothesized that carbon dioxide acted as the source of carbon, forming organic compounds from which living things developed. P. Becquerel (1924) apparently shared this view, pointing out that carbon dioxide was the first carbon compound to exist on earth, and that organic substances arose by combining with it water and minerals, using some of the energy sources mentioned above. The proposal is that the prerequisite for the development of simple organic substances was the production of formaldehyde from carbon dioxide as a consequence of the effect of solar energy.

A few investigators have also demonstrated that formic acid and formaldehyde can be produced from a mixture of water and carbon dioxide in silent discharges (Oparin 1957). Under the presumed constituents of an early atmosphere, Groth and Suess (1938; see Fox and Dose 1972, p. 67) obtained formaldehyde and glyoxal as the major products from the same mixture, using ultraviolet light as the energy source on aqueous solutions containing ferrous ion in equilibrium with an atmosphere of carbon dioxide and hydrogen. By irradiation with alpha-particles, Calvin and associates (Garrison et al. 1951) demonstrated the production of formic acid, formaldehyde, and succinic acid. (That the presence of ferrous ion in the early geological evolution was important for the prebiosynthesis, as was speculated by W. Francis [1925], seems to be supported by the above experiment.) We can see that formic acid and formaldehyde, viewed as simple compounds, having undergone prebiotic synthesis, can be produced under different conditions. These results well support the previously held views relating to the beginning of prebiotic synthesis.

Another aspect of earth's original carbon compounds is in the formation of hydrocarbons. In his book *The Origin of Life on the Earth,* Oparin showed how evolution of simple carbon compounds into hydrocarbons led to the formation of proteinlike compounds. These compounds can become colloids through undergoing gradual differentiation of their internal organization. For example, methane, the simplest hydrocarbon known to exist in the primitive atmosphere of the earth, can be produced by various reactions that occurred during the geological formation of primeval earth. According to Nekbrasov (1955; see Oparin 1957, p. 167), each carbide of a number of metals, such as aluminum and beryllium, when reacted with water, gave rise to methane, in addition to acetylene and ethylene, depending on the particular carbide being involved. Some of the hydrocarbons so formed dispersed into the atmosphere; the rest underwent various reactions within the lithosphere.

From the above-mentioned simple compounds and molecules, further reactions can take place. Knowing that organic chemistry is essentially the chemistry of hydrocarbon derivatives, one can imagine that by further reaction and condensation these derivatives could give rise to more complicated compounds on the surface of the earth when it was still devoid of life. The entire second stage and the further development of matter were inherent in the original hydrocarbons themselves. For instance, methane reacts with ammonia to form methylamine and hydrocyanic acid, and acetylene reacts with water to produce heterocyclic compounds (Noyes and Leighton 1967). As will be shown later, these are the compounds necessary for the production of amino acids and nucleotides.

FORMATION OF MICROMOLECULES

In this section, laboratory evidence is given for the production of slightly more complicated compounds than those discussed in the previous section. These more complicated compounds are required for syntheses of the two major macromolecular systems, polyamino acids and polynucleotides, for the development of life. In addition, evidence is shown for production of porphyrin, the basic compound for the development of photosynthesis.

Amino Acids

One significant advance in the study of prebiotic evolution is the laboratory production of amino acids under primitive earth conditions. It will be seen that, as one type of basic compound, amino acids are as important for the development of primordial life as they are in the present living systems. Miller (1953) produced amino acids experimentally on the basis of evidence regarding the composition of the atmosphere of primeval earth. He passed electrical discharges through a mixture of CH_4, NH_3, H_2, and H_2O, producing large amounts of amino acids, along with glycolic, lactic, formic, acetic, and propionic acids. He explained this phenomenon as a result of reactions between the free radicals of hydrocyanic acid, amines, aldehydes, alcohols, and ions as they exist in silent discharges, and he drew the following set of equations for the production of amino acids:

$$RCHO + NH_3 + HCN \rightarrow RCH(NH_2)CN + H_2O$$
$$RCH(NH_2)CN + 2H_2O \rightarrow RCH(NH_2)COOH + NH_3$$

and for hydroxy acids:

$$RCHO + HCN \rightarrow RCH(OH)CN$$
$$RCH(OH)CN + 2H_2O \rightarrow RCHOHCOOH + NH_3$$

Miller's results were completely confirmed later by Russian workers. Table 1.2 shows the percentage of amino acids recovered under heating and sparking conditions.

TABLE 1.2 Yields from Sparking a Mixture of CH_4, NH_3, H_2O, and H_2
(710 mg of Carbon Added as CH_4)

Compound	Yield (moles $\times$ 10^5)
Glycine	63
Glycolic acid	56
Sarcosine	5
Alanine	34
Lactic acid	31
N-Methylalanine	1
α-Amino-n-butyric acid	5
α-Aminoisobutyric acid	0.1
α-Hydroxybutyric acid	5
β-Alanine	15
Succinic acid	4
Aspartic acid	0.4
Glutamic acid	0.6
Iminodiacetic acid	5.5
Iminoacetic-propionic acid	1.5
Formic acid	233
Acetic acid	15
Propionic acid	13
Urea	2.0
N-Methyl urea	1.5

SOURCE: Reprinted, by permission, from Calvin, 1969.

Saccharides

Because of the importance of DNA as the information carrier in contemporary organisms, experiments dealing with the origin of life have focused on the formation of its constituents: the purines, the pyrimidines, and the pentoses. Besides pentose, sugars—as one group of carbohydrates—are needed for cellular structure as well as for sources of chemical energy in present organisms, and so they would presumably have been important for primordial forms of life.

Formaldehyde, considered to be readily available in primitive earth, can be easily condensed into sugars, such as fructose, cellobiose, sorbose, xylose, and glycolaldehyde (fig. 1.3), in the presence of alkaline catalysts in aqueous solution. By formaldehyde condensation, A. von Butlerow demonstrated sugar formation early in 1861 (see Fox and Dose 1972). Since then a number of investigators have produced many kinds of monosaccharides, thereby confirming the results of previous experimentation. The reaction proceeds readily in the presence of alkaline

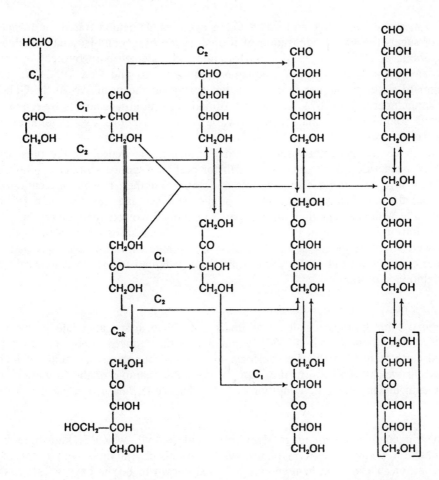

Fig. 1.3. Formation of monosaccharides from formaldehyde by base catalyses. (Reprinted, by permission, from Pfeil and Ruckert, 1961.)

earth hydroxides or weak basic hydroxides of lead and tin, as indicated by Loew's experiment (see Fox and Dose 1972). Among the sugars, it is known that pentose and hexose are produced in preference to others because of their stabilizing ring structure.

With respect to the methods of monosaccharide production through base-induced condensation, there have been objections to high concentrations of formaldehyde as unrealistic, as well as to the alkaline condition; Abelson (1966) claimed that earth water has never been highly alkaline. Therefore, investigators have sought evidence of the formation of prebiotic monosaccharides through physical means as possible alternatives. Ponnamperuma (1965) pointed out that sugars such as ribose and deoxyribose can be produced directly by electron irradiation of methane, ammonia, and water. Other experiments demonstrated the formation of sugars through condensation of formaldehyde by gamma-irradiation (Ponnamperuma and Mariner 1963), and formation of 2-deoxyribose by heating glyceraldehyde and acetaldehyde in aqueous solution (Oro and Cox 1962). In an attempt to simulate the conditions of a hot spring on the primitive earth, Gabel and Ponnamperuma (1967) showed the formation of monosaccharides by refluxed aqueous solutions of formaldehyde in various concentrations over kaolinite (a pure white clay).

The above experimental results should be sufficient evidence for the presence of monosaccharides on the primitive earth for prebiotic synthesis under conditions that may have been present and suitable one way or another. It may be mentioned also that the stabilizing ring structures of the pentoses and hexoses seem to be compatible with the fact that these sugars are more involved than others in cell replication and metabolism in present life. This is because of the possibility that they were more abundant at the beginning of organic evolution and thus had a greater chance than did others to be utilized and adapted to the systems involved in prebiotic evolution.

Fatty Acids

Although some have speculated that lipids must have arisen at a late period of chemical evolution, Oparin (1957) claimed that the reduced conditions on primeval earth were favorable for the formation of hydrophobic compounds of high molecular weight. To confirm this idea, he cited the example of the formation of petroleum under anaerobic conditions at considerable depths in the earth.

Porphyrin

Porphyrin combined with magnesium forms chlorophyll, which is known to be essential for photosynthesis in plants, while iron-porphyrin links with a specific protein molecule to form hemoglobin, which is known to be the oxygen carrier in vertebrates. Porphyrin is important in biological evolution, not only because it is essential for the present life of individuals in both kingdoms, but because it is involved in photosynthesis, which is the major source of free oxygen contribution on earth. Porphyrin is essential, at least, for the evolution of eukaryotes from prokaryotes. According to evidence obtained from fossils, some prophyrin complexes date at least as far back as two billion years, coinciding with the periods of BIF. This provides further evidence suggesting that photosynthesis is the origin of free oxygen.

Porphyrins may be regarded as derivatives of the fundamental substance, porphin, which consists of four pyrrolelike rings linked by four CH groups in a ring system (fig. 1.4). Porphyrins found in nature are compounds in which side chains are substituted for the eight hydrogen atoms of porphin at the numbered positions as shown in figure 1.4.

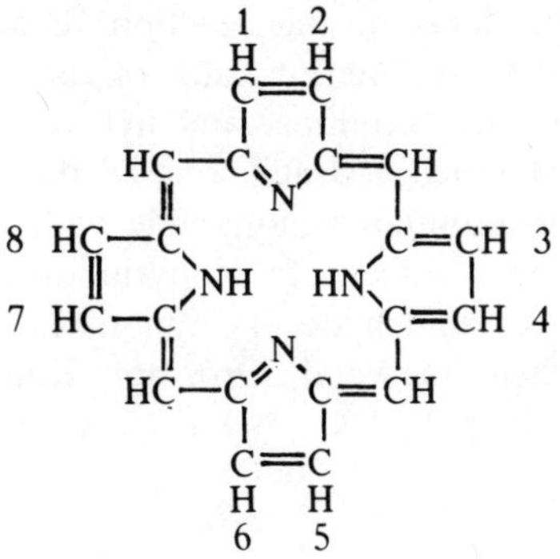

Porphin ($C_{20}H_{14}N_4$) 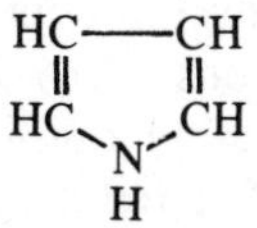   Pyrrole

Fig. 1.4. The chemical structure of pyrrole and porphin.

A number of experiments bear on prebiotic synthesis of porphyrins. Pyrroles, the probable precursors of porphyrins, must be formed at first during the stage of chemical evolution, since the synthesizing of pyrroles was achieved from acetylene and hydrogen cyanide by Meyer (1913; see Fox and Dose 1972). By condensation of pyrrole with benzaldehyde by ultraviolet irradiation, Szutka, Hazel, and McNabb (1959) and Szutka (1966) promoted the formation of porphyrin. They also obtained porphyrinlike substances through gamma-irradiation of mixtures of pyrrole, benzaldehyhe, pyridine, and zinc acetate. Aqueous alpha-aminolevulinic acid, the intermediate in the biosynthesis of porphyrins, has been condensed to porphobilinogen, a biological precursor of porphyrin, under alkaline and anaerobic conditions (Scott 1956). There is other experimental evidence, but the above results may be sufficient to substantiate the possibility of prebiotic synthesis of porphyrins from simple chemical molecules that, as previously discussed, presumably existed at that time.

Up to this point in our discussion of chemical evolution, we have seen chemical synthesis involving the use of existing simple molecules and simple compounds, such as H_2, H_2O, CO_2, and N_2, present in the primitive atmosphere, to form slightly more complicated molecules, such as methane, formaldehyde, sugars, and fatty acids. We also see chemical evolution beginning to involve the utilization of the Mg and Fe present in the lithosphere.

Purines and Pyrimidines

Purines and pyrimidines are so-called nucleotide bases, the building blocks of nucleic acids, just as the amino acids are for proteins. Thus, before discussing the polynucleotides, we need to examine the feasibility of the prebiotic synthesis of purines and pyrimidines.

Under conditions similar to those of primitive earth, Oro and Kimball (1962) produced adenine (one of the purines) by reacting a mixture of HCN, NH_3, and H_2O, which were abundantly present in the primeval soup. Figure 1.5 shows the individual steps and the intermediate products of the reaction. From 4-amino-imidazole-5-carboxamide (AICAI), one of the intermediates of the reaction, the production of other purines, such as guanine, xanthine, and hypoxanthine (Kliss and Mathews 1962), can be achieved (fig. 1.6). Uracil, one of the pyrimidines, was produced by the condensation of beta-aminopropionamide with urea in ammoniacal solutions (Oro 1963). Thymine, another of the pyrimidines, has been produced by methylation of uracil with formaldehyde in ammoniacal solution, in the presence of a reducing agent (Stephen-Sherwood, Oro, and Kimball 1971).

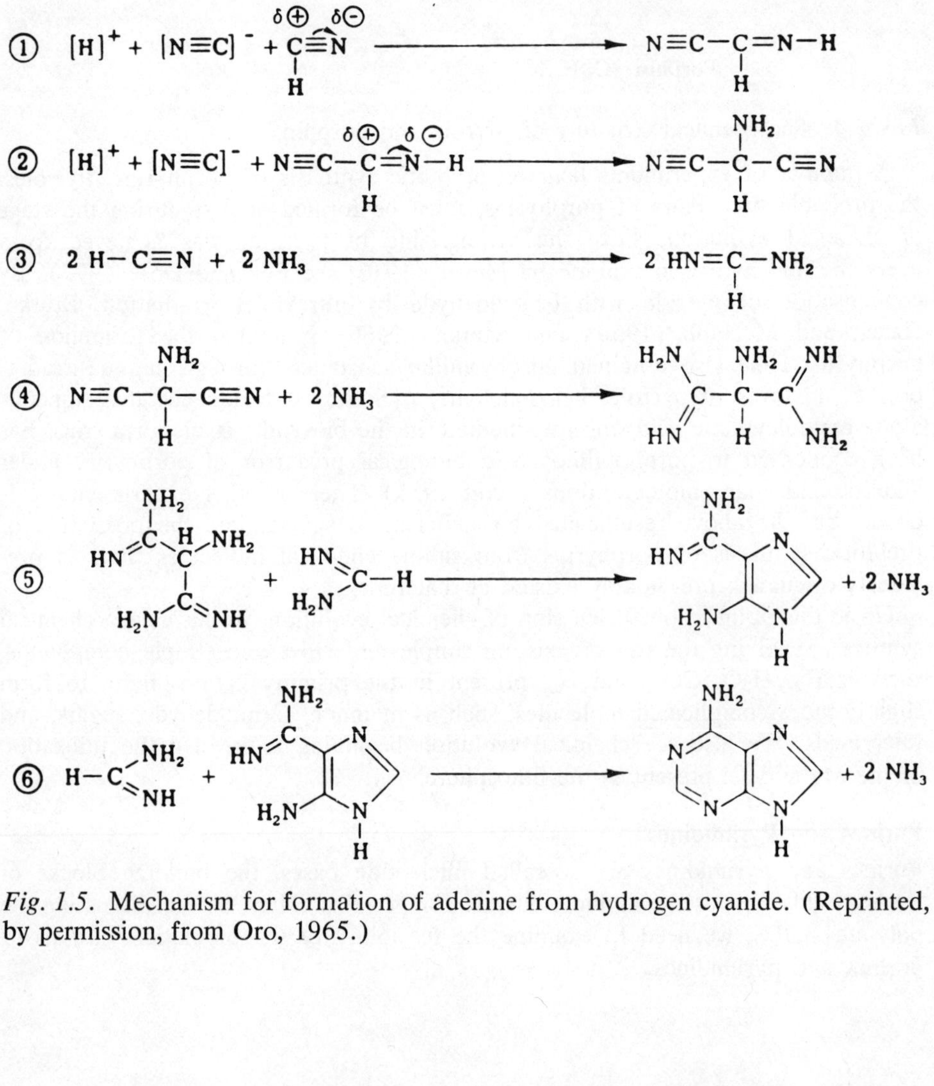

Fig. 1.5. Mechanism for formation of adenine from hydrogen cyanide. (Reprinted, by permission, from Oro, 1965.)

Recently, beta-cytidylic acid, one of the single nucleotide units of a typical nucleic acid, was produced in relatively high yield by Sanchez, Ferris, and Orgel (1966), whose experiment was done under presumably prebiotic conditions. The experiment demonstrates the prebiotic synthesis of the nucleosides and nucleotides, and the possible natural processes leading to their abiogenesis.

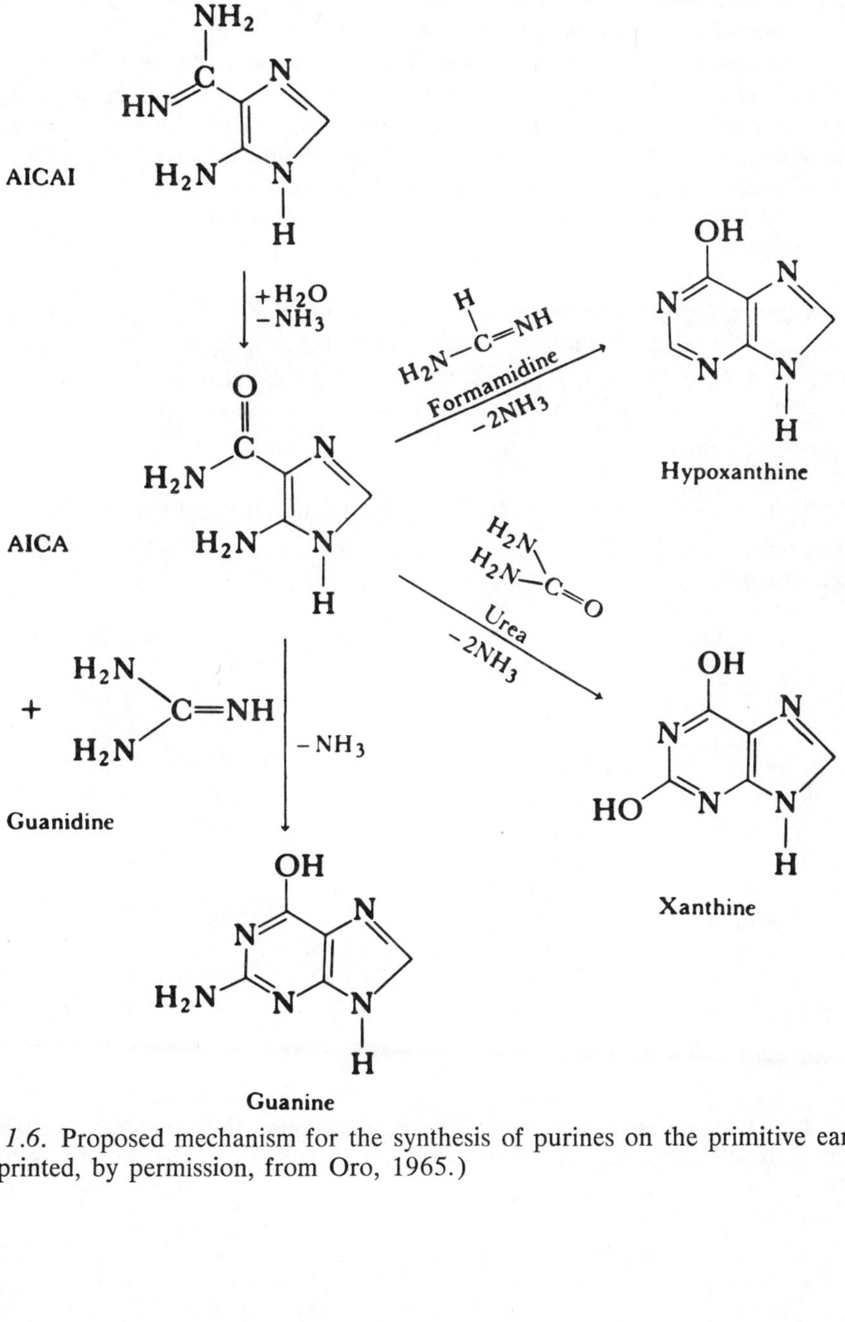

Fig. 1.6. Proposed mechanism for the synthesis of purines on the primitive earth. (Reprinted, by permission, from Oro, 1965.)

Nucleoside Phosphates

It is well known that, as building blocks for DNA (see chap. 2) and as an energy intermediary, nucleoside phosphates, the products of phosphorylation of nucleosides, are important compounds in modern organisms. With the recent discovery of cyclic phosphorylated nucleosides in gene transcription and translation (see chap. 5), additional functions or roles of the phosphorylated nucleosides have been recognized. Such phosphorylated compounds could also be important in prebiotic syntheses, not only because they were needed as building materials for the nucleic acid, but also perhaps as a chemical energy source for the formation of large molecules at a later stage of chemical evolution. Up to this point, all the energy sources we have discussed for prebiotic synthesis have been physical. It would be convenient to have different forms of energy, such as adenosine diphosphate (ADP) and adenosine triphosphate (ATP), available. Indeed, experimental production of nucleoside phosphates under presumed primeval earth conditions have recently been reported.

The general mechanism of phosphorylation, elucidated and exploited in a wide variety of laboratory conditions, is believed to be the displacement of the conjugates of the phosphoryl reagents (Rabinowitz, Chang, and Ponnamperuma 1968). The chemical reaction is illustrated in figure 1.7, where X represents

$$
\begin{array}{cccc}
O & NR & O & CH_2 \\
\| & \| & \| & \| \\
\end{array}
$$
$-OP(OR)_2$, $-OCR$, $-OCR$, $-OCOR$, $-NR_2$, halogen, and so forth;

R represents H, aryl, alkyl, or aralkyl; and N represents nucleosides or some other sugar compounds.

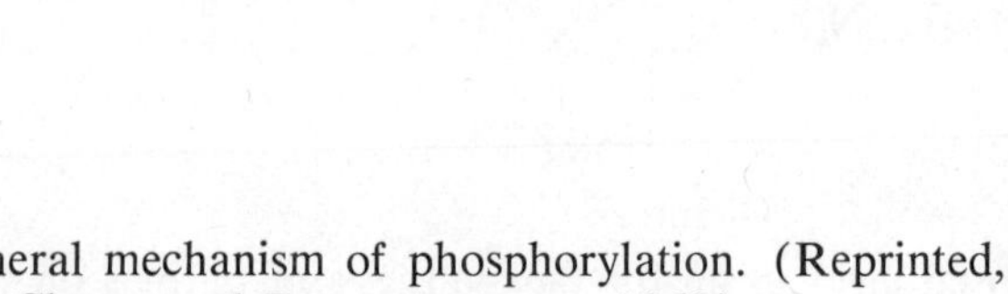

Fig. 1.7. The general mechanism of phosphorylation. (Reprinted, by permission, from Rabinowitz, Chang, and Ponnamperuma, 1968).

Under conditions assumed to have been present on the primitive earth, phosphorylation of glucose with *N*-cyanoguanidine and orthophosphoric acid (Steinman, Lemmon, and Calvin 1964) and of nucleosides with polyphosphoric acids (Waehneldt and Fox 1967) have been described.

Abiogenic nonenzymatic production of nucleoside phosphates and related molecules, under simulated primitive earth conditions, was reported by Ponnamperuma, Sagan, and Mariner (1963) (table 1.3). Starting with adenine, adenosine, adeno-

TABLE 1.3. Percentages of Nucleoside Synthesis under Presumably Primitive Earth Conditions

Experiment	Adenosine	AMP	ADP	ATP	A4P
1.					
(i) Adenine-^{14}C + ribose					
(ii) Adenine-^{14}C + ribose + phosphoric acid	+ (0.01%)	−	−	−	−
(iii) Adenine-^{14}C + ribose + ethyl metaphosphate	+ (0.01%)	+ (0.08%)	+ (0.06%)	+ (0.05%)	+ (0.04%)
2.					
(i) Adenosine-^{14}C + phosphoric acid		−	−	−	−
(ii) Adenosine-^{14}C + ethyl metaphosphate		+ (0.5%)	+ (0.2%)	+ (0.1%)	+
3.					
(i) AMP-^{14}C + phosphoric acid			−	−	−
(ii) AMP-^{14}C + ethyl metaphosphate			+ (3%)	+ (0.3%)	+ (0.1%)
4.					
(i) ADP + phosphoric acid				−	−
(ii) ADP + ethyl metaphosphate				+	+

Figures in brackets show conversion as percentage of starting material.

With the techniques used in this experiment, the lower limit of detectability was 0.001 per cent.

In experiment 4, no quantitative estimates were performed, as unlabeled ADP was used. The ATP in this case was located by shadowgrams.

SOURCE: Reprinted, by permission, from C. Ponnamperuma, C. Sagan, and R. Mariner, *Nature* 199 (1963): 222-26.

sine monophosphate (AMP), or ADP in separate experiments, they observed the production of adenosine, and adenosine tetraphosphate (A4P). The question arises why ATP, not the triphosphates of other purines and pyrmidines, was produced in the primitive earth and why it is utilized today as the primary biological energy intermediary. Ponnamperuma, Sagan, and Mariner (1963) gave the following explanation:

First, in primitive earth simulation experiments under reducing conditions with low hydrogen content, adenine is produced in far greater yield than are other purines and pyrimidines. Secondly, no biological purine or pyrimidine has a larger absorption cross section between 2,400 and 2,900 Å. Thirdly, adenine is among the most stable of such molecules under ultraviolet irradiation. Finally, the ultraviolet excitation energy is readily transferred, especially by π electrons, along the conjugated double bonds of the molecule; the excited states are very long-lived, and thereby serve to provide bond energies for higher synthetic reaction. All but the first of these properties of adenine derive from the fact that it has the greatest resonance energy of all the biochemical purines and pyrimidines.

The greater abundance and energy of adenine, compared with other purines and pyrimidines, led Ponnamperuma, Sagan, and Mariner (1963) to conclude that molecules ideally suited for the origin of life were preferentially produced in primitive times. The synthesis of adenosine 2′, 3′, and 5′- phosphates by simple heating of an aqueous solution of adenosine and linear polyphosphate salts ($Na_4P_2O_7$ and $Na_5P_3O_{10}$ has been demonstrated recently (fig. 1.8).

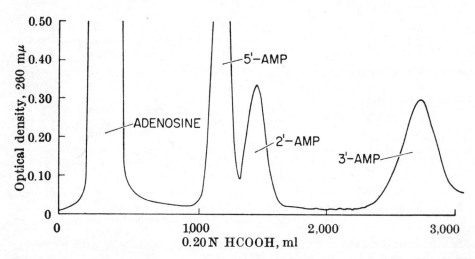

Fig. 1.8. Fractionation of the products of the phosphorylation of adenosine with $Na_5P_3O_{10}$. The reaction was run at 100°C in a sealed tube in 0.75N NH_4OH. (Reprinted, by permission, from A. Schwartz and C. Ponnamperuma, *Nature* 218 (1968):443.)

In connection with phosphorylation, a question that remains is how phosphates became available. The concentration of dissolved inorganic phosphate in the oceans is at present in the range of $10^{-5}M$ to $10^{-6}M$ (Gulick 1955). Such a low concentration is believed to be due to the insolubility of minerals of the apatite groups, particularly hydroxyl apatite, $Ca_5(OH)(PO_4)_3$. Thus a number of authors have suggested that the insolubility of the most abundant phosphate mineral constitutes a difficulty for theories of prebiotic phosphorylation. However, the problem may not be as serious as it seems. Miller and Parris (1964) have found that pyrophosphate is formed on the surface of apatite as a result of reaction with cyanate ion. Lohrmann and Orgel (1971) have shown that in the presence of urea and ammonium chloride, at temperatures ranging from 85°C to 100°C, hydroxyl apatite will phosphorylate nucleosides and nucleotides, although slowly. It was reported recently that struvite ($MgNH_4PO_46H_2O$) may be another phosphorylating mineral. Handschuh and Orgel (1973) have shown that when struvite is heated with urea at 65°C, magnesium pyrophosphate is produced. If nucleotides are also present in this reaction, nucleoside pyrophosphates are formed. These authors have speculated that the concentration of ammonia in evaporating tide pools may have been $10^{-2}M$ or higher in the primitive earth, basing their theory on the concentration of about $10^{-3}M$ to $10^{-2}M$ present in primitive oceans (Bada and Miller 1968). Under this condition, struvite would have precipitated rather than hydroxyl apatite. It is believed that the evaporation of prebiotic lakes or tide pools could have led to the formation of solid films containing struvite, urea, and other organic compounds, including nucleosides and nucleotides (Lohrmann and Orgel 1971).

In connection with the formation of struvite, which contains NH_4^+, Handschuh and Orgel hypothesize the evolution of a shift from NH_4^+ to K^+ in contemporary organisms. The isomorphism of many NH_4^+ and K^+ salts is well recognized. It is also known that NH_4^+ can replace K^+ in protein synthesis, although NH_4^+ is toxic to most contemporary cells. When the atmosphere became oxidizing and ammonia began to disappear from the surface of the earth, it is possible that a continuous transition from a biological system rich in NH_4^+ to one containing K^+ could have taken place, as substantial amounts of K^+ can replace NH_4^+ isomorphously in struvite. This shift, if it occurred, would have begun about three billion years ago, when the earth began to accumulate oxygen.

Formation of Macromolecules

The most important functions carried on in a living cell are rooted in two types of macromolecules, proteins and nucleic acids. Proteins are an assembly of amino acids, and nucleic acids are an assembly of nucleosides. These two different macromolecules are more complex than the cellulose, chitin, lipids, and so forth found in living organisms. But such complexity does not imply that their evolutionary precursors were also complex. In this section, as we examine polyamino acids and polynucleotides assembled from their monomers under simulated geological conditions, we will extend our discussion from simple and small molecules to large and complex ones.

Polyamino Acids

Speculation on the possibility of polymerizing amino acids in prebiotic evolution has been stimulated since the industrial production of nylon by W. H. Carothers

(1936). The molecular logic of the chain initiation of amino acids for the primordial synthesis of peptides has been discussed by Oparin (1957). Ehrensvärd (1962; see Fox and Dose 1972) suggested that the geological locale for such synthesis would have been near volcanic regions.

Both small and large peptide molecules (50 to 100 amino acids) have been produced experimentally by amino acid condensation. All types of amino acids commonly found in protein can be incorporated. Necessary conditions for the condensation are that the amino acids be reacted on in an initially dry state and at a temperature far above the boiling point of water (Fox and Yuyama 1963). These conditions are believed to have been prevalent in the primitive earth. Such heat-synthesized polypeptides have practically all the properties of the contemporary proteins produced by organisms, with portions containing the 18 to 20 amino acids common to proteins (Fox and Wood 1968). Some of them have enzymelike activities, catalyzing classical biochemical reactions such as hydrolysis, decarboxylation, and transamination. Their activities are weaker than those of enzymes synthesized by living organisms. They lack some of the sophistication and complexity of contemporary proteins, which apparently are the result of millions of years of evolution.

Fox and Wood pointed out that macromolecular preparations are relatively uniform and ordered, not disordered polymers, as one might anticipate as a result of the pyrolytic thermal process. Apparently the steric and electronic individuality of each amino acid results in a selective coupling at each stage of polymerization. Large and complex molecules (proteinoids) with hormonal and nutritive qualities, in addition to more versatile enzymatic properties, have been discussed by Fox and Dose (1972). Notice that the ordered polymeric structures, together with some enzymatic and hormonal activities of the macromolecules, are an indication of the presence of genetic information. This information may be crude, but it was originated entirely by the molecules themselves.

Polynucleotides

Now we come to one of the most important groups of macromolecules, polynucleotides. Naturally occurring polynucleotides in contemporary organisms are characterized by 3'- and 5'-phosphate ester linkage between the two adjacent nucleotides. In the nucleic acid, known as the information carrier, there are two molecules of polynucleotides with cross-linkages between them (see fig. 2.2 for molecular structure). The number of nucleotides per molecule varies among organisms, with a few thousand in viruses and a few million in bacteria. Laboratory attempts of synthesizing polynucleotides under simulated geological conditions have had some preliminary success, although none have really managed to produce molecules with the same intrastrand and interstrand linkages as in the naturally occurring nucleic acid.

Schramm and his colleagues (1962; see Fox and Dose 1972, p. 181; Schramm 1965) achieved a polymerization of nucleotides from polymetaphosphate ethyl ester (PMP). The polymers show some relationship to the corresponding naturally occurring polynucleotides. These experiments were criticized, however, by Fox and Dose (1972) on the basis that the type of condensing agent used lacked geological relevance. In a typical experiment in which adenylic acid was heated in the presence of PMP, Schramm (1965) claimed that the polymers obtained by using 3'-nucleotides as the reactant contained the 3'- and 5'-phosphate linkage. He further claimed that his synthetic polymer of uridylic acid possessed the same coding ability as polyuridylic acid prepared by phosphorylase. According to Fox and Dose (1972), many investigators were not able to confirm his results. By ultraviolet irradiation, Oro, Kimball, and McReynolds (1969) polymerized nucleotides in an aqueous solution. They tentatively considered that the products resembled the naturally occurring polynucleotide, except, as Fox and Dose pointed out, they did not have the unusual cross-linkages present in the biological nucleic acid.

Under different conditions, Ibanez, Kimball, and Oro (1971) showed that deoxyribonucleotides, such as thymidine 5'-monophosphate, could be condensed to oligodeoxyribonucleotides in the presence of cyanamide. Oligomers up to four nucleotide units in length could be formed in an aqueous solution at a neutral pH. They also showed that pentanucleotides are produced in the presence of a clay, montmorillonite; thus a clay surface allows for elongation of a nucleotide chain. Oligonucleotides containing up to eight or nine nucleotide residues have been produced in recent work (Pongs and Ts'o 1971; Navyvary and Nagpal 1972).

Generally speaking, the above experimental demonstration of the production of different nucleotide polymers, although not identical with the biological polymers, is still meaningful in an evolutionary context. Since evolution implies change in time and space, one may not expect the polymers initiated in the stage of prebiotic evolution to be identical with the modern type. Some of their functional and structural similarities to the biogenetic nucleotides are of sufficient value for speculation that at least some processes of nucleotide polymerization could have occurred on the primeval earth. In regard to the phosphate ester linkage, it is possible that various linkages for the polynucleotides could have been present before the first organisms. If so, why was the particular 3'- and 5'-linkage chosen? Prebiotic selection will be discussed briefly in connection with optical asymmetry.

Development of a Multimolecular System

At the turn of the century, it was observed by many investigators that hydrophilic colloid solutions can separate into two layers, one rich in colloidal substances and another almost free from them. Bungenberg de Jong called this phenomenon "coacervation" and the colloid-rich layer was referred to as a "coacervate." In many cases, the coacervate did not occur as a continuous layer but appeared in the form of very small droplets floating in the liquid. In protein coacervates, de Jong observed that the diameters of the droplets ranged from 2μ to 670μ. He studied the formation of coacervates with solutions of gelatin and gum arabic.

Coacervates can be produced by mixing proteins or nucleic acids with various polyoses, such as amylophosphophoric acid or araban, and lecithin, or other lipids, in addition to other substances. De Jong believed that electrostatic and hydration forces play an important role in the formation of this system.

Based on his observation of the formation of droplets with components of gelatin, gum arabic, and RNA (fig. 1.9), Oparin (1957) pointed out that when

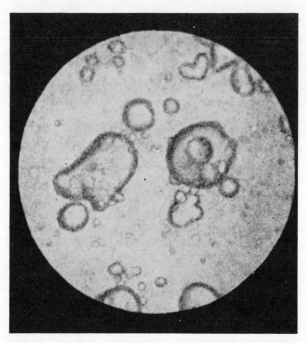

Fig. 1.9. Oparin's coacervates ($\times$ 300). (Reprinted, by permission, from Oparin, 1957.)

a droplet is composed of proteins or other substances in a medium, a continuous migration of protein, lipids, or polysaccharides to the surface takes place. Thus viscosity and electricity are built up in the droplet and its resistance to mechanical deformation is increased. The stability of the droplet depends on the surface from which it separates: the higher the surface tension, the greater the stability. In the primeval ocean, various organic compounds of high molecular weight, in the form of polymers of carbohydrates, amino acids, or nucleotides, may not have been different from those present today. These polymers have a marked tendency to form intermolecular aggregates without requiring special conditions, and these aggregates remain to interact with their surrounding medium, leading to further development.

Oparin pointed out that the phenomenon of coacervation is a powerful means of concentrating compounds of high molecular weight, such as polypeptides and polynucleotides, and it is an important one in the process of evolution of organic substances. Thus, he hypothesized, organic molecules were distributed throughout the homogeneous solution of the hydrosphere. Such a solution in the primitive ocean is referred to as the "primeval broth." The molecules in it united at particular points to form clumps or coacervates. They separated from the medium because of their different chemical properties. This system led to the development of an organized metabolism by the interaction of coacervates with the external medium, as well as by the gradual internal organization, which was a process of prebiotic evolution. Eventually the simplest living body could thus be formed.

Oparin's theory of coacervate formation with proteinoids or similar polymers has been tested. Heating proteinoids with water and allowing the hot solution to cool resulted in the formation of microspheres less than 2μ in diameter (fig. 1.10) that were as stable as true cells and that could be sedimented with a centrifuge without disruption. Further, they did not break down when allowed to stand. According to Fox (1967), the microspheres have a number of properties in common with cellular systems found in contemporary organisms, such as stability in standing and centrifugation, uniformity of size, stainability, size changes in hypertonic and hypotonic solutions, and simulation of cell division. Another interesting feature is the formation of budlike appendages and the induction sep-

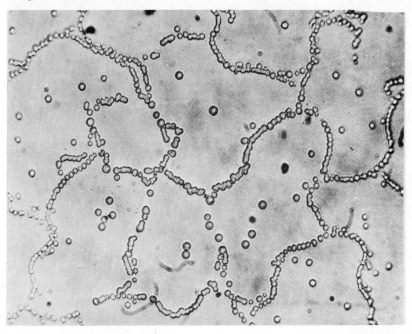

Fig. 1.10. Microspheres arise from proteinoid with associations resembling algae. Larger spheres are approximately 10μ in diameter. (Reprinted, by permission, from S. W. Fox and S. Yuyama, *Ann. N.Y. Acad. Sci.* 108(1963):487–94.)

tate fission (separation of two liquids by a dividing wall) by raising the pH of the suspension by one to two units (fig. 1.11). Analysis of the boundary (wall of the microspheres) material revealed only small differences in amino acid composition from the remaining polymer, and electron microscopic studies revealed that the boundaries were a double-layer structure. Experiments have shown that contemporary enzymes can catalyze reactions at higher rates when included within the boundaries than when outside. This is another major advance in the study of prebiotic synthesis.

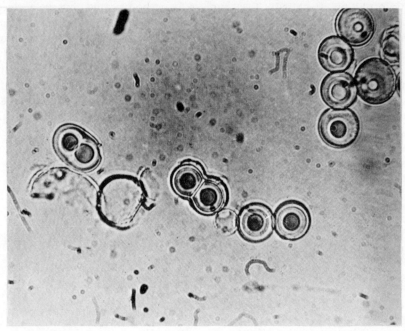

Fig. 1.11. Microspheres arise from proteinoid with double boundaries following increase in pH. Larger spheres are approximately 10μ in diameter. (Reprinted, by permission, from S. W. Fox and S. Yuyama, *Ann. N.Y. Acad. Sci.* 108(1963): 487–94.)

The tendency for the protein molecules to aggregate in order has been observed as a coincidence by Francis O. Schmitt (1951), when he studied the ultrastructure of collagens. He found that the separated collagen molecules, if allowed to precipitate from an aqueous solution, can form microfibrils of uniform diameter with cross-striations at regular intervals. Schmitt's findings stimulated George Wald (1954) to elaborate that:

> To form an organism, molecules must enter into intricate designs and
> connections. . . . For a time this problem of molecular arrangement
> seemed to present an almost insuperable obstacle in the way of imagining

spontaneous origin of life. . . . The change in view has come about because we now realize that it is not altogether necessary to bring order into this situation; a great deal of order is implicit in the molecules themselves. . . . We have therefore a genuine basis for the view that the molecules of our oceanic broth will not only come together spontaneously to form aggregates but in doing so will spontaneously achieve various types and degrees of order. . . . What it means is that, given the right molecules, one does not have to do everything for them, they do a great deal for themselves.

The microspheres of the thermally prepared proteinoids are another example.

The growing chain reaction of amino acids during thermal condensation is distinctly selective (Fox and Dose 1972), and the above example of the reconstitution of collagen molecules indicates that selective assembly occurs even among larger and more complex molecules. Such nonrandom reactions suggest that genetic information developed during chemical synthesis. It is on this basis, together with the ease with which proteinoids are formed as compared to that of polynucleotides, that Fox and Dose (1972) strongly support Oparin's view (1957) that the first organisms must have been developed primarily from proteinoids. Arguments against this assumption question the accuracy of protein replication without a code relationship with the nucleotides having been established.

EVOLUTION OF GENETIC APPARATUS

All contemporary organisms, from microbes to mammals, consist of two basic molecules: nucleic acids and proteins. The nucleic acids perpetuate the code for achieving genetic continuity, and the proteins, as catalysts, make it possible for the system to use the chemicals in their environments to carry on the genetic information. We have discussed the evolution of proteins and polynucleotides in the early development of life. How did this cooperative and well-organized system of nucleic acids and proteins, which may be called the genetic apparatus, evolve? Up to the present, there has been no good evidence available to provide an answer to this question. However, some hypotheses and some of the problems involved are worth discussing.

Laboratory evidence has shown that polypeptides, as well as polynucleotides, can be formed under primitive earth conditions. The polypeptides form spontaneously under less restrictive conditions than the polynucleotides. Furthermore, the sugar components of the nucleic acids, ribose and deoxyribose, are much less stable in aqueous solution than are amino acids. For these reasons, together with the demonstration of laboratory production of proteinoids, it seems that the prebiotic polymerization of amino acids may have preceded the appearance of nucleic acids.

The primeval soup contained a sufficient variety of amino acids so that, with electrical discharges or light as a source of energy, condensation of polypeptides could easily proceed spontaneously. In the process of condensation, the property of each amino acid would have had some influence on the choice of a specific type of the next to be connected in the polypeptide sequence (Harada and Fox 1960). There is a possibility, therefore, that there exists enough order in polymerization so that the amino acid sequence in a polypeptide is not completely a random event. It is also possible that some components could have possessed catalytic activity (Fox and Krampitz 1964) and that the pairing of certain amino acids was favored. A fairly high level of biological organization in such a system could have emerged through evolution. According to Polanyi (1968), the nucleotide base sequence can function as a code "only if its order is not due to the forces of potential energy. It must be as physically indeterminate as the sequence of words is on a printed page. . . . It is this physical indeterminacy of the sequence [that] thereby enables it to have a meaning—a meaning that has a mathematically determinate information content equal to the numerical improbability of the arrangement." But how far could a system of proteins have developed toward life without the presence of nucleic acids?

The difficulty would be encountered in the process of replication, which is a well-defined and fairly long polypeptide sequence requiring new enzymes. Each enzyme in itself would represent a well-defined polypeptide chain and would, in turn, require a whole further series of enzymes. As pointed out by Orgel (1968), it is highly questionable that a self-replicating unit lacking such elements could exist and be able to evolve. The possibility of unit-by-unit replication of polypeptides, or complementary replication (where positively charged and negatively charged amino acids attract each other, and neutral amino acids go together) has been investigated and considered unfeasible (Orgel 1968); and even if this were possible, in one way or the other, there has as yet been no lead as to how the transition from self-replicating proteins to a nucleic acid-protein system could have been made.

Orgel (1968) speculated that accurate replication of polynucleotides without enzymes is apparently possible because of the structural characteristics inherent in the bases themselves. A polynucleotide chain, once formed, can act as a template to orient mononucleotides, and the base-pairing specificity resides in the bases themselves. Thus the complementary template-directed synthesis would be sufficiently accurate without polymerizing enzymes. Supposing that in the primeval soup the components of mononucleotides for building polymers were freely available, then natural selection would favor the accumulation of those sequences best able to replicate. If a given ancestral polynucleotide had been formed, from which a number of descendants arose, some sequences might replicate more rapidly than others under different environmental conditions. Due to possible errors of replication, different families of molecules with slight variations in the base sequence could be produced, and certain sequences, over wide geographical regions, could become more dominant than others. Thus, a fairly accurate nonenzymic template synthesis could be well established, although a system so developed would not function in the environment except by depleting the supply of available monomers. Orgel (1968) described such a system as natural selection without function, in contrast to the system found in contemporary organisms, in which the polynucleotides affect the chemicals in their environment toward the development of metabolism: natural selection with function. But how would transition from a biological system without function to one with function occur? As yet no experimental evidence has demonstrated any feasible process.

It is unlikely that the genetic code was suddenly developed in a completed stage; rather, it is believed that there may have been a stereochemical correlation between certain polynucleotide sequences and certain available amino acids (or groups of amino acids). Therefore the whole system could subsequently perfect itself in a long evolutionary process. In the early stages, some polynucleotides may have been able to catalyze chemical reactions, although such activity would subsequently be taken over by the more versatile polypeptides. It has been reported (Fox and Krampitz 1964) that thermally polymerizing nucleotides, as well as RNA and DNA, bind with polyamino acids when the polynucleotides contain a sufficient proportion of basic units. Crick (1968) speculated that the evolution of nucleic acid began at a stage when few amino acids were coded, either because the mechanism was imprecise or because only a few triplets were used, and that later there was more precise recognition and the introduction of amino acids.

The successful replication of nucleic acid molecules outside a living cell by Mills, Kramer, and Spiegelman (1973) provides us with an opportunity to explore Darwinian selection in prebiotic time. The experiment began with the isolation of a specific enzyme called RNA replicase from *Escherichia coli* infected by an RNA bacteriophage. With RNA as the instructive agent, the enzyme preparation could mediate a virtually autocatalytic synthesis of biologically competent and infectious RNA. (Normally, bacteriophage or virus cannot replicate outside a living cell.) From mutants produced through numerous transfers of RNA molecules replicating under restricted conditions, Kacian et al. (1972) were able to isolate one RNA molecule that contained 218 nucleotides. Its primary, secondary, and tertiary structures[1] were determined through very elaborate procedures (Mills, Kramer, and Spiegelman 1973). The two complementary RNA strands are shown in figure 1.12.

It should be noticed immediately that there are "hairpin loops" stemming from each strand, formed by pairing complementary regions in the same strand. Intra-strand complementation increases the morphological complexity of the RNA molecule, and generates the possibility of forming unique secondary and tertiary structures. It should be noticed that the helix portion of each hairpin in the plus strand (the original strand from which the replication started) has an identical counter-

1. In the RNA molecule, primary structure refers to the nucleotide sequence regardless of its spatial arrangement; secondary structure refers to the local spatial arrangement of a segment of the nucleotide sequence; and tertiary structure refers to the arrangement of atoms of the RNA molecule in space involving interaction between segments of nucleotides.

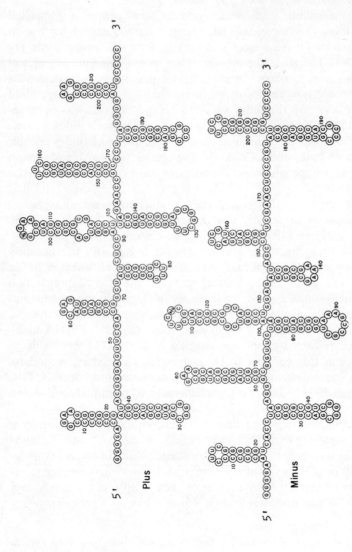

Fig. 1.12. The complete sequence of the complementary strands of an RNA molecule derived from *E. coli* bacteriophage through a large number of in vitro transfers under specified conditions. This RNA molecule consists of 218 nucleotides, much reduced from the original bacteriophage RNA. Note that, in addition to the complementation between the two strands of nucleotides from the 5' end of one strand with those from the 3' end of the other strand, there is also intrastrand complementation. This results in the formation of the hairpin structures and the complex morphology of the RNA, and is assumed to confer fitness; it is also of significance in prebiotic evolution. (Reprinted, by permission, from D. R. Mills, F. R. Kramer, and S. Spiegelman, *Science* 180(1973):916–27, © 1973 by the American Association for the Advancement of Science.)

part in the complementary minus strand. For example, the helix of nucleotides 148 to 170 of the plus strand is identical to the helix of nucleotides 49 to 71 of the minus strand. In this sense, one of the two strands forms a mirror image to the other.

From these observations, Mills, Kramer, and Spiegelman (1973) argued that if a given variant strand is superior because of a particular sequence arrangement, optimization of the selective advantage would encourage the same sequence in the minus strand. This can be assured only if the plus strand contains the intrastrand complementary sequence. As a consequence of such complementation, new dimensions are added to the molecular structure. Such molecular morphology is referred to as the phenotype of the RNA itself. If nucleic acid molecules were formed in precellular stages of organic evolution, natural selection would act directly on such phenotypes. However, once cells evolve, natural selection does not operate on the nucleotide base sequence directly but through their translated proteins (gene products). Under these circumstances, the secondary and tertiary structures of the genetic materials become irrelevant, as in the present living organisms.

EVOLUTION OF OPTICAL ASYMMETRY

Chemical evolution may be viewed as a continuous interaction between molecules that have the properties of self-reaction, self-ordering, self-assembly, and self-replication. In the course of chemical evolution, prevailing environments favored the survival of certain molecules and certain reactions by others, because of particular chemical structures and compositions. This may be called prebiotic selection. The phenomenon of optical asymmetry in the carbon compounds, as observed in present living systems, is believed to be rooted far back in prebiotic evolution and has aroused much speculation.

Let us review some of the physical and chemical properties of the optical isomers before discussing evolutionary aspects. The atoms in molecules are arranged not on a plane, but in space. Increases in the complexity of the molecules leads to the formation of asymmetrical configurations. The most common and most thoroughly studied case of optical asymmetry is the carbon atom when it is united through its four valences to other atoms. Two different molecules contain exactly the same groups but are arranged differently in space; for example, a particular group is on the right in one form and on the left in another. Organic compounds containing asymmetric carbon atoms have the capacity of rotating the plane of polarized light, and are said to be optically active. If it rotates the polarized light clockwise, it is called dextro (D); otherwise, levo (L). A mixture of both, resulting in no rotation, is said to be racemic.

When organic compounds, such as the amino acids, are synthesized under ordinary laboratory conditions, a mixture of both forms is always obtained. The proteins of all living organisms are built, however, with L-amino acids. This is a distinguishing characteristic of such systems, and a sure sign of life. How did the organisms make the choice and why?

There has been much discussion and many assumptions regarding the causes. One assumption is that natural light is circularly polarized to a slight degree, causing selectively stereospecific synthesis or selectively stereospecific destruction of one member of a pair of racemates. Another assumption is that the electrons from radioactive beta-decay are predominantly left-handed. These polarized radiations may cause stereospecific destruction of isomers. There are other theoretical explanations, none of which have obtained sufficient experimental support (see Fox and Dose 1972 for a more detailed discussion).

Relevant experiments in stereospecific synthesis of polyamino acids showed that D- or L-polymers generally propagate much faster and become much longer and more stable than DL-polymers (Blout, Doty, and Yang 1957; Blout and Idelson 1956). The rate of polymerization of the mixture was about one-twentieth that of the L or the D form. Furthermore, the DL polymers were found to be less stable than polymers of either one form. In the case of nucleic acids, asymmetry is due to the sugar residue, or D-pentose, which leads to the formation of a right-handed helix. Two such helixes can be united so that a given purine in one always forms a pair of hydrogen bonds with a given pyrimidine in the other. No double helix of any substantial length could be formed if both D and L forms of pentose were used. We can see now why organisms chose one specific configuration rather than the mixtures.

It is not at all clear, however, why the L-amino acids in proteins and the D-pentose in nucleic acids are used instead of the opposite configuration in each case. Wald (1957) once discussed this matter with Albert Einstein, who said: "You know, I used to wonder how it comes about that the electron is negative. Negative-positive—these are perfectly symmetric in physics. There is no reason whatever to prefer one to the other. Then why is the electron negative? I thought about this a long time, and at last all I could think was 'It won in the fight!' " Wald then said, "That's just what I think of the L-amino acids. They won in the fight."

Unless there are physical and chemical reasons for the presence of either amount in primeval earth, the conferring of any specific evolutionary advantage, even though small, could lead to the eventual victory of one form over the other in time. Such a possibility would be comparable with one of two allelic gene forms conferring a selective advantage to 1% or even as small as 0.1% over the other; after a large number of generations, it can become fixed in the population. If, on the other hand, there are no physical and chemical reasons for us to believe that there would be a selective advantage of one form over the other, then the only explanation for the occurrence is chance. Such an explanation would correspond to the theory of random genetic drift in biotic populations, as will be discussed in chapter 7. It will be seen that random drift is affected by sample size, applying to the prevalence of certain specific configurations of molecules in prebiotic synthesis. But the effect of a specific chemical configuration on further synthesis appears to be distinct and definite. When the polymerization of molecules with one type of configuration is once started, it becomes "a one-way street."

Catalysts of a certain configuration will be used for all time to come. Gaffron (1960) drew an analogy to the effect that if the mechanics depend on asymmetric parts, the best way is to use parts of one type only. In biotic populations, after one gene allele is fixed, it may affect the selection of other genes in other loci. The joining effect of all genes for a phenotypic expression is generally assumed, but such effect is not so distinct and clear as that of specific molecular configuration.

In conclusion, the basis for the establishment of the L-amino acids in protein synthesis and the D-pentose in the nucleic acid in the present living systems is not clear. However, the use of one specific configuration rather than a racemic mixture definitely confers a greater advantage for the production of more and highly specific genetic information. As a distinctive characteristic, the presence through all living organisms of the same configuration, for all types of biological substances, suggests that the choice was made very early, perhaps before the first living organisms developed. This conclusion, in turn, favors the monophyletic theory of evolution.

SUMMARY

Since the turn of the century, scientific thinking about the origin of life has crystallized into the idea that life must have arisen from lifeless matter. Experimental approaches to the question have been accelerated in the past twenty years or so. From geological and laboratory evidence, prebiotic evolution may be viewed as going through the different stages of chemical syntheses, from simple cosmic molecules to complex molecules to macromolecules. By self-assembly of the macromolecules, some organic structures possessing enzymatic and hormonal activities and metabolism could have formed. Because of the nonrandomness of the order of the polyamino acids, it is believed that some genetic information, although crude, should be present in them.

It was questioned, however, whether these forms could replicate precisely without the polynucleotides as the code. Thus a genetic apparatus is proposed, and in it the code relationship has to be gradually evolved for precise replication. This seems to be the last step toward the end of the long journey of prebiotic evolution.

the evolution of genes

In the early studies of Mendelian inheritance, the term *gene* was given to the unit affecting inherited characteristics. Although there can be no precise definition, *gene* refers to a biological substance in the germ cells that specifically affects the organism that carries it. Today, the gene is defined operationally as a specific sequence of nucleotides in the DNA molecule (or the RNA molecule in viruses) that determines the sequence of amino acids in a certain polypeptide (or the sequence of nucleotides in a certain RNA molecule) and also regulates the transcription of other genes.

Let us suppose that the first organisms developed through a long period of chemical evolution that took place more than three billion years ago. The code relationship between polynucleotide and polyamino acids having been established, this organism would have had only one gene, consisting of one thousand nucleotide bases, and would have performed basically the same function as genes do in present living systems. It is estimated that the genome of a man consists of 4×10^9 nucleotide bases. In the course of three billion years, there has been roughly a 4×6 order of magnitude increase in DNA or genetic materials. Through what process was this large amount of DNA produced? We see size, shape, and physiological differences among living things now on earth. Through what process must the large number of nucleotides have been arranged to bring about the vast variations that resulted in, for example, a man or a mouse? Since genetic information resides in DNA, the whole story of the evolution of genes is essentially that of the growth and alteration of nucleotide sequences.

We speak of the evolution of genes rather than of the evolution of DNA for the following reasons: (1) It would be most difficult to examine nucleotide sequence changes directly. Evidence of such changes can be seen in the biological effects they produce. (2) It is these very biological effects that determine the survival or fitness of an organism and through which selection operates. The characteristics that constitute the biological uniqueness of a species or of an individual originate in the hereditary unit, the gene.

GENE CONCEPT

Our concept of the nature of genes has undergone very drastic revision in the last fifty years. The term *gene* was proposed by Johannsen in 1907 and gained wide currency following the *Drosophila melanogaster* work by T. H. Morgan. According to Morgan (1917), a gene is a certain amount of material in the chromosome that may separate from the chromosome in which it lies and be replaced by a corresponding part (and none other) of the homologous chromosome. To the pioneers of Mendelian genetics, the gene was an indivisible hereditary unit within which crossing-over or other forms of chromosome rearrangement could not occur.

The gene concept was later found inadequate when a curious phenomenon was observed in *Drosophila*, that is, the phenotype of a gene varies depending on the neighboring genes on the same chromosome. For example, in double heterozygotes when two very closely linked genes exist on one chromosome and their normal alleles exist on the other, a normal phenotype is produced. But when one of the mutant genes is present on one chromosome and the other mutant gene on the other homologous chromosome, a mutant phenotype is produced. This phenomenon is known as "position effect" (Sturtevant 1925). It contradicted the then-existing gene concept representing a gene as a chromosome segment functioning as a unit but physically divisible (for example, crossing-over occurs within a gene). Benzer (1960) called such a genetic unit a cistron, and described it as a genetic map segment having a unitary function.

Evidence that genetic information is carried in the DNA molecule first came from work in bacteria and virus. Avery, MacLeod, and McCarty (1944) showed that genetic traits can be transferred from one strain of pneumococcus to another by pure DNA. Later, Hershey and Chase (1952) demonstrated that DNA is the primary chemical material of the virus particle. In tobacco mosaic virus, it was RNA that carried the genetic information. Of course, the clearest demonstration of the chemical structure of genetic material is the well-known DNA double-helix model of Watson and Crick (1953). Further evidence equating DNA and the gene came after the discovery of the double-helix model, from proof of *colinearity*: mutational alterations of single bases in DNA of bacterial virus resulting in colinear changes of amino acids in proteins (Yanofsky et al. 1963).

Chemically a gene is part of a molecule of DNA, made up of two strands, each of which is composed of combinations of four basic nucleotides: two purines—adenine and guanine—and two pyrimidines—thymine and cytosine, usually represented by A, G, T, and C, respectively (fig. 2.1). Linked to a sugar and a phosphoric acid molecule (fig. 2.2), the two strands are wound in a double helix in which the phosphate-sugar groups form the backbone, the purine (A, G) and the pyrimidine (C, T) bases facing inward with their rings at right angles to the axis of the helix. The diameter of the helix is about 20 Å; a full turn is 34 Å, which accommodates 10 nucleotide units. A polynucleotide, such as the DNA molecule of a bacterial virus, consists of 300,000 nucleotide units with a molecular weight of 10^8 (Pauling and Corey 1956). The neatness of fit is provided by the exact complementary configuration of the purines and pyrimidines, so that the As of one strand are always opposite the Ts of the other; the same is true for the Gs and Cs—sharing hydrogen atoms (hydrogen bonds) to stabilize the configuration. These four base units are arranged in various combinations to code genetic information (fig. 2.3). Such a chemically unique structure accounts for the four essential properties of genes: specificity, replication, mutation, and biological function.

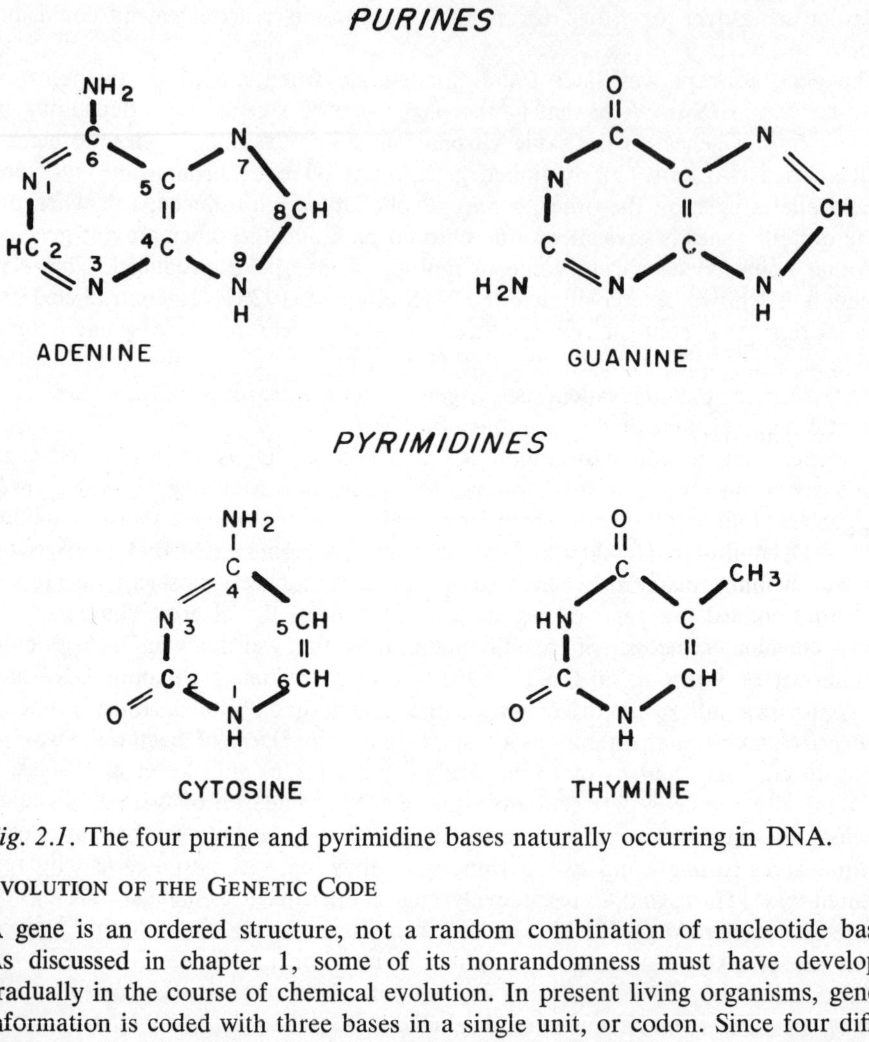

PURINES

ADENINE   GUANINE

PYRIMIDINES

CYTOSINE   THYMINE

Fig. 2.1. The four purine and pyrimidine bases naturally occurring in DNA.

EVOLUTION OF THE GENETIC CODE

A gene is an ordered structure, not a random combination of nucleotide bases. As discussed in chapter 1, some of its nonrandomness must have developed gradually in the course of chemical evolution. In present living organisms, genetic information is coded with three bases in a single unit, or codon. Since four different bases of two purines and two pyrimidines are found in the nucleic acids of contemporary organisms, sixty-four triplets are possible. Except for triplets UAA and UAG, and probably UGA, which are terminator codons giving signals for the termination of the polypeptide translation, each of the remaining codons codes one amino acid, as shown in table 2.1. These conclusions were based on the results of studies on *E. coli*, but it is assumed that they are applicable to all organisms (Stent 1971, p. 555).

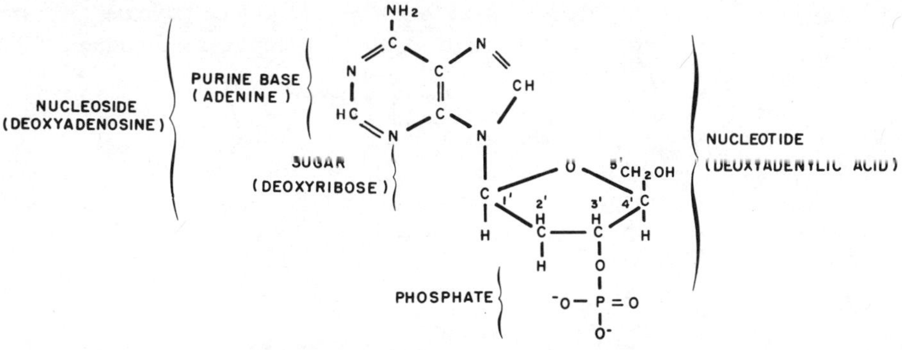

Fig. 2.2. Organization of phosphate, sugar, and purine or pyrimidine in the DNA nucleotide. A purine or pyrimidine linked with the deoxyribose forms a nucleoside (thymidine, deoxyadenosine, deoxyguanosine, and deoxycytidine). A nucleoside linked with a phosphate forms a phosphoric acid ester called nucleotide (thymidylic acid, deoxyadenylic acid, deoxyguanylic acid, and deoxycytidylic acid or thymidine (3' or 5') monophosphate, deoxyadenosine (3' or 5') monophosphate, and so on). Adenine is used here as the nucleotide base.

The following questions arise: Why does the codon operate on a triplet basis? Why does the code happen to be universal? How did the relationship between the codon and amino acids develop? This last question bears on the hypothesis of separate evolution for polynucleotides and polyamino acids and the development of a coding relationship.

One concept accounting for universality as well as specificity in coding is the stereochemical theory of Woese (1967). He suggested that the amino acids, due mainly to their side chain differences, are in some way stereochemically related to their respective codons. Crick (1968) proposed the "frozen accident" theory, which states: "that code is universal because at the present time any change would be lethal or at least very strongly selected against." Because the code determines the proteins, most of which are highly evolved, any change of the code would cause changes in many proteins, which an organism would be unable to tolerate or to survive. Thus, Crick pointed out, to account for its being the same in all organisms one needs to accept the monophyletic theory of evolution, that is, all life evolved from a single organism.

There have been suggestions that the primitive code was not triplet in nature, that the bases were read one at a time, evolving to two and then to the present triplet system. Crick (1968) denied such a possibility as a violation of the principle of continuity, that is, a change of codon size would make all previous genetic information nonsense, and furthermore would be lethal to the organism.

Crick believes instead that the primitive code was a triplet code, but with only the first two bases being read at the time when the code began functioning. This differs considerably from a two-base code. This supposition seems to be supported by the characteristics of the codons. For example, in eight out of sixteen cases

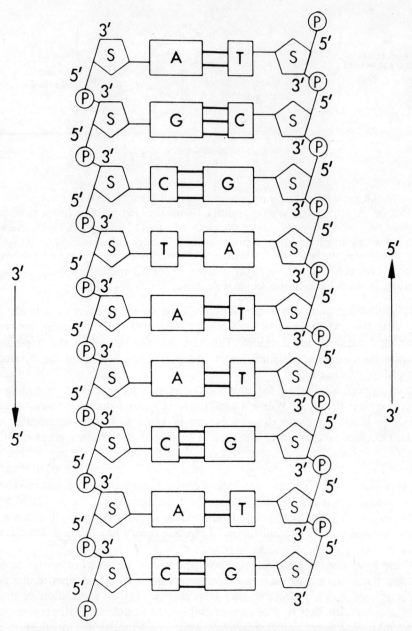

Fig. 2.3. The DNA double helix. The two strands are not identical but complementary. They are held together through hydrogen bonding of (complementary)
bases, A-T, and G-C. The 3′:5′-phosphodiester linkages are in reverse order in the
two chains. (Reprinted, by permission, from DeBusk. © 1968 by A. Gib DeBusk.)

TABLE 2.1 The Genetic Code

5' Terminal	Middle Nucleotide				3' Terminal
	U	C	A	G	
U	Phe	Ser	Tyr	Cys	U
	Phe	Ser	Tyr	Cys	C
	Leu	Ser	N2 (Ocher)	N3	A
	Leu	Ser	N1 (Amber)	Trp	G
C	Leu	Pro	His	Arg	U
	Leu	Pro	His	Arg	C
	Leu	Pro	Gln	Arg	A
	Leu	Pro	Gln	Arg	G
A	Ile	Thr	Asn	Ser	U
	Ile	Thr	Asn	Ser	C
	Ile	Thr	Lys	Arg	A
	Met	Thr	Lys	Arg	G
G	Val	Ala	Asp	Gly	U
	Val	Ala	Asp	Gly	C
	Val	Ala	Glu	Gly	A
	Val	Ala	Glu	Gly	G

SOURCE: Reprinted, by permission, from A. Garen, *Science* 160 (1968): 149-59. © 1968 by the American Association for the Advancement of Science.

NOTE: Each amino acid listed is coded for by an RNA triplet (codon). N1 (amber), N2 (ocher), and N3 designate the three nonsense triplets, UAG, UAA, and UGA. RNA contains uracil (U) instead of thymine (T) which is contained in DNA. Therefore, U is used in the table.

the base in the first two positions determines the specific amino acids, and the third position contributes no effect on the selection of the amino acids. It is also true that in codons with U and C at the third position, the bases at the first and second positions determine the amino acids. For all codons with G and C in the first two positions, the codons with the same initial doublets code the same amino acids; the same is true for codons with C in the middle.

Both Crick and Woese tended to favor the theory that polypeptide synthesis, as guided by the genetic code, was rather imprecise when the relationship was first established. The primitive system might have been coded according to classes of amino acids based on their general differences and similarities in structure and other chemical properties. For instance, all codons with U in the second position stand for hydrophobic amino acids; those with A in the middle, all the acidic residues; and those with U, the basic residues. Crick further speculated that the primitive code involved only a few amino acids, such as glycine, alanine, serine, and aspartic acid. In view of the rarity of the codons for methionine and tryptophan, he believed them to be later additions. He also suggested the possibility that only two of the bases (adenine and inosine) existed at first, instead of the four found in contemporary organisms. Inosine could have arisen from adenine merely by deamination, with the possibility that both were available on primitive earth.

GENE DUPLICATION

Gene duplication is defined in a chromosome as the occurrence of an additional and identical chromosome segment, which originally consists of part of a gene, a whole gene, or a number of genes. Chemically it is a repetition of the corresponding nucleotide base sequence. Such a description of gene duplication would put one nucleotide base and the complete nucleotide sequence of a chromosome as lower and upper limits. A duplication of the whole chromosome is called polyploidization, a matter that will be treated as chromosomal mutation in chapter 4.

Gene duplicating is one of the major processes for increasing genome size and thus genetic information. Its far-reaching evolutionary consequences concern adaptation and speciation; and although its evolutionary impact may not be fully appreciated at this point, a number of examples will show gene duplication to be a common mutational process, affecting the various biological systems of different species, and taking place from the past to the present time.

As a mutational process, the importance of gene duplication was not fully appreciated until Ingram (1963) showed that genes for hemoglobin synthesis were originated by duplication. This work led investigators to examine and reexamine many genetic systems whose origins had been both complex and puzzling. In gene evolution, the present reflects its past. It is in this context that the impact of gene duplication in evolution should be recognized. L. C. Dunn (1956) once remarked, during his work on tail mutants in mice, that when our genetic techniques become more refined we may find that practically every genetic locus is a complex one, that is, one with more than two alleles, and some, if not all, with repeated nucleotide base sequences, originated by a gene duplication. If so, it may be possible that with further advances in both genetic and biochemical techniques we may be able to trace for most genes the ancestral and homologous genes that resulted from duplication.

Genes Affecting Gross Morphology

Perhaps Tice (1914) was the first to become aware of gene duplication when he observed reduced-eye mutants in *Drosophila*, called bar-eye (B). It was a phenomenon later recognized by Sturtevant and Morgan (1923) as a novel type of mutation resulting from unequal crossing-over. The two genes present in the parental chromosomes emerged from meiosis in a single chromosome (bar-double), while the other chromosome resulting from the crossing-over lacked any (bar-reverted). Two bar genes in the same chromosome ($BB/B+$) caused greater reduction in the size of the eye than two bar genes in separate ones (B/B). (The intensification of action is called position effect.) In the chromosomal preparation of the mutant flies, the duplication in the BB chromosome was later confirmed. Bridges (1936) found that an extra, short section of chromosome band was present in excess of the normal complement, the insertion point of this duplication being in the bulbous "turnip" segment (fig. 2.4).

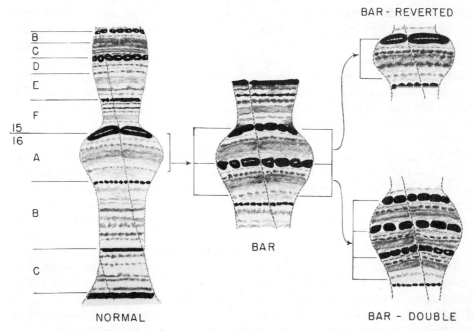

Fig. 2.4. Portions of X chromosomes from salivary glands of *D. melanogaster* showing the bar region. Note the segment that is duplicated in bar and triplicated in bar-double. (Reprinted, by permission, from C. B. Bridges, *Science* 83(1936): 210–11.)

This classical example of duplication shows the dosage effects of the duplicated bar-eye gene that had not yet undergone the further mutation, observed in most natural populations, to be cited later. It also illustrates the fact that since the duplication is of a chromosome band, it can be assumed that other genes next to the *B* gene must also have duplicated. Their dosage effects, short of gross morphological changes or those reaching beyond the normal physiological range for pathology, have not been detected. That a gene duplication without further mutation often escaped detection is perhaps the general case, rather than the exception. We should also recognize that such duplication causes an increase of at least many thousands of nucleotides.

In plants, several mutant series reflect functional complexities and suggest an origin of gene duplication. For example, the *A* (anthocyanin) locus in corn is known to affect the relative proportion of brown and purple pigments in different parts of the plant. Although the *A* locus was formerly considered to be a single locus involving a series of multiple alleles, Laughnan (1949), using genetic markers, was able to demonstrate that the origin of certain mutants was associated with crossing-over within the *A* locus, with at least two adjacent loci, or subloci, involved. Because both regions apparently controlled pigment formation, Laughnan believed that the locus represented repeated loci, derived by duplication from a common ancestral locus, and since modified. The *R* series of genes in corn, also affecting the color of both seed and plants, and similar to the *A* series in high mutability, could also have resulted from gene duplication. The foregoing examples illustrate (1) gene duplication possibly followed by further mutation or (2) partial duplication. From the point of view of variation in gross morphological characters, however, these different possibilities cannot be separated.

Genes Affecting Amino Acid Synthesis

In microorganisms such as bacteria, it was reported by Demerec and Hartman (1959) that genes involved in synthesis of the same amino acids are often closely linked. In a histidine cluster in *Salmonella typhimurium,* nine genes were implicated in histidine synthesis, originating with ATP and phosphoribosylpyrophosphate (Ames and Hartman 1963), as shown in figure 2.5. Some of these genes are located in the same sequence as their enzymes, participating in the synthesis. Their close linkage was discovered by means of transductional analysis.[1]

The natural question that arises is: How do we know that these genes were derived through gene duplication? First, the chance of forming such a gene cluster by mutation and translocation of individual genes from other chromosomal regions is very small. Second, since they react on similar substrates it is reasonable to assume that the enzymes must have rather similar chemical structures. This implies that genes that code for the enzymes are partially identical in their nucleotide base sequences. The rationale underlying this line of argument is that protein hormones with similar functions have identical amino acid sequences of large sections of their polypeptides (see chap. 5). Therefore, the best explanation for the formation of closely linked histidine genes with similar functions would involve gene duplication followed by mutation involving nucleotide base substitution. Horowitz (1965) later reached the same conclusion. On the basis of further experiments, an identical conclusion was reached regarding a small cluster of genes involved in the synthesis of leucine (Margolin 1963) and tryptophan in *E. coli* (Somerville and Yanofsky 1964).

In bacteriophages, Epstein et al. (1963) mapped about one-half of the genome of bacteriophage T4D, where gene order was found to correlate with genetic function. In *Neurospora* and yeast, however, the number of functionally related genes

1. Transduction is the transport of genetic materials by bacteriophage or virus from one bacterium to another (see chap. 3). Analysis is based on the fact that the bacteriophage can carry only a limited amount of genetic material from a donor to a recipient bacterium. It is assumed that two genes can be transported together from the donor only when they are closely linked. Thus, simultaneous transduction of two genetic markers is used as a measure of close linkage: the greater the frequency of simultaneous transduction, the closer the linkage of the two marker genes. From these studies, a cluster of sites affecting a similar biochemical system was established in *S. typhimurium;* these sites were then equated with genetic loci and hence called genes.

was less than in bacteria and bacteriophages. Since many chromosomes are present in these organisms, it was speculated that genes which originated by duplication may have been incorporated into different chromosomes and are hence difficult to identify (Horowitz 1965).

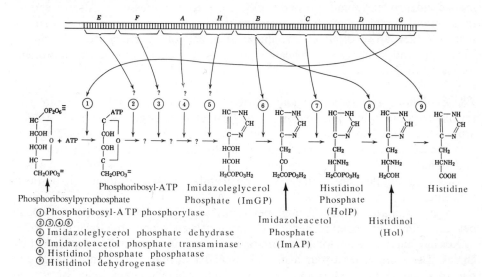

Fig. 2.5. Relationship between the genetic organization of the histidine controlling region in *S. typhimurium* and the sequence of reactions controlled by the histidine loci, *E, F, A*, and so on. Each is responsible for the production of a specific enzyme as shown. (Reprinted, by permission, from Wagner and Mitchell, 1955.)

If we accept the above interpretation, this would be chemical evidence of gene duplication. Each of the duplicated genes has gone through some mutational changes, performing specific functions but affecting the same chemical system. They are present next to each other, as detected by the method of transduction, an ingenious method for an exclusive analysis of all genes in a short chromosome region.

Amino Acid Sequence Analyses

Since each amino acid of a protein is coded by a codon, the nucleotide base sequence of a gene can be constructed with reference to the amino acid sequence. Gene homology or relative identity in nucleotide base sequence can be so established. In contemporary organisms, amino acid sequence is determined by many proteins. The degree of identity between certain proteins is so great that it is beyond a doubt that the genes originated by duplication.

In general it may be expected that the longer the time lapse after duplication, the less the degree of identity between genes due to subsequent occurrence of mutations. Depending on the functions to be evolved, duplicated genes may retain a certain amount of identity for a long time, in order to produce proteins that can be mutually recognized for cooperative or similar functioning. However, if one of the two genes must gradually evolve to undertake a different function, both mutate independently and are under different selection pressures. After a certain time period, the degree of identity between them tends to become so small that the fact that they arose originally by duplication cannot be recognized. It will be shown presently that gene duplication involves a number of important proteins necessary to the vertebrates.

The Evolution of Hemoglobin and Myoglobin Genes. The study of amino acid sequences in the polypeptide chains of the hemoglobins, begun by Ingram (1963), a pioneer in this field, contributed significantly to genetics. Besides providing certain clinical applications, it revealed the evolution of globins through two basic mutational processes: gene duplication and nucleotide base substitution. We will discuss gene duplication at this point; nucleotide base substitution will be dealt with later.

An evolutionary tree of hemoglobin genes (fig. 2.6) based on the homology between human hemoglobin peptide chains has been proposed (Baglioni 1967).

Normal human hemoglobin is a tetramer containing two pairs of polypeptide chains. There are at least four hemoglobins (Hb) in man, each consisting of one pair of alpha peptide chains and one pair of either beta, gamma, delta, or epsilon chains:

Hb-A (adult hemoglobin)	α_2	β_2
Hb-F (fetal hemoglobin)	α_2	γ_2
Hb-A$_2$ (minor adult hemoglobin)	α_2	δ_2
Hb-E (embryonic hemoglobin)	α_2	ϵ_2

Hb-E is found only in embryos (Huehns et al. 1964) and Hb-F gradually disappears after birth. All other hemoglobin polypeptide chains are present in adult humans. The alpha chain consists of 141 amino acids; the beta chain and others, 146 each.

Genes coding the various hemoglobin polypeptide chains are believed to have been derived by repeated duplications from a primitive gene (Ingram 1963; Fitch 1966; Baglioni 1967). The peptide chain of the primitive globin was about half the length of the present hemoglobin chains. Based on a comparison of every thirty amino sequences between the alpha and beta chains, using a digital computer, Fitch concluded that the ancestral protein was lengthened one way or another by sixty-six amino acids. Subsequently the portion of the genes coding for the sequence of sixty-six amino acids was doubled. Later another gene duplication, followed by translocation, gave rise to the genes for the alpha and beta chains (Fitch 1966) (fig. 2.7). The presence of these globin chains in both mammals and reptiles suggests that all of these duplications occurred during or prior to the reptile stage in vertebrate evolution.

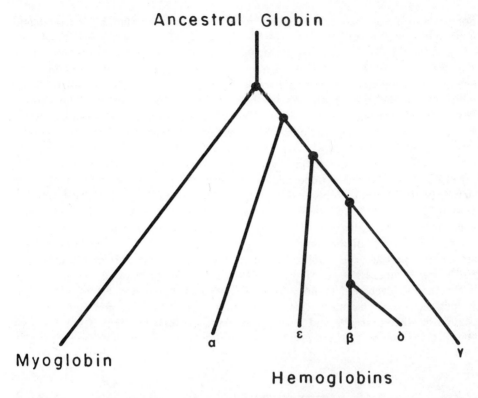

Fig. 2.6. Evolutionary tree of hemoglobin genes. Each branching point in the tree represents a hemoglobin gene duplication. (Reprinted, by permission, from Baglioni. © 1967 by The Johns Hopkins University Press.)

A protein closely related to hemoglobin is myoglobin, a ferrous protoporphyrin globin complex present in the muscle that acts as a store of oxygen. It has a tertiary structure similar to that of hemoglobin (Perutz, Kendrew, and Watson 1965), with two types of peptide chains present in almost all vertebrates. The monomeric myoglobin appeared similar to the alpha hemoglobin polypeptide chain (Braunitzer 1966), suggesting duplication from the alpha chain.

There is a high degree of homology between the beta and delta polypeptide chains, the genes for which are closely linked, suggesting that they are the result of a recent duplication.

Evidently, deletion and duplication of hemoglobin genes are continually taking place in the present human population. In one mutation described by Baglioni in 1962, called hemoglobin Lepore, two cases were reported, Lepore Boston and Lepore Hollandia, which differed in the length of the peptide chains deleted. The peptide chain, consisting partly of an NH_2-terminal delta chain sequence and partly of a COOH-terminal beta chain sequence (fig. 2.8), resulted from a deletion and fusion of the delta and beta genes caused by presumably unequal and nonhomologous crossing-over.

The Evolution of Haptoglobins. Haptoglobins are serum proteins binding free hemoglobins, and they consist of two polypeptide chains: alpha and beta. With minor exceptions, no variation in the beta chain has been found among people. However, as common types in human populations, three different alpha chains are found in human populations. These chains are determined by one genetic locus with three alleles: Hp^{1Fa}, Hp^{1Sa}, and Hp^{2a}. (The Hp^{1Fa} and Hp^{1Sa} are subtypes of HP^{1a}, with alpha chains exhibiting different rates of migration in starch gel electrophoresis. These three alleles are also called Hp^{1F}, Hp^{1S}, and Hp^2 for simplicity; Hp^1 without subtype designations refers to both types.) The genes coexist in a population and thus constitute polymorphisms. The gene frequencies vary among populations (fig. 2.9), indicating evolution in process.

From the results of the amino acid sequence analysis, it is believed that Hp^{2a} arose from partial gene duplication. The partial sequences of amino acids for Hp^{1Sa}, Hp^{1Fa}, and Hp^{2a} are shown in figure 2.10. The polypeptide chain of Hp^2 is shorter by seventeen amino acids than those of Hp^{1Sa} and Hp^{1Fa} combined. Among the seventeen amino acids, three belong to the COOH terminal of the polypeptide specified by Hp^{1Fa} and fourteen belong to the NH_2 terminal of the polypeptide specified by Hp^{1Sa}. These results have led Smithies, Connell, and Dixon (1962)

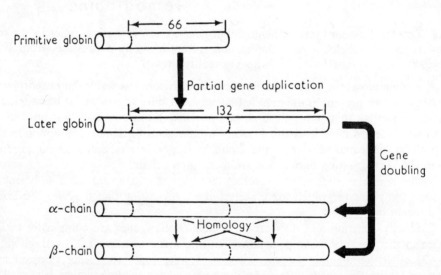

Fig. 2.7. A possible process of hemoglobin gene duplication. Numbers (66 and 132) represent amino acid residues. (Reprinted, by permission, from W. M. Fitch, *J. Mol. Biol.* 16(1966):9–16.)

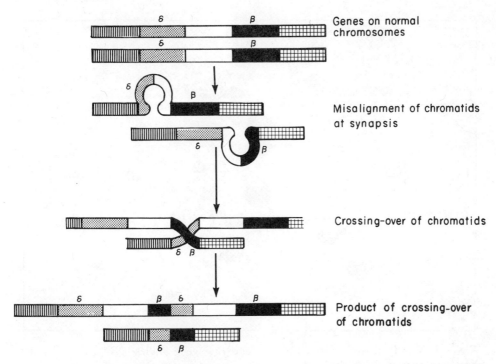

Fig. 2.8. Possible origin of Lepore gene by aberrant pairing of beta and delta hemoglobin genes. These genes are closely linked, as shown by a space between them. At synapsis, misalignment of the chromatids occurs. There is a large amount of homology in the nucleotide base sequence between the beta and delta genes. This is probably one reason for the aberrant pairing. The aberrant pairing leads to the unequal crossing-over, the production of partial duplication, and partial deletion of the beta and delta genes. The chromosome at the bottom is assumed to be the one carried by the Lepore individuals. The one directly above it may produce a normal condition or may be lethal.

to conclude that a rare and possibly unique genetic event occurred in the evolution of the haptoglobin gene, so that a partial gene duplication due to nonhomologous crossing-over should be considered. The polypeptide chain for the Hp^{2J}, triple the size of that for Hp^1, is assumed to have been produced by unequal crossing-over of Hp^2 (Smithies 1964).

The function of haptoglobins is to form a tight bond with free hemoglobin for the conservation of iron in the body. The binding capacity differs in a ratio of roughly $Hp^1/Hp^1 : Hp^2/Hp^1 : Hp^2/Hp^2 = 135:110:85$. The hemoglobin and haptoglobin complex appears to be a substrate for the action of alpha-methenyl oxygenase to degradate the heme ring to biliverdin. Homozygotes for the Hp^2 were found to be more active for this reaction than the other two genotypes, but there was no discernible effect on health.

Eskimos
Lapps
Athabascans
Europeans
Iranians
Chinese
Apaches
Punjabis
Japanese
Nigerians
Polynesians
Congolese
Papuans
Paraujano
Tamils
Bushmen
Peruvian indians
Aborigines
Haptoglobin Genes
Hp¹
Hp²

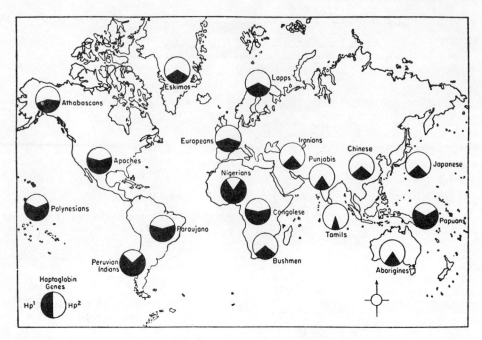

Fig. 2.9. World distribution of the Hp^1 and Hp^2 gene alleles. (Reprinted, by permission, from R. L. Kirk, 1968.)

The Evolution of Immunoglobulins. Immunoglobulins comprise a large and complex family of evolutionary-related proteins. The immunoglobulins within a single organism are extraordinarily heterogeneous. So far there has been no satisfactory theory to account for the heterogeneity. However, it is generally agreed that gene duplication is involved in the evolution of these proteins and that this evolution may have taken place before the evolution of the vertebrates.

Immunoglobulins comprise at least four classes of protein molecules: IgA, IgG, IgM, and IgD, each defined by the presence in the protein molecule of a specific heavy chain: alpha, gamma, mu, and delta, respectively. They are associated with

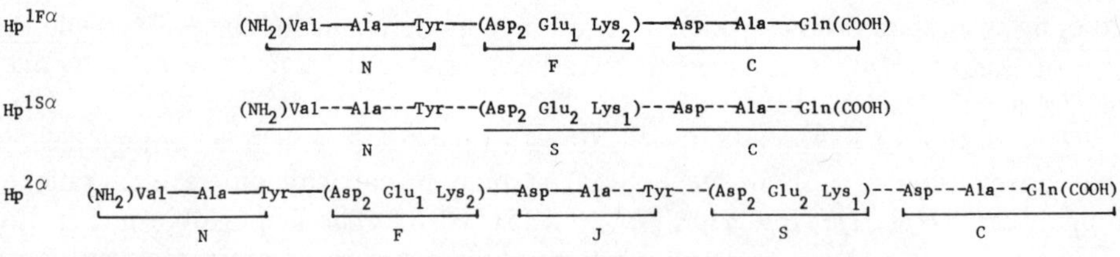

Hp1Fα
(NH2)Val—Ala—Tyr—(Asp2 Glu1 Lys2)—Asp—Ala—Gln(COOH)
N F C
Hp1Sα
(NH2)Val---Ala---Tyr---(Asp2 Glu2 Lys1)---Asp---Ala---Gln(COOH)
N S C
Hp2α
(NH2)Val—Ala—Tyr—(Asp2 Glu1 Lys2)—Asp—Ala---Tyr---(Asp2 Glu2 Lys1)---Asp---Ala---Gln(COOH)
N F J S C

Fig. 2.10. The structures of the $Hp^{1F\alpha}$, $Hp^{1S\alpha}$, and $Hp^{2\alpha}$ polypeptides and the relationships between the chymotryptic peptides (N, C, and J isolated from Hp^{2a} for amino acid sequence determinations. (Reprinted, by permission, from O. Smithies, *Cold Spring Harbor Symp. Quant. Biol.* 29(1964):309–19. © 1964.)

the light chain, κ and λ, common to all classes of immunoglobulins. Homologous chains are present, with minor differences, in all mammals investigated thus far.

Chemical analyses suggest that the structure of immunoglobulin molecules is organized into a series of domains having different functions (Edelman 1972; Feldmann and Nossall 1972) (fig. 2.11). In the light chain, the COOH terminal half (unstriped part, fig. 2.11) is constant (except that amino acid interchanges from valine to leucine are found at position 191 in the κ human light chains, and lysine to arginine at position 190 in the λ chain). Thus this region is referred to as the constant part of the light chain, C_L. In the NH$_2$ terminal half of the light chain, amino acid variations are found in a large number of positions for the same or different individuals of a species; this section is referred to as the variable half of the light chain, V_L. In the heavy chain, the pyrrolidine carboxylic acid (PCA) terminal part is variable, and is therefore referred to as V_H. In addition, three constant regions are present in the heavy chain, $C_H{}^1$, $C_H{}^2$, and $C_H{}^3$, as shown in figure 2.11. Each variable and constant region contains one disulfide.

Antigen-combining sites are present in the variable parts of both light and heavy chains, the "Fab" regions. Other biological properties, such as complement fixa-

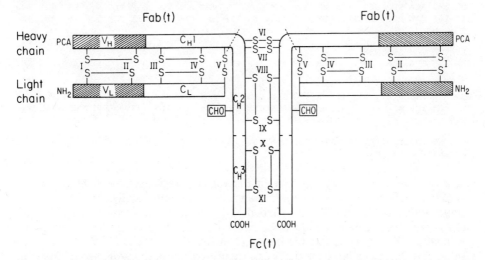

Fig. 2.11. An immunoglobulin molecule IgG. Half cystinyl residues are numbered I to XI beginning from the NH$_2$ terminal; C represents constant regions and V variable regions; numbers I to V designate corresponding residues in light and heavy chains; PCA, pyrrolidone carboxylic acid; CHO, carbohydrate. Fab (t) and Fc (t) refer to fragments produced by trypsin, which cleaves the heavy chain (dashed lines above half-cystinyl residues VI). V_H and V_L are homologous variable regions of heavy and light chains; $C_H{}^1$, $C_H{}^2$, and $C_H{}^3$ are homologous regions of the heavy chain; and C_L refers to the constant region of the light chain which is homologous to $C_H{}^1$, $C_H{}^2$, and $C_H{}^3$. (Reprinted, by permission, from Edelman, 1972.)

tion and cell fixation, are modulated by the constant part of the heavy chains, referred to as the "Fc" region. This reaction occurs after binding of the antigen by the Fab portion of the antibody molecule.

The V_H and V_L regions are homologous to each other, as shown by a large number of identical amino acids in the polypeptide chains (fig. 2.12), but they are not obviously homologous to C_H or C_L. More homology exists among the constant regions, C_L, C_H^1, C_H^2, and C_H^3 (fig. 2.13).

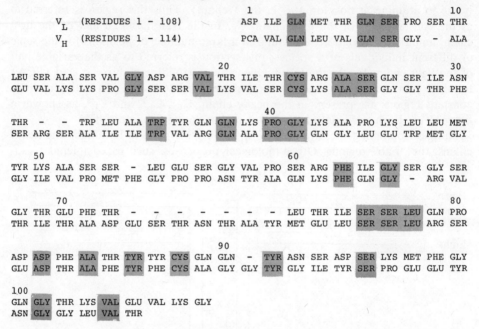

Fig. 2.12. The amino acid sequence of the variable region of the light chain, V_L, and of the heavy chain, V_H, from a human immunoglobulin molecule. The identical residues between V_L and V_H are shaded. The alignment is made so that a maximum number of identical amino acids can be obtained. The hyphens indicate the continuation of the sequence; for instance, in the second line, GLY is the eighth residue and ALA the ninth of the V_H. (Reprinted, by permission, from Edelman, 1972.)

It has been proposed that the original immunoglobulin gene coded for approximately 100 amino acid residues (Hill et al. 1968; Singer and Doolittle 1966). Duplication of this gene about one hundred million years ago (veterbrates emerging) (Wang and Fudenberg 1974) gave rise to the ancestral constant light chain gene; further duplication and mutation formed the present heavy chain genes. Because there is no clear evidence of homology between V and C regions, it is questionable whether the V and C regions evolved from the same gene. Thus, an alternative hypothesis proposes that two precursor genes, one for V and one for C, were brought together because of the selective advantages of combining their functions in a single protein molecule (Edelman 1972).

		110								120		
C_L	(RESIDUES 109–214)	THR VAL ALA ALA	PRO SER VAL PHE	ILE	PHE PRO PRO	SER
C_H1	(RESIDUES 119–220)	SER THR LYS GLY	PRO SER VAL PHE	PRO	LEU ALA PRO	SER
C_H2	(RESIDUES 234–341)	LEU LEU GLY GLY	PRO SER VAL PHE	LEU	PHE PRO PRO	LYS
C_H3	(RESIDUES 342–446)	GLN PRO ARG GLU	PRO GLN VAL TYR	THR	LEU PRO PRO	SER

		130		
ASP GLU GLN – –	LEU LYS SER GLY THR ALA SER VAL VAL CYS LEU LEU ASN ASN PHE
SER LYS SER – –	THR SER GLY GLY THR ALA ALA LEU GLY CYS LEU VAL LYS ASP TYR
PRO LYS ASP THR LEU	MET ILE SER ARG THR PRO GLU VAL THR CYS VAL VAL VAL ASP VAL
ARG GLU GLU – –	MET THR LYS ASN GLN VAL SER LEU THR CYS LEU VAL LYS GLY PHE

140		150		
TYR PRO ARG GLU ALA	LYS VAL – – GLN TRP LYS VAL ASP ASN ALA LEU GLN SER GLY
PHE PRO GLU PRO VAL	THR VAL – – SER TRP ASN SER – GLY ALA LEU THR SER GLY
SER HIS GLU ASP PRO	GLN VAL LYS PHE ASN TRP TYR VAL ASP GLY – VAL GLN VAL HIS
TYR PRO SER ASP ILE	ALA VAL – – GLU TRP GLU SER ASN ASP – GLY GLU PRO GLU

160		170		
ASN SER GLN GLU SER	VAL THR GLU GLN ASP SER LYS ASP SER THR TYR SER LEU SER SER
– VAL HIS THR PHE	PRO ALA VAL LEU GLN SER – SER GLY LEU TYR SER LEU SER SER
ASN ALA LYS THR LYS	PRO ARG GLU GLN GLN TYR – ASP SER THR TYR ARG VAL VAL SER
ASN TYR LYS THR THR	PRO PRO VAL LEU ASP SER – ASP GLY SER PHE PHE LEU TYR SER

180		190		
THR LEU THR LEU SER	LYS ALA ASP TYR GLU LYS HIS LYS VAL TYR ALA CYS GLU VAL THR
VAL VAL THR VAL PRO	SER SER SER LEU GLY THR GLN – THR TYR ILE CYS ASN VAL ASN
VAL LEU THR VAL LEU	HIS GLN ASN TRP LEU ASP GLY LYS GLU TYR LYS CYS LYS VAL SER
LYS LEU THR VAL ASP	LYS SER ARG TRP GLN GLU GLY ASN VAL PHE SER CYS SER VAL MET

200		210		
HIS GLN GLY LEU SER	SER PRO VAL THR – LYS SER PHE – – ASN ARG GLY GLU CYS
HIS LYS PRO SER ASN	THR LYS VAL – ASP LYS ARG VAL – – GLU PRO LYS SER CYS
ASN LYS ALA LEU PRO	ALA PRO ILE – GLU LYS THR ILE SER LYS ALA LYS GLY
HIS GLU ALA LEU HIS	ASN HIS TYR THR GLN LYS SER LEU SER LEU SER PRO GLY

Fig. 2.13. The amino acid sequence of the constant region of the light chain, C_L, and those of the heavy chain, C_H^1, C_H^2, and C_H^3, from a human immunoglobulin molecule. (Reprinted, by permission, from Edelman, 1972.)

It is worthwhile to mention at this point that, as a generally accepted theory, there is one gene for one peptide chain. The structure of the immunoglobulins is perhaps the first instance of one polypeptide chain being produced by more than one gene.

The Evolution of Genes Affecting Enzymes

In some cases, two or more forms of an enzyme are present in the same individual. These enzymes are called isozymes. As gene products, each polypeptide chain is affected by a single locus. In some, because of their similarity in chemical, physical, and biological properties, investigators believe that the responsible genes are originally derived by duplication. Lactic dehydrogenase (LDH) is an example. In both vertebrates and invertebrates, LDH is a tetrameric molecule formed by associating four identical or similar protein molecules. It has a molecular weight of 135,000, catalyzing the oxidation of lactic acid to pyruvic acid. In the mammalian and avian species studied, LDH is determined by two separate loci, one codes for subunit A and the other for B (Markert 1964). Four subunits of one type or both are associated, forming five different tetrameric molecules—A_4, A_3B_1, A_2B_2, A_1B_3, and B_4—appearing as five distinct bands on starch gel electrophoresis (fig. 2.14). According to Ohno, Wolf, and Atkin (1968), the two genetic loci appeared to arise by gene duplication and the gene products remained mutually recognizable and free associating.

Among teleost fish, Heterosomata have only a single locus coding for LDH subunits in somatic tissues (Salthe, Chilson, and Kaplan 1965; Markert and Faulhauber 1965); duplication of the LDH locus has apparently not taken place in this group. It was found, however, in the hagfish (*Eptatretus stoutii*) that two loci code for the LDH subunits in somatic tissues, but the subunits of LDH are not associated, and consequently only two molecular forms, A_4 and B_4, are found (Ohno, Wolf, and Atkin 1968). A similar form of autotetramer was noted in thirteen of thirty teleost fish studied by Markert and Faulhauber (1965). Ohno, Wolf, and Atkin suggest that in vertebrate evolution the first duplication of a gene locus for LDH subunits may have occurred in fish at the jawless stage.

Note that the process of duplication of genes coding the A and B molecular forms of LDH may be similar to that of hemoglobin genes. In the association of protein molecules, however, all or some tetrameric combinations were observed for LDH, whereas in the hemoglobin molecules only one form for each combination was found.

DNA Reassociation

Up to this point, we have seen examples of gene duplication for *specific loci* in different species. In some instances, evidence is sharper and clearer than in others. We must now examine another field of research products, one that considers the question of gene duplication in the long path of organic evolution and sheds light not on specific genes but on all genes in the genome.

This is the study of DNA reassociation. It is based on the fact that when double-stranded DNA is dissociated into single strands, these strands reassociate to the double-stranded form by incubation under appropriate conditions (Marmur and Lane 1960). Those strands that have more identical sequences associate faster and the association is tighter than in those with less identical sequences. The ex-

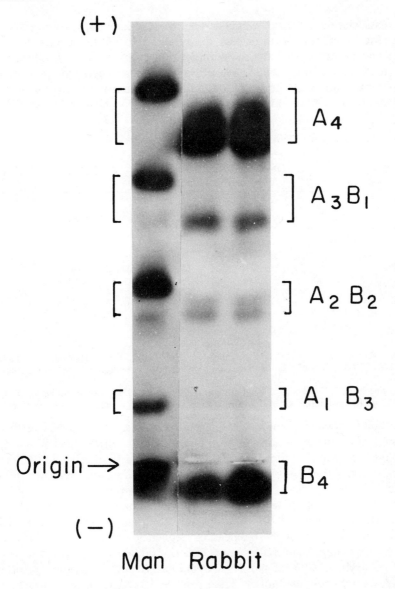

(+)

A_4

A_3B_1

A_2B_2

A_1B_3

Origin →

B_4

(−)

Man   Rabbit

Fig. 2.14. Lactic dehydrogenase from erythrocytes of rabbits and man as revealed by starch gel electrophoresis. The enzyme consists of two basic subunits, A and B, each determined by a gene. The subunits associate to form tetrameres with all four possible units so that five tetrameres, as represented by the five bands, are formed. Note difference in electrophoretic mobility between the human and rabbit forms, and in the relative amount between the tetrameres.

tent of reassociation between strands derived from different species correlates with phylogenetic order, that is, the closer the relationship of the species, the greater the rate and percentage of reassociation; within species, the percentage is even higher (table 2.2). These results have led investigators to conclude that the extent of DNA reassociation can be used as a measure of recurring nucleotide sequences arising from a large-scale duplication of genes (Britten and Kohne 1968).

TABLE 2.2. Binding of Fragments of DNA, Human and Mouse Origin, by Single-stranded DNA of Different Species

DNA in Agar	DNA (μg/gm of Agar)	Labeled DNA Bound (%)	
		C^{14}	P^{32}
Human	650	18	5
Mouse	1,020	6	22
Rhesus monkey	450	14	8
Rat	350	3	14
Hamster	370	3	12
Guinea pig	280	3	3
Rabbit	390	3	3
Beef	720	5	4
Salmon	600	1.5	1.5
E. coli	400	0.4	0.4
None	0	0.4	0.4

SOURCE: Reprinted, by permission, from B. H. Hoyer, B. J. McCarthy, and E. T. Bolton, *Science* 144 (1965): 959-67. © 1965 by the American Association for the Advancement of Science.

NOTE: The human DNA was C^{14}-labeled, and the mouse DNA was P^{32}-labeled. They were mixed with agar containing DNA of different species or agars lacking DNA, and incubated.

Repetition-frequency of the nucleotide base sequence for the DNA of the same organism, such as the mouse, is shown in figure 2.15. As Britten and Kohne (1968) pointed out, the width of the peaks in the figure is partly due to the difficulty of resolving reassociation rates that differ by less than a factor of ten, but the curve is correct in a broad aspect. The large peak at the left represents the mouse satellite DNA (the small discrete band well resolved from the broad principal band, when the DNA is fractionated by equilibrium centrifugation in concentrated CaC1 solutions). It shows the highest repeating frequencies and accounts for 10% of the DNA. Repeating frequencies with 1,000 to 100,000 copies account for 15% of the DNA. About 70% of the DNA is unique. Repetition-frequency spectra of calf DNA do not show the large isolated peak of 10^6 copies as does mouse DNA, but a large, broad peak is present in the region of 10^4 to 10^5 and accounts for 40% of the DNA.

Britten and Kohne (1968) reported that in higher organisms more than one-third of DNA comprises sequences that recur anywhere from 10^3 to 10^6 times per cell. They concluded that the genetic materials, instead of being a collection of randomly derived genes, consist mainly of families of sequences that display the similarities of a common origin. Notwithstanding a certain amount of controversy

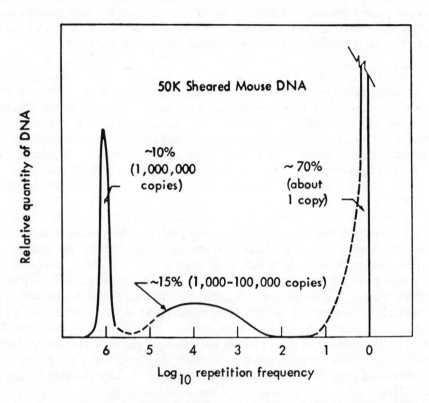

Fig. 2.15. Spectrogram of the frequency of repetition of nucleotide sequences in DNA of the mouse. Relative quantity of DNA plotted against the logarithm of the repetition frequency. These data are derived from measurements of the quantity and rate of reassociation of fractions separated on hydroxyapatite. Dashed segments of the curve represent regions of considerable uncertainty. (Reprinted, by permission, from R. J. Britten and D. E. Kohne, *Science* 161(1968):529–40. © 1968 by the American Association for the Advancement of Science.)

regarding the sources of DNA and the need for more refined studies, general results seem to indicate the presence of a large fraction of repeating sequences of DNA in the cells of all species.

How did these similarities arise? The best explanation seems to be that they were the result of gene duplication. Thus it is very likely that the genes derived from a common ancestor gene may perform different functions and may be too different to identify through the analyses of phenotype in contemporary organisms. They have not, however, completely lost their identities in the nucleotide base sequence revealed through DNA association studies. If we accepted the monophyletic theory of gene evolution, or an even less rigid assumption that all genes

now present in living organisms were derived from a few present in the first organisms, the single copy of genes accounting for two-thirds of the total number must
also have derived from gene duplication but have since lost their homology or
identity with their duplicated copies because of subsequent mutations.

GENE DELETION

The classical technique of studying gene deletion or deficiency is to observe
chromomeres, knobs, or other landmarks that may be missing in the aberrant
homologous chromosome. An unpaired loop, for example, is characteristic of
deficient heterozygotes, either in meiotic cells at synapsis or in somatic cells if
chromosome pairing occurs. When a segment of a chromosome is deleted, the
genes present in the missing section also are absent. When a number of genes in
the section play a vital role in physiological processes, lethality may result from
their deletion.

On the molecular genetic basis, a mutant gene with part of its DNA sequence
missing ordinarily codes for an abnormal polypeptide with unusual structure and
aberrant function, the nature of the abnormality depending on the number of
nucleotide base pairs missing, as well as on their positions in the sequence. If the
missing base pairs are three or a multiple of three, beginning from the first pair of
the triplet, the polypeptide produced will be short by exactly the amino acid residues normally coded by the deleted base pairs. If the deletion begins from the
second or third pair of the triplet, the translation will be different thereafter because, since the number of base pairs is not exactly three or a multiple of three,
translation of the code cannot proceed normally; the base sequence will be cor-
rectly translated prior to deletion and misread thereafter. Such deletion is said to
involve "frame shifts," which may affect the neighboring genes and cause pleiotropic effects (Garen 1968).

Individuals having Hb Freiburg, in which the amino acid, valine, at position 23
of the beta polypeptide chain of human hemoglobin was missing (Jones et al.
1966), suffer from mild anemia even where normal hemoglobin is found. It is
therefore assumed that the individuals were heterozygotes, and the normal adult
hemoglobin was produced by the normal Hb-β allele. With Hb Gun Hill (Bradley,
Wohl, and Rieder 1967; Rieder and Bradley 1968), in which deletion also occurred in the Hb-β gene, with five amino acid residues at positions 91 to 95,
92 to 96, or 93 to 97 in the beta hemoglobin chain missing, chronic hemolytic
disease results and, since normal adult hemoglobin was found, presumably these
individuals are also heterozygotes.

Among gross morphological mutations, some may be due to gene deletion,
either in part or as a whole, and their exact nature is not known. When there is
gene duplication in one chromosome, often there is simultaneous gene deletion in
its homologous partner (fig. 2.8). This may have occurred with complex loci such
as the A and R loci in corn discussed previously. Each may have involved duplication in some alleles, deletion in others, and some alleles with both deleted and

duplicated nucleotide sequences. Incidentally, the presence of both colored and colorless patches in the corn mutation may be comparable to the case of deletion in *Hb-β*, where both normal and abnormal hemoglobins were observed.

The harmful effects of gene deletion on an organism vary according to the physiological importance controlled by the lost genes. On the other hand, during evolution there must have been genes that were useful at earlier stages but became useless later. Since it would be a nuisance to carry these genes generation after generation, gene deletion would thus become a useful device, one that might actually have been occurring undetected. When a gene plays no vital function, its presence or elimination cannot be phenotypically observed.

MECHANISMS OF GENE DELETION AND DUPLICATION

Having surveyed a number of cases of gene duplication and deletion, we can now examine various theories on the mechanisms involved. The processes take place in the course of nuclear division, a biological phenomenon the physical and chemical basis of which is not clearly understood. Therefore, the mechanisms herewith proposed, as based on observed cytological and molecular evidence, are considered to be the most reliable, although certainly there may be others, some unknown, some merely accidental, as in cases of minute deletions and duplications. (Instances involving large portions of a chromosome are treated as chromosomal mutations or rearrangements and will be discussed in chap. 4. Since a gene is part of a chromosome, one cannot, strictly speaking, draw sharp lines between gene duplications and deletions from corresponding types of chromosomal mutations.)

Chromosome Breakage and Cross-reunion

Breakage sometimes occurs simultaneously in two homologous chromosomes but at nonhomologous points. Thus when cross-reunion of two broken ends takes place, one chromosome will be a deletion type, resulting from the connection of the two short chromosome segments, and the other will be a duplication, resulting from the union of the two long segments. Depending on the distance between the two breakage points, duplication and deletion can vary from a portion of a gene to a number of genes (Harris 1970) (fig. 2.16). Such reciprocal translocation (one segment of a chromosome transferred to another chromosome; detailed description given in chap. 4) between two homologous chromosomes, presumably from simultaneous breakage, is followed by aberrant reunions (Court-Brown and Smith 1969).

Homologous Pairing but Unequal Crossing-over

During cell division, homologous pairing of chromosomes normally occurs. Thus when homologous partners are present, chromosome pairing is highly specific, except where structural heterozygosity restricts the opportunity for the association of homologous regions, so that there is a form of nonspecific affinity to associate small, adjacent, nonhomologous segments (McClintock 1933).

Some time ago it was suggested (Darlington 1935) that the longitudinal congruence of chromosome homologues could stem from an affinity between relatively few pairing sites followed by the zipperlike completion of synapsis in hitherto unpaired segments. This view is evidently supported by a recent study of bacteriophage induction that indicated that as few as twelve nucleotides are sufficient for such a recognition site (Thomas 1966).

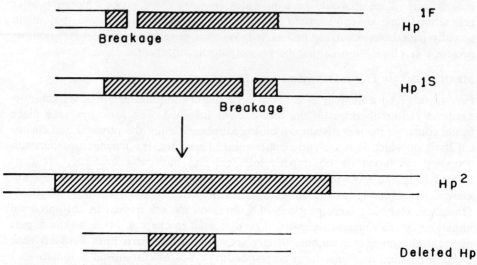

Fig. 2.16. Postulated formation of gene duplication and deletion as a result of breaks in the two homologous chromosomes; followed by a reunion, using the haptoglobin genes as an example.

When a gene contains repeated nucleotide sequences, unequal crossing-over may occur between its parts. In other words, chromosome pairing itself is homologous (fig. 2.8), but the exchanged parts of a gene are not. In heterozygotes, the number of nucleotide base pairs for the two gene alleles may not always be equal. Thus synapsis cannot always be perfect. Unequal crossing-over can occur over a wide range of sites, and the duplicated and deleted regions may be large or small, depending on the distance between the two sites. The classical case of bar locus duplication in *Drosophila* provides good cytological evidence for two different events of duplication: a novel event of unequal crossing-over and a case of homologous pairing but unequal crossing-over. Such duplications may involve other loci in the immediate neighborhood of the bar locus, rather than a single locus; nevertheless, it serves as an interesting example of duplication and deletion.

Once unequal crossing-over has occurred, it may recur frequently, depending on the number of repeated sequences built into the chromosome, since the processes of duplication and unequal crossing-over are synergistic. There could be many mutant genes with different repeated sequences, many of which might be eliminated by natural selection for various reasons because frame shift and chain terminating types could be lethal.

In general, a locus having repeated nucleotide sequences would have a higher rate of mutation than one without these sequences, because of greater chance of incorrect matching during homologous chromosome pairing. Such mutation would be primarily of the deletion and duplication types seen in the specific mutation rate of a specific locus listed for different species. In their work with yeast, Magni and Sora (1969) (table 2.3) obtained results that show evidence for the relative rate of gene duplication and deletion compared to that of other types of mutations. They compared the mutation rates during mitosis and meiosis (corresponding, respectively, to replicative and recombinational cell cycles of reproduction). Their results clearly indicate significantly higher mutation rates during meiosis than during mitosis. It is assumed that during meiosis an unequal exchange between homologous chromosomal regions produces deletion and duplication types of mutation. This mechanism cannot operate during the replicative mitotic cycle.

TABLE 2.3 Mutation Rates during Mitosis and Meiosis in Yeast, *Saccharomyces cerevisiae*

Mutation	Mutation Rates		Me/Mi
	Mitosis	Meiosis	
Prototrophy → auxotrophy	$< 5 \times 10^{-6}$	3.5×10^{-4}	> 70
thr4-1 → *Thr*	0.32×10^{-9}	9.9×10^{-9}	30
hi1-1 → *Hi*	1.65×10^{-9}	2.4×10^{-8}	14.5
s(ar4-17) → $>$ *S(ar4-17)*	8.2×10^{-8}	1.3×10^{-6}	16.3

SOURCE: Reprinted from G. E. Magni and S. Sora, 1969, by permission of J. & A. Churchill and the Ciba Foundation, London.
NOTE: Prototrophy: exogenous growth factor dependent; auxotrophy: exogenous growth factor independent; in general, wild-type genes are exogenous growth factor independent and are dominant; *thr4-1*: a recessive mutation of threonine dependence; *Thr*: the dominant gene of threonine independence; *hi1-1*: a recessive mutation of histidine dependence; *Hi*: the dominant gene of histidine independence; s → S: mutation from inactive nonsense suppressors to the alleles active in the mutant *ar4-17* of the ocher type.

To summarize, chemical, cytological, and morphological evidence all point to different types of gene duplication and deletion. The duplicated segments may be contained in the same chromosome immediately adjacent to the original segment or separated by other segments. They may possibly be incorporated into a nonhomologous chromosome by later translocation. There is good evidence that many complex loci contain repeated nucleotide base sequences resulting from the gene duplication that has recurred through the course of organic evolution. It is a most important, and perhaps the only, biological mechanism for increasing genome size and adding genetic information. It would be difficult indeed to imagine any other mechanism as effective and convenient.

NUCLEOTIDE BASE SUBSTITUTION

Mutation by nucleotide base substitution may be considered as a trial and error device for improving the fitness of organisms, and it is of great evolutionary importance. A large proportion of all mutations are probably due to this cause. This can be ascertained only by chemical analysis of the amino acid sequences of the protein molecules coded by the genes involved.

Gene Evolution within and between Phyletic Lines

Comparisons of the amino acid sequences of the polypeptide chains for structural genes, within and between species, have afforded much insight into the genic evolutionary changes that have taken place from the remote past to the present. Just as the gross morphological characters can be compared between living species, so phylogenic relationships of polypeptides can be analyzed between species. Furthermore, from differences and similarities in the amino acid sequences of closely related proteins, inferences can be drawn concerning ancestral genes. Only a limited number of protein molecules have been analyzed so far, but the list has grown rapidly in the past decade. A few are discussed in the following paragraphs to illustrate their significance in the study of evolution.

Let us begin with cytochrome c, an enzyme located in the mitochrondria of all aerobic cells. It accepts electrons from cytochrome b and transfers them to cytochrome oxidase. Cytochrome c is widely distributed in nature and readily extractable from biological materials. Its rate of mutation appears to be optimal for phylogenetic studies of distantly related species. There are known amino acid sequences of cytochrome c from a variety of organisms. All the cytochrome c obtained from animals, plants, and yeasts are homologous proteins, with the amino acid sequences in certain sections identical for any two species (table 2.4). Examining the differences in the number of amino acids between species, in the table one can see the evolutionary changes of the cytochrome c gene in both the remote and the more recent past. Between the human cytochrome c and the wheat cytochrome c, the amino acids are identical in 61 out of 104 positions. It is remarkable that these two proteins, one from a plant and one from one of the most advanced mammals, are so similar. It has always been assumed that the plant kingdom and the animal kingdom descended from a common origin, but we have no conclusive genetic evidence of this; the discovery of the homology of the cytochrome c gene, universally present through all of the species examined, provides perhaps the best evidence to date (and also, in effect, for the monophyletic theory of evolution).

The universal presence of the cytochrome c gene indicates that this gene has been in existence for at least two billion years, that is, as far back as the divergence of the animal and plant kingdoms, and that its evolution along phyletic lines follows the relationships between the species very closely. If we assume that *Homo sapiens* derived from an ape or some monkeylike creature prior to the Miocene age, and that these monkeys or apes were genetically very much like present primates, one may be surprised that in this long time period there was only one amino acid change in the human system as against that of the rhesus monkey.

There are minor discrepancies in the number of amino acid differences as between the phylogenetic relationships of the species. For instance, the wheat cyto-

TABLE 2.4 Matrix of Numbers of Amino Acid Differences in Cytochrome c between Species

Species	Human	Rhesus monkey	Horse	Pig, bovine, sheep	Dog	Gray whale	Rabbit	Kangaroo	Chicken, turkey	Penguin	Pekin duck	Pigeon	Snapping turtle	Rattlesnake	Bullfrog	Tuna fish	Dogfish	Lamprey	Fruit fly	Screwworm fly	Silkworm moth	Tobacco horn worm moth	Wheat	N. crassa	Baker's yeast
Rhesus monkey	1																								
Horse	12	11																							
Pig, bovine, sheep	10	9	3																						
Dog	11	10	6	3																					
Gray whale	10	10	5	3	5																				
Rabbit	9	8	6	5	3	2																			
Kangaroo	10	11	7	6	7	6	9																		
Chicken, turkey	13	12	11	9	9	8	6	12																	
Penguin	13	12	12	10	10	8	8	10	2																
Pekin duck	11	10	10	8	8	6	6	10	3	3															
Pigeon	12	11	11	9	9	8	7	11	4	4	3														
Snapping turtle	15	14	11	9	9	8	8	11	8	8	7	8													
Rattlesnake	14	15	22	20	21	19	18	21	19	20	17	18	8												
Bullfrog	18	17	14	11	11	11	11	13	11	12	11	12	10	22											
Tuna fish	21	21	19	17	18	17	17	18	17	18	17	18	18	24	15										
Dogfish	24	23	17	16	17	16	16	20	19	20	17	19	19	26	20	20									
Lamprey	20	20	16	14	14	15	14	17	18	19	18	19	19	27	21	26	17								
Fruit fly	29	28	24	24	23	23	23	26	25	26	24	25	24	31	22	25	26	29							
Screwworm fly	27	26	22	22	21	22	21	24	23	24	22	24	22	29	25	24	26	29	2						
Silkworm moth	31	30	29	27	25	27	27	28	27	28	26	27	27	31	29	32	32	31	15	14					
Tobacco horn worm moth	31	30	28	27	25	27	28	28	28	28	27	27	26	33	30	32	30	31	12	15	5				
Wheat	43	43	46	45	44	44	44	47	46	46	46	46	46	46	48	49	51	47	45	45	42	46			
N. crassa	48	46	46	46	46	46	46	49	47	47	46	46	49	49	48	49	48	51	51	51	41	47	54		
Baker's yeast	45	45	46	45	45	45	46	46	45	45	46	46	46	49	48	47	47	48	49	45	45	47	47	41	
Candida krusei	51	50	51	50	49	50	51	51	50	50	50	50	52	52	52	51	47	47	47	46	50	50	50	54	28

SOURCE: Reprinted, by permission, from M. O. Dayhoff, 1969.

NOTE: This is based on a total of 104 amino acid residues, as found in most species. The highest total number of residues is 112, observed in wheat.

chrome c resembles that of the primates more than that of any other species examined. The number of variant residues between primates and wheat is forty-three, whereas between wheat and all the remaining species it ranges from forty-four to fifty-one. Such discrepancies or particularities apply also to some other proteins, such as insulin. Insulin is composed of two polypeptide chains, A and B; A consists of twenty-one residues and B of thirty. Insulin differences and similarities are generally in line with the phylogenic relationships between species, except for a large difference in the amino acid sequence between guinea pigs and all other species (table 2.5). There seems no good explanation for such a strange deviation except to say that genetic drift and different selection pressures in various environmental niches could be causal factors. Such odd phenomena suggest that in studying genetic relationships between species or populations it is always wise to include as many genetic variants as possible.

TABLE 2.5 Matrix of Numbers of Amino Acid Differences in Insulin A and B between Species

	Human	Rat 1	Rat 2	Guinea pig	Rabbit	Elephant	Horse	Pig	Bovine	Sheep	Sei whale	Chicken	Bonito	Toadfish 1	Toadfish 2	Angler fish
Human		3	4	18	1	2	2	1	3	4	3	7	11	17	19	*9*
Rat 1			1	17	2	5	4	3	5	6	5	9	13	18	20	*9*
Rat 2				18	3	6	5	4	6	7	6	10	14	19	20	*10*
Guinea pig					18	17	17	18	17	16	16	19	21	22	24	*14*
Rabbit						3	2	1	3	4	3	7	11	17	19	*9*
Elephant							2	3	3	2	4	7	11	17	19	*9*
Horse								1	3	2	3	6	11	17	19	*9*
Pig									2	3	2	6	11	17	19	*9*
Bovine										1	1	6	11	17	19	*9*
Sheep											2	6	11	17	19	*9*
Sei whale												5	11	17	19	*9*
Chicken													8	15	17	*8*
Bonito														12	14	*8*
Toadfish 1															3	*3*
Toadfish 2																*3*
Angler fish																

SOURCE: Reprinted, by permission, from M. O. Dayhoff, 1969.
NOTE: Matrix elements are italicized when the sequence of a chain has not been completely determined.

We have already discussed hemoglobin genes from the standpoint of gene duplication. As we compare the number of amino acid residue differences between the polypeptide chains within and between species, we find additional evolution significance. The differences in the number of amino acids between the alpha and beta chains within a species are greater than those differences for the same polypeptide chains between species (table 2.6). Why did they come out this way and what does it mean? First, genes for the alpha and beta hemoglobins must have been derived through gene duplication and must have been present before the vertebrates diverged, since both exist in contemporary vertebrates. Obviously, then, a number of amino acid differences between them must have already evolved by the time the lines diverged. But the species difference in the number of amino acids for a specific gene evolved only since the species in question were separated.

Second, within the same species the beta and alpha hemoglobins, and hence their genes, function differently. Thus they are under different selection pressures in comparison with the same gene in different species.

TABLE 2.6 Matrix of Numbers of Amino Acid Differences in Alpha and Beta Hemoglobins and Myoglobin between Species

	α Human	α Rhesus monkey	α Mouse	α Rabbit	α Horse	α Bovine	α Carp	β Human	β Rhesus monkey	δ Human	β Rabbit	β Bovine	β B Sheep	β C Sheep	γ Human	Myoglobin horse	Myoglobin sperm whale
α Human		4	16	25	18	17	71	84	84	85	87	87	84	85	89	115	115
α Rhesus monkey			16	25	16	16	71	84	84	87	86	87	85	86	89	116	116
α Mouse				27	22	19	71	84	86	87	86	87	83	85	87	117	117
α Rabbit					25	25	74	88	86	90	91	92	89	88	90	119	120
α Horse						18	70	86	85	87	87	84	83	84	87	117	118
α Bovine							68	85	84	87	85	87	84	84	88	114	114
α Carp								88	87	90	90	88	86	87	87	119	121
β Human									8	10	14	25	26	32	39	116	117
β Rhesus monkey										11	16	27	27	33	37	117	118
δ Human											19	27	28	34	41	117	118
β Rabbit												30	28	36	40	116	117
β Bovine													12	24	40	119	120
β B Sheep														21	41	116	119
β C Sheep															41	122	122
γ Human																120	121
Myoglobin horse																	19
Myoglobin sperm whale																	

SOURCE: Reprinted, by permission, from M. O. Dayhoff, 1969.

Genes that originated by duplication, evolving side by side in the same phyletic lineage and resulting in different functions, such as the alpha and beta hemoglobin genes in the *same species*, may be said to be paralogous (Fitch and Margoliash 1970). Genes performing the same function in *different* species, such as the alpha *or* the beta hemoglobin gene, are said to be *orthologous*. As a general rule, then, one may assume that the differences in terms of number of nucleotide base differences between paralogous genes are greater than between orthologous genes.

Paralogous gene evolution in the pituitary hormones may serve as a better example. Between the human pituitary growth hormone (GH) and the pituitary lactogenic hormone (LGH) of sheep, the human growth hormone consists of 188 amino acid residues and that of the sheep lactogenic hormone 198. Three homologous segments have been found in the molecules, comprising approximately 45% of either peptide chain (Bewley and Li 1970). Adrenocorticotropin (ACTH) and melanocyte-stimulating hormone (MSH) are also homologous within and between species, their function and chemical structure being similar to those between the hemoglobin and the myoglobin. These genes, coding these molecules, although derived from a common origin, have apparently evolved some functional differences. For instance, LGH is believed to affect somatic growth as well as lactation, while GH affects somatic growth only. Similarly, ACTH has a melanotropic effect in addition to adrenal cortical stimulation.

Paralogous gene evolution, which concerns differentiation and morphogenesis, and thus ontogeny, will be further discussed in chapter 5. Orthologous gene evolution, which concerns adaptation and speciation, and thus phyletic evolution, will be scrutinized in chapter 7. As a genetic measurement, nucleotide base substitution—one of the mutational processes—adds another dimension to our knowledge of organic evolution.

Gene Mutation Rate

Mechanisms involved in mutation, such as nucleotide base substitution, are imperfectly understood. Many factors cause mutation, and the incorporation of a wrong nucleotide base could occur only by random error of DNA replication. Experiments have demonstrated the frequency of error in nucleotide replication. Incorporation of G in the enzymatic replication of dAT polymer (double-stranded copolymer composed of alternating deoxyadenylate and deoxythymidylate) occurred at less than one residue per 28,000 to 580,000 adenine and thymine nucleotides polymerized (Trautner, Swartz, and Kornberg 1962). Assuming that a gene consists of 1,000 such nucleotide pairs, the error of replication would then be equivalent to a mutation rate of 1/28 to 1/580 per gene. During the synthesis of polymer dG (homopolymer of deoxyguanylate) on a dC (homopolymer of deoxycytidylate) template by DNA polymerase of T4 bacteriophage, T was incorporated instead of G at a rate of 1/100,000 to 1/1,000,000 (Hall and Lehman 1968), giving an equivalent mutation rate of 1/100 to 1/1,000 on the same basis of 1,000 nucleotide pairs for a gene.

When the rate of nucleotide base changes is calculated from the observed number of amino acid changes in the protein molecules, a correction for synonymous codons is needed. Since there is a three-way possibility of mutation for each base pair, each codon may mutate in any of nine ways. Of the 549 possible single substitutions (61×9, the value 61 obtained from the total of 64 codons minus 3 nonsense codons), 134 are synonymous codons (see table 2.1). Therefore, in each heritable change in the genetic material, there is only an average of (549 —

134)/549 or 0.76 changes in the peptide chains of the protein molecules, or for one amino acid change there is a 1.3 (1/0.76) triplet nucleotide base change.

Many physical and chemical mutagens can cause mutations by incorporation of various base analogues into DNA or by incorporation of a wrong nucleotide base. A full discussion of this subject is outside the scope of this book, but it may be worthwhile to note that some physical mutagens, such as different types of radiation present in the natural environment through the whole course of evolution, have specific effects on certain nucleotide bases (table 2.7) (Wacker and Chandra 1969). Such effects apply not only to DNA bases as cited in the table, but also RNA bases.

TABLE 2.7 Sensitivity of Nucleotide Bases to Different Radiations during Transcription of DNA

	Purines		Pyrimidines	
Radiation	A	G	T	C
Ultraviolet			+	(+)
Visible light + thiopyronin		+		
+ acetone			+	
Neutrons (thermal)	+	(+)		
X rays			+	+

SOURCE: Reprinted from A. Wacker and P. Chandra, 1969, by permission of J. & A. Churchill and the Ciba Foundation, London.
NOTE: Strong sensitivity, +; weak sensitivity, (+).

In breeding experiments, it is difficult to pinpoint the exact nature of mutation, for example, whether it is the result of nucleotide base substitution, duplication, deletion, or chromosomal aberration. Estimates of mutation rates tend to be too low because many mutations go undetected. In general, information obtained from lower organisms is apt to be more accurate than that obtained from higher organisms because of larger sample sizes and because the types of mutations are better known.

In light of the above information, we can examine some statistics on the mutation rate of laboratory populations. In studies of the spontaneous mutation rate of *D. melanogaster,* the average lethal mutation rate was 3×10^{-6} per locus, 10^{-2} per genome (Muller 1962), and 10^{-3} per X chromosome (Demerec 1937) (table 2.8). It should be understood that a large percentage of mutants causing slightly lowered fitness, that is, most of the mutations resulting from nucleotide base substitutions, were not included. The implication is that for every lethal mutation there are about nine that are nonlethal (Whitfield, Martin, and Ames 1966; King and Jukes 1969). If Muller's estimate represents only mutations of the chain terminating type, plus a roughly equal number of the substitutional type that produce nonfunctional proteins, there were nine times more nonlethal mutations that were not included, and the actual mutation rate for all genes with different degrees of fitness would be about ten times greater, roughly 3×10^{-5} per locus. The polygenic mutation rate estimated by Mukai (1964) in *Drosophila* may very likely be close to the true rate. He found that the rate of spontaneous mutation for the whole genome for slightly deleterious mutations is at least twenty to thirty times as great as that for recessive lethal mutations, including homozygotes with fitness greater than 98% of normal.

TABLE 2.8 Frequency of Spontaneous Lethals in X Chromosomes of
         Wild-Type Stocks of *D. melanogaster*

Stocks	No. of X Chromosomes Examined	No. of Lethals Observed	Lethals (%)
Oregon-R	3,049	2	0.066 ± 0.03
Swedish-b	1,627	3	0.18 ± 0.07
California-c	708	2	0.28 ± 0.13
Huntsville, Tex.	938	...	...
Urbana, Ill.	1,016	1	
Canton, Ohio	922	...	...
Amherst, Mass.	572	1	
Woodbury, N.J.	1,159	1	
Tuscaloosa, Ala.	545	1	
Lausanne	955	2	0.21 ± 0.09
Seto, Japan	1,236	...	...
Kyoto	875	1	
Total	13,602	14	0.10 ± 0.02

SOURCE: Reprinted, by permission, from M. Demerec, *Genetics* 22 (1937):467-83.

Some estimated mutation rates at specific coat-color loci in mice are as follows: an overall spontaneous mutation rate for five specific loci of 8.9×10^{-6}, based on about ten million mice (Schlager and Dickie 1967); one of 7.5×10^{-6} for seven specific loci (Russell 1963); and another of 10^{-5} for seven specific loci (Carter, Lyon, and Phillips 1958). These estimates, reported by different laboratories, are in close agreement and roughly on the order of 10^{-5}, a value generally used as a standard mutation rate per locus per gamete.

The chemical nature of specific locus mutations as found in mice is unknown. If we assume that each of these mutations is due to a specific amino acid change in the polypeptide sequence, the mutation is then the result of a specific nucleotide base substitution. If we further assume that the error of base substitution occurs with equal likelihood for any nucleotide base, there are three possible errors of substitution per base pair, and nine possible errors per codon. For a gene with, say, 100 codons, we find 900 possible errors of substitution. The observed specific locus mutation rate thus represents only 1 out of 3 per base, 1 out of 9 per codon, and 1 out of 900 per gene. The substitutional type of mutation with any base at any one position of a gene can be represented as the observed rate multiplied by 3 on the per nucleotide basis, 9 on the per codon basis, and 900 on the per locus basis. Each may be further multiplied by 1.3 as a correction for synonymous changes. Taking the value of 10^{-5} to 10^{-6} as the observed rate of specific locus mutation, and the factor of 1,000 as a rough correction for errors in substitution at any base of the DNA, we arrive at the value of 10^{-2} to 10^{-3}, closely comparable with the mutation rate observed in vitro as discussed above. Then it can be concluded that the other mutations are either undetectable or lethal. The question that remains is: How realistic is the assumption that specific locus mutation results from replacement of a specific amino acid? This question cannot be answered until the proteins are discovered and the amino acids are determined. This information will probably not be available for some time.

A list of mutation rates at specific loci for maize, *Drosophila,* and mice is given in table 2.9. The rates vary between genes and within and between species. Although the exact nature of the mutations cannot be ascertained, it is my opinion that some of them, especially those with high rates, were possibly due to intragenic crossing-over, as discussed previously. This opinion may not be shared by others.

Certain genes called "mutators" increase the mutation rate of other genes. They have been studied in *Drosophila* (Demerec 1937), maize (McClintock 1951), and several strains of bacteria and bacteriophage (Gibson, Scheppe, and Cox 1970).

TABLE 2.9 Spontaneous Mutation Rates at Specific loci

Organism	Locus	No. of Gametes	Mutation Rates per 10^6 Gametes	References
Maize	+ → *wx* (wax endosperm)	1,503,744	0	Stadler (1942)
	+ → *pr* (red aleurone)	647,102	11	Stadler (1942)
	+ → *sh* (shrunken)	2,469,285	1	Stadler (1942)
	+ → *su* (sugar endosperm)	1,678,731	2	Stadler (1942)
	+ → *i*	265,391	106*	Stadler (1942)
	+ → *r* (colorless authocyanin)	554,786	492*	Stadler (1942)
D. melano-	white and zesty region	10,799	164*	Green (1962)
gaster	(one to others)	(flies)		
	+ → *B* (bar)	4,122	7.3*	Green (1962)
		(flies)		
	+ → *ct* (cut)	60,000	1.5	Muller (1950)
	+ → *y* (yellow)	70,000	0.29	Ives (1950)
	+ → *lz* (lozenge)	70,000	0.29	Ives (1950)
Mouse	H-2 histocompatibility	1,936	39.0*	Stimpfling & Richardson
	Brachyury (*T*)	(mice)		(1965)
		11,164	18.9*	Dunn (1956)
	+ → A^{iy} (intermediate yellow)	33,679	29.7	Schlager & Dickie (1967)
	+ → *b* (brown)	514,558	3.9	Schlager & Dickie (1967)
	+ → *c, c^p* (albino, platinum)	196,594	10.2	Schlager & Dickie (1967)
	+ → d^l, d^s (dilute lethal, slight dilute)	479,777	12.5	Schlager & Dickie (1967)
	+ → *ln* (leaden)	125,725	8.0	Schlager & Dickie (1967)

*Mutation possibly involving unequal crossing-over.

One, the Treffers mutator gene (*mut T*) of *E. coli,* increases the mutation rate by at least 100-fold at most, if not all, genetic loci. E. C. Cox and C. Yanofsky (1967) found that when *E. coli* of the *mut* T strain is repeatedly subcultured, the *mut T* gene produces a trend toward DNA of a guanine-cytosine content higher than that in the original stock. The gene favors the substitution A-T → C-G. The effect may be mediated through an altered DNA polymerase so that A- and T-rich codons are changed to G- and C-rich codons. Cox and Yanofsky also noted, in bacteriophage T$_4$, a bias in the opposite direction, G-C → A-T.

In summary, the mutation rates cited for different species may only represent the minimum levels. They are all based on laboratory evidence. In natural populations, accurate estimations of mutation rates are difficult, although they can be extrapolated, based on some assumptions. These will be discussed in chapter 6. What can be observed in natural populations is the rate of mutation and fixation combined.

Rate of Nucleotide Base Substitution

Analyses of amino acid sequences of protein molecules are also useful in estimating rates of nucleotide base substitution on an evolutionary time basis, by observing the number of differences in amino acids between two species and the time of their separation. Note that the substitution rate so estimated represents the coincidence of the mutation rate and the rate of complete fixation of a gene.

A distinction needs to be made between fixation rate and substitution rate. In natural populations, the fixation rate of a gene is the amount of time elapsing from its first appearance to its complete replacement of the other genes in the population. The substitution rate represents the total amount of time required for both the occurrence of mutation and fixation within a population.

King and Jukes (1969) calculated the average rates of nucleotide base substitution per codon per year (table 2.10) according to the number of amino acid differences between species for a number of proteins. The calculation is based on seventy-five million years for the divergence of all major euplacental orders from a common ancestor (G. G. Simpson, see Smith and Margoliash 1964). The rate ranges from 3.3×10^{-10} per codon per year for insulin A and B, to 42.9×10^{-10} for fibrinopeptide A. Fitch and Margoliash (1970), assuming that the Artiodactyl-Perissodactyl[2] ancestor existed about sixty million years ago, arrived at an estimated substitution rate of 5×10^{-9} per codon per year. (They did not mention what proteins were used in making the estimation.) Corbin and Uzzell (see Fitch and Margoliash 1970), using their own data as well as those of others, arrived at an average of 16×10^{-9} substitutions per codon per year for those codons for which they assumed the replacements to be "neutral." Kimura (1968) estimated molecular evolution in vertebrates at the rate of one amino acid substitution every

---

2. Artiodactyl, an order of ungulate mammals, even-toed, such as cattle, sheep; Perissodactyl, an order of nonruminant ungulate mammals, odd-toed, such as horses, rhinoceroses.

1.8 years, based on comparisons of the beta chains of horse and human hemoglobins, cytochrome c, and triosphosphate dehydrogenase. On the basis of an average of 100 codons for a gene and 1.3×10^7 (or $[4 \times 10^9]/300$) for an organism, the substitution rate would be 0.48×10^{-9} per codon per year.[3]

TABLE 2.10 Rates of Amino Acid Substitutions in Mammalian Evolution

Protein	Total Number of Comparisons of Amino Acids	Observed Number of Amino Acid Differences	Observed Number of Differences per Codon	10^{-10} Substitutions per Codon per Year*
1. Insulin A and B	510	24	0.047	3.3
2. Cytochrome c	1,040	63	0.061	4.2
3. Hemoglobin, alpha chain	432	58	0.137	9.9
4. Hemoglobin, beta chain	438	63	0.144	10.3
5. Ribonuclease	124	40	0.323	25.3
6. Immoglobulin light chain (constant half)	102	40	0.392	33.2
7. Fibrinopeptide A	160	76	0.475	42.9
8. Bovine hemoglobin fetal chain	438	97	0.221	22.9†
9. Guinea pig insulin	255	86	0.337	53.1‡

SOURCE: Reprinted, by permission, from J. L. King and T. H. Jukes, *Science* 164 (1969): 788–98. © 1969 by the American Association for the Advancement of Science.
*The estimate for time elapsed since the divergence of euplacental mammalian orders is taken as seventy-five million years. The average rate of evolution for the seven protein species represented by entries 1 through 7 is 16×10^{-10} substitutions per codon per year.
†Bovine line of descent only.
‡Guinea pig line of descent only.

To evaluate the rate of gene evolution within a species, Fitch and Neel (1969) estimated substitution rates for six blood group loci in three South and Central American Indian tribes. Using a separation time of 15,000 years between the tribes, they estimated an approximate 0.16 gene substitution in any one tribe, which amounts to $0.16/(6 \times 15,000) = 1.8 \times 10^{-6}$ per locus per year. If we assume 1,000 nucleotide pairs for each gene, we obtain an average substitutional rate of 5.5×10^{-9} per codon per year. This value comes out very close to the estimate of Fitch and Margoliash.

3. Kimura's calculation is based on the average of 100 amino acids of a protein. He took Muller's estimate of 4×10^9 as the number of nucleotide base pairs for all the nuclear genes, and used 1.2 as a correction for synonymous mutations for which 1.3 is more accurate. When 1.3 is used as the correction factor, the average substitutional rate for any gene of the whole genome is

$$28 \times 10^6 \text{ year} \div \left(\frac{4 \times 10^9}{300} \right) \div 1.3 = 1.6 \text{ years.}$$

The rate of substitution per codon per year would be
$$1/[1.6(4 \times 10^9)/3] = 0.48 \times 10^{-9}.$$

For appreciation of the difference in rate of evolution between genes, the percentage differences in amino acids were plotted against time for the phyletic lines for three proteins: fibrinopeptides, hemoglobin, and cytochrome c (fig. 2.17). First, it should be noted that the percent of amino acid changes is approximately linear with respect to time for each protein. This suggests that the evolution rate for each protein is uniform through time. Second, the difference in slope clearly shows the striking differences in the rate of evolution of the genes for the proteins. Whereas twenty million years are required to produce a 1% amino acid change in cytochrome c of two phyletic lines, for the same percent change it takes a little less than six million years in hemoglobins, and about one million years in the fibrinopeptides. What is the explanation for the rate differences between these proteins?

We have discussed the mutation rates for different genes. Since genes consist of different numbers of nucleotides, the larger the number of nucleotides, the greater the chance of error of base substitution during replication. Differences in mutation rate among genes do not necessarily imply differences in rate of nucleotide base substitution. So far there has been no evidence showing that error of substitution (mutation) is more frequent in one specific base than in others. The differences in the rate of evolution between the genes are clearly due to differences in natural selection pressures.

Selection pressure concerns both the physical property of the individual amino acid and the physiological role that the whole protein molecule plays. Alterations in the proteins must be of a nature that improves, or at least does not impair, their function. There are sites in the polypeptide chains where no amino acid substitutions have been found, and others where changes have been limited to the polar types, hydrophobic or hydrophilic, aromatic or aliphatic types. Smith (1966) called the latter type of substitution "conservative," that is, limited to amino acids with certain specific properties. Any mutations other than those will produce harmful effects and will be rapidly eliminated.

As far as the whole range of protein molecules is concerned, selective pressure seems to be correlated with their specific functions and chemical mechanisms. For instance, the fibrinopeptides in the blood are merely serving as spacers that prevent the fibrinogen from adopting the fibrin configuration before the clotting process begins. They seem to have no other requirements as long as they can be removed by an enzyme when the time comes for the blood to clot. The hemoglobin molecules have more constraints than the fibrinopeptides. Each molecule embodies four heme groups and is responsible for binding and releasing oxygen. A random mutation of the gene is therefore expected to be more harmful in hemoglobin than in fibrinopeptides. Cytochrome c is a small molecule and interacts over a large part of its surface with a molecular complex larger than itself; thus, a perfect match over a large portion of its surface is required. This is probably why the property of positive and negative charges of cytochrome c has been so faithfully preserved throughout the long history of evolution. Histone IV, one of the basic proteins binding to DNA, is apparently the most conservative, since a difference of only 2 out of 102 amino acids was found between molecules of pea seedlings and those of the calf thymus. Since histone IV controls the DNA replicating process (see chap. 4), the center of genetic information, high sensitivity to the substitution of any of its amino acids is to be expected. The evolution of protein molecules is somewhat like that of certain specific organs, with shape, form, and function all of importance. In a sense, they must preserve their basic function while being continually modified for adaptation to their internal environment.

SUMMARY

The discovery of the genetic code, the amino acid sequence analysis of proteins, and DNA association studies, as the outgrowth of the double helix of Watson and Crick, have led to an insight into gene structure and gene mutation. What I hope to have made clear is that the evolution of genes, from their primitive form to the present stage, depends essentially on gene duplication and nucleotide base substitution. The former is produced by errors in chromosome pairing or unequal crossing-over, and the latter by errors in nucleotide replication. Duplication, which may be in a unit of a part of a gene, a whole gene, or a number of genes, is a biological device to provide "genetic space." Only in this space can new genetic information, provided by alteration of nucleotide base pairs, be added. Bridges (1936) saw an evolutionary significance of gene duplication in his time and stated:

> In my first report on duplications at the 1918 meeting of the A.A.A.S.
> [American Association for the Advancement of Science], I emphasized
> the point that the main interest(s) in duplications lay in their
> offering a method for evolutionary increase in lengths of chromosomes
> with identical genes which could subsequently mutate separately
> and diversify their effects. The present demonstration that certain sections
> of normal chromosomes have actually been built up in blocks through
> such "repeats" goes far toward explaining species initiation.

In *Mathematical Challenges to the Neo-Darwinian Interpretation of Evolution* (1967), Eden and others claimed that biologists had not yet supplied enough information to make neo-Darwinian evolution adequate as a scientific theory. It appears to me that confusion arose from some misconceptions involving mutation processes. Gene duplication, one of the fundamental events in mutation that adds genetic information, was not taken into consideration. Of course, symbiosis and chromosome mutation are other shortcuts for gaining genetic information on a large scale. Nature operates in the shortest way possible.

Perhaps one of the most interesting facts revealed in protein analysis is the gene homology between species. These homology studies, referred to as molecular taxonomy, become a new approach to phylogeny, parallel and supplementary to classical taxonomy.

Primitive genes and proteins presumably no longer exist and most contemporary genes have lost their original identity through evolution. We can trace the origins

of some homologous genes, but not the nonhomologous genes. Nevertheless, overall evidence and theory seem to support a hypothesis of the monophyletic origin of
genes, which in turn fits well with the monophyletic theory of the origin of species.

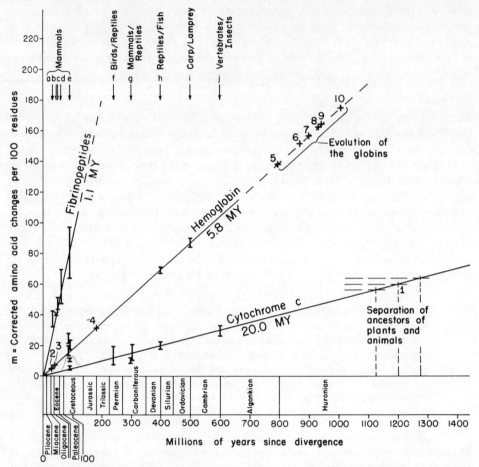

Fig. 2.17. Rates of evolution of proteins may be inferred by plotting average differences in amino acid sequences between species on two sides of an evolutionary
branch point that can be dated; for example, the branch point between fish and
reptiles or between reptiles and mammals. The average differences (vertical axis)
have been corrected to allow for the occurrence of more than one mutation at a
given amino acid site. The length of the vertical data bars indicates the experimental scatter. Times since the divergence of two lines of organisms from a common branch point (horizontal axis) have been obtained from the geological record.
The rate of change is proportional to the steepness of the curve. It can be represented by a number called the unit evolutionary period, which is the time required
for the amino acid sequence of a protein to change by 1% after two evolutionary

lines have diverged. For fibrinopeptides this period is about 1.1 million years (MY), hemoglobin 5.8 MY, and cytochrome c 20.0 MY. Point 1 represents a date of 1,200 $\pm$ 75 MY for the separation of plants and animals, based on a linear extrapolation of the cytochrome curve. Points 2–10 refer to events in the development of the globin family the δ/β separation is at point 3, γ/β is at 4, and so on. Under mammals, a = divergence of deer and cattle, b = divergence of dog and cat families, c = divergence of camel, llama, and vicuna from deer and cattle, d = separation of swine from other artiodactyls, and e = differentiation of various orders of mammals. (Reprinted, by permission, from R. E. Dickerson, *J. Mol. Evol.* 1(1971):26–45.)

symbiosis

Supposing mutation to have occurred in two different organisms, one having acquired a new gene at one locus and another at a different locus, both genes would be useful, and together would contribute greater fitness to an organism than either one could alone. How can they be brought into a single complex organism? Obviously, this can be accomplished by sexual reproduction through the union of two gametes which act as gene carriers. But sexual reproduction as a way of perpetuating life was not evolved until long after the first organisms appeared. There may have been, then, another means by which a similar purpose was achieved: symbiosis.

Symbiosis as a cooperative way of life allows one organism to borrow genetic information from another, not only as it involves one gene, but all the genes present in one member and not in the other. Between two organisms of different species, obviously a large number of genes must be different. Results from cell and microbial research suggest that, as variant forms of association, symbiosis could have occurred at various levels, including that of genome integration and intracellular incorporation of chromosomes.

As symbionts (the smaller of two symbiotic members), some organisms, apparently having chosen this mode of life in the early history of evolution, still retain it. Symbiosis as a biological process of adding to and expanding genetic information can be traced in the cell formation of present life.

INTERCELLULAR SYMBIOSIS

The classical and perhaps the best known example of symbiosis is provided by the lichens (Fink 1935), a group of plants unique in being a combination of two different groups: algae and fungi. Most of the fungi are members of the Ascomycetes, but the algae belong either to the Cyanophyta, or blue green algae, or the Chlorophyta, or green algae. The shape of the lichens depends on the specific combination of the alga and the fungus; for example, where the association involves the blue green algae, the shape of the lichen is usually determined by that alga, and the fungal threads are usually within its gelatinous sheath (fig. 3.1). However, in the case of an association with the green algae of the Chlorophyta, the fungus determines the shape of the lichen. In any case, the association is an intercellular or intertissue type of combination.

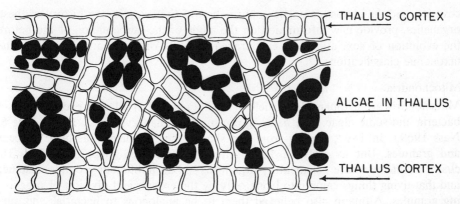

THALLUS CORTEX

ALGAE IN THALLUS

THALLUS CORTEX

Fig. 3.1. Section of thallus of a lichen (*Leptogium cyanescens*) showing inter-cellular symbiosis. The fungus produces a definite layer of cells on either side of the thallus, and between the cortex the threads of fungus are intermixed with those of the alga.

Symbiosis in lichens stems from the fact that while fungi absorb water from the atmosphere, algae manufacture food by photosynthesis; in addition there exists an exchange of complex organic compounds between them. Such association has made the lichen extremely adaptive, occupying niches tolerated by no other plant. For instance, lichen is the most dominant form of vegetation in the alpine and arctic tundras.

There is no fossil record of the antiquity of lichens, but these symbiotic organisms must have evolved after the appearance of their components. It is assumed that the earliest lichens were a loose association of fungal hyphae with free-living algae. It is believed that some lichens are still on this border line, and that evolution from a looser to a more definite structure is continually in process.

INTRACELLULAR SYMBIOSIS

Surprisingly, cells of the ancestors of present higher organisms may have once harbored a body or bodies of free-living organisms like bacteria. These bacterialike organisms somehow entered the host cells, having adapted to intracellular living, and presently passed through the gametes as cellular organelles, identified as plastids and mitochondria ever since their discovery at the end of the last century. It is only in the present decade that evidence from the morphology, chemistry, and physiology of these organelles has provided so vivid a picture indicating that they were bacterialike organisms and that they entered the cells of the ancestors of the present life as symbionts. Through the long evolutionary journey, they have neither completely merged with the nuclei of the host cells nor lost their own identity. As a result of certain adaptive evolutionary changes made in both the host and the symbiont, however, one can function and live well without the other. Such discoveries, and the almost complete confirmation of the evolutionary history of these organelles, provide new information on the mode of the perpetuation of life before the evolution of sex, and have encouraged students of systematics to reexamine taxonomic classifications.

Mitochondria

Mitochondria are cytoplasmic particles (fig. 3.2) found in all cells except those of bacteria and some algae. The name, given by a German cytologist, Benda (see S. Nass 1969), in 1897, derives from the shapes they frequently assume: threads and granules. But as early as 1890 Altmann discovered these organelles. He claimed that they were "elementary organisms" in an indifferent ground substance and that living things came into being only by the growth and division of preexisting granules. Altmann also believed them to be analogous to bacteria, and suggested that the bioblast could even exist as a free-living form in some cases when it was not part of a bacterial colony within the cell.

Working from Altmann's granulum theory, Mereschovsky (1910) hypothesized that the eukaryotic cell was derived from a symbiotic association of two types of ancestral microorganisms: nonnucleated amoebaplasm and the bacterialike biococci, which evolved into the cell nucleus. Minchin (see Wilson 1928) claimed that the eukaryotic cell evolved from biococci composed of chromatin and that the cytoplasm was a secondary product in which the biococci were suspended.

Portier (1918) held that mitochondria were symbiotic bacteria, capable of being cultivated in vitro. Wallin (1927) suggested that bacteria evolved into eukaryotic cells via varying degrees of symbiotic association. Regarding speculations on the origin of the eunucleate cell, Wilson (1928) suggested that: "To many, no doubt, such speculations may appear too fantastic for present mention in polite biological society; nevertheless it is within the range of possibility that they may someday call for more serious consideration."

Not until the middle 1950s were mitochondria recognized as prime centers of oxidative metabolism, playing an important role in aerobic respiration. Mitochondrial enzymes are involved in the Krebs cycle and its associated electron transport system, which produces ATP, a high-grade universal fuel necessary to the life process. The mitochondrial enzymes also catalyze a wide array of such chemical reactions as concentrating ions and synthesizing proteins.

Evidence linking mitochondria and the inheritance of certain characteristics was obtained from a study of mutants of *Neurospora,* a fungus (Mitchell and Mitchell 1952). The "poky" mutants, showing a pattern of slow growth not affected by supplementing the medium with growth factors, are inherited only through the female parent. (The male *Neurospora* parent contributes only the nucleus.) One poky strain showing deficiencies in respiratory capacity had cytochromes b and $a + a^3$ missing. Such deficiencies result from mitochondrial gene mutation.

All the plant and animal mitochondria so far examined possess DNA, and protein biosynthesis in isolated mitochondria, relatively independent of the nucleus,

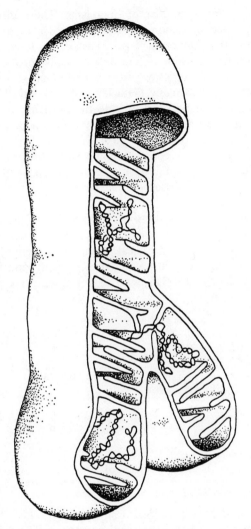

Fig. 3.2. Possible arrangement of circular DNA with a mitochondrion. A branched mitochondrion typical of L cells (mouse fibroblasts) and many other cell types emphasizes the polymorphous structure of these organelles. The number of nucleoids containing DNA molecules is variable. The DNA molecules may be attached to portions of the membranes. The DNA circles (half-length 2μ to 2.5μ) may be coiled or folded inside the matrix compartments, which are about 0.5μ in diameter. The twists or supercoils shown on the DNA do not necessarily reflect their occurrence or number in vivo. (Reprinted, by permission, from M. M. K. Nass, *Science* 165(1969):25–35. © 1969 by the American Association for the Advancement of Science.)

has been demonstrated for a number of organisms. These discoveries have provided the information that mitochondria are involved in cytoplasmic inheritance, a genetic phenomenon that has puzzled geneticists for a long time.

Mitochondria replicate themselves quite independently of the nucleus, new ones being derived from the old by fission after a period of growth, as observed in *Neurospora* with electron microscopy (Hawley and Wagner 1967). Mitochondrial protein is synthesized both inside and outside the mitochondria.

Isolated mitochondria can perform most of the fundamental biochemical cycles necessary for life, including oxidative phosphorylation and protein, lipid, and nucleic acid synthesis; like many endosymbionts and parasites, however, they are incapable of synthesizing all of the enzymes they contain, and are dependent on their eukaryotic[1] hosts for some essential materials.

Evidence for an Evolutionary Relationship between Mitochondria and Bacteria

1. *Size and shape.* Bacteria and mitochondria are similar in shape and size (table 3.1) and may have like cytochemical properties, as demonstrated by Bourne and Tewari (1964), through staining with Janus green B, and by Regaud (1919; see S. Nass 1969) and Knaysi (1951), who disclosed the presence of phospholipid and protein.

TABLE 3.1 Size and Shape of Bacteria and Mitochondria

Mass	Shape	Width (μ)	Diameter (μ)	Length (μ)
Bacteria 1.1 × 10⁻¹² gm,	Mostly rods	0.2–2.0	. . .	0.3–5 (up to 50 reported)
for example, *E. coli*	Rods	0.4–0.7	. . .	1.0–3.0
	Spherical	. . .	0.8–2.5	. . .
Mitochondria 1.3 × 10⁻¹³ gm,	Mostly rods	0.2–0.5	. . .	3–5 (up to 40 reported)
for example, rat liver, mitochondria	Rods	0.5	. . .	3.3
	Spherical	. . .	0.2–1	. . .

SOURCE: Reprinted, by permission, from S. Nass, *Intern. Rev. Cytol.* 25(1969): 55-129.

2. *Membranes.* The difference between mitochondria and bacteria (prokaryotes[1]) is in the composition of their external envelopes. The cell wall of bacteria is composed of mucopolysaccharides or lipopolysaccharides and peptides, while sialic acid was found in some mitochondrial preparations and was thought to be a vestige of bacterial cell-wall mucopolysaccharides. M.M.K. Nass (1969) believed that the outer layers of both mitochondria and prokaryotes were either secondarily

1. The major distinction between prokaryotes and eukaryotes is in the nuclear structure and cell division process. The former have a single linear structure of DNA molecules and the nuclear division is invariably direct. The genome of the eukaryotes contains two or more DNA molecules associated with basic proteins, with complicated cell division processes. Bacteria and blue green algae are prokaryotes; all higher organisms are eukaryotes. See chapter 4 for further description.

derived or nonessential to the vital functions of these organisms and organelles and were, hence, no barrier to a consideration of their relationship. However, the ultrastructure, molecular composition, and specific functions of the bacterial membrane and the mitochondrial inner membrane are more closely related than are the bacterial and other membranes found in the eukaryotic cell.

Most characteristic of the mitochondrial membrane is the indentation (fig. 3.2) that penetrates into the matrix of the organelle, similar to the infolding of a bacterial plasma membrane. Another is a stalked 90-Å diameter subunit in the mitochondria similar to that found in the bacterial membrane. All major components of the respiratory chain are present in isolated mitochondrial and bacterial membranes. Stanier's (1954) suggestion that the bacterial cell membrane was the functional equivalent of that of the mitochondrion was later validated (Marr 1960; Lascelles 1965).

The presence of an ion-regulating mechanism within a cell organelle can be taken as presumptive evidence that the organelle was originally a free-living organism. Ion accumulation is very similar in bacteria and mitochondria (see S. Nass 1969). It is clear that at the present point in evolution many differences separate mitochondrial and bacterial functions, but equally striking is the fact that so many similarities still exist between them even after a presumably long period of independent evolution, indicating that the membranes are homologous in structure.

3. *DNA.* Chevremont, Chevremont-Comhaire, and Baeckeland (1959) and Chevremont (1963) were the first to present evidence for the existence of mitochondrial DNA. Many more recent studies still require confirmation, but certain generalizations can now be made: (*a*) a circular configuration appears to be almost universal; (*b*), the contour length of approximately 5μ, equivalent to the molecular weight of 10×10^6 daltons per monomeric unit, is constant; (*c*), although the DNA content per mitochondrion (2×10^{-17} gm to 5×10^{-16} gm) does not reflect the broad range of values found for nuclear DNA of eukaryotes (7×10^{-14} gm to 7×10^{-10} gm), about 2% to 5% of that found in bacteria, the much smaller quantity of DNA in mitochondria may not be taken as evidence against an evolutionary relationship between mitochondria and bacteria, since there are 1,000-fold differences in the DNA content of some eukaryotes (see table 4.1); (*d*) unlike nuclear DNA there is no apparent relationship between the base contents of mitochondrial DNA from closely related organisms; (*e*) the mean guanine cytosine (GC) content of mitochondria from various organisms of between 20% to 50% is close in range to that found in bacterial species, but greater than that observed for all eukaryotes except protozoa and the unicellular plants, such as algae, fungi, and lichens, in the division of protophytes.

Plastids

Plastids are discrete cytoplasmic organelles found in the cells of higher plants and certain unicellular organisms. Just as bacteria gave rise to mitochondria, it is believed that, through symbiosis, plastids in the higher plant cells were derived from one group of blue green algae. Before discussing the possible evidence for such a relationship, it may be helpful to review briefly the function and structure of the plastids, as well as the characteristics and classifications of the algae.

Plastids often occur as free spherical or disklike bodies in the cytoplasm and are ordinarily grouped into two classes: leukoplasts (colorless) and chromoplasts (pigmented). Leukoplasts, usually found in plant tissue not exposed to light, have as their function the formation and storage of starch granules or oil droplets, and are not associated with photosynthesis. Chromoplasts may or may not be photosynthetic. Those that are photosynthetic are called chloroplasts, and they vary considerably in size, shape, and pigment content. Some, containing chlorophyll, appear green; others, where the chlorophyll is masked by carotenoids or by phycobilin proteins (proteins combined with pyrrole derivatives related to bile pigments), appear brown or red, respectively. Any pigments present help the chloroplasts to absorb light for the photosynthesis.

Chloroplasts are the most prevalent chromoplasts occurring in plant cells, with sizes ranging from 3μ to 7μ. Each is constructed with numerous nongreen granular materials called stroma, in which are embedded minute disklike units called grana, consisting of parallel chlorophyll-containing laminations of lamellae (fig. 3.3).

Isolated chloroplasts, when suspended in a defined medium, divide (fig. 3.3) (Ridley and Leech 1970). Chloroplasts from young expanding leaves of *Vicia faba* were observed in the process: they remained "dumbbell" shaped for several hours; then the central constriction became tubelike, and within an hour the two halves were completely separate. The configuration of the dividing plastids closely resembles the appearance of intracellularly dividing algal chloroplasts of the genus *Sphacelaria* (small, feathery, brown algae).

Algae

Algae consist of a large group of simple, free-living chlorophyllous plants of the subkingdom Thallophyta, separate from the rest of the thallophytes, which are fungi, by reason of their possession of chlorophyll, which gives them the ability to manufacture their own food by photosynthesis. Algae are chiefly water plants, varying in size from unicellular microscopic organisms to seaweeds several hundred feet long, and only slightly differentiated in tissue. They lack the characteristic complex vascular systems, such as xylem and phloem, and other tissues of higher plants, as well a true stems, roots, and leaves.

Reproduction processes in algae are exceedingly varied. Certain primitive species reproduce only asexually, by cell division, or fission, or a primitive type of sexual isogamy (union of morphologically similar cells). Higher forms reproduce by heterogamy, the union of dissimilar sexual cells, such as sperm and eggs. Blue green algae are simple unicelluar colonial organisms, lacking differentiated nuclei and plastids in their protoplasm and reproducing by fission. A bluish pigment is present in addition to chlorophyll. Blue green algae are of the class Cyanophyceae, division Schizophyta of the subkingdom Thallophyta and they reproduce by cell division. Green algae are of the division of Chlorophyta including the true green

algae and the stoneworts. Red algae, of the division Rhodophyta of the subkingdom Thallophyta, often resemble certain blue green algae in color. Some show a
complex branched structure; the most advanced are found in tropical and semi-
tropical seas. Red algae are sexually reproducing organisms, but the spermatia are
nonmotile and bring about fertilization by drifting to the female reproductive

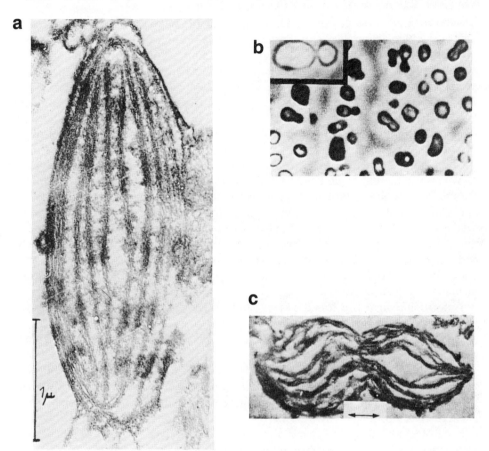

Fig. 3.3. a, Electron micrograph of cross section of a plastid in the brown alga
Fucus, showing parallel chlorophyll-containing laminations of lamellae, embedded
in the stroma. (Reprinted from D. von Wettstein, 1961, in *The Encyclopedia of
the Biological Sciences,* Peter Gray, ed. © 1961 and 1970 by Litton Educational
Pub., Inc., with permission from the Van Nostrand Reinhold Company.) *b* and *c,*
Chloroplasts isolated from leaves of spinach, *Vicia faba,* and suspended in a
medium, (*b*) showing dumbbell-shaped chloroplasts (×880) and (*c*) a dividing
chloroplast. (Reprinted, by permission, from S. M. Ridley and R. M. Leech,
Nature 227(1970):463–65.)

organ. Brown algae are of the division Phaeophyta and include the Isogeneratae. They have flattened, many-branched stalks and typically are anchored by a hold-fast. Isogeneratae are ancestral to the Heterogeneratae (kelp and Cyclosporeae). Among the above-mentioned algae, only blue green algae are prokaryotic organisms.

Symbiotic Relationships between Plastids and Algae

The very simple structure of the chloroplasts in red algae is virtually identical to that of the blue green algae. Furthermore, as one group of free-living prokaryotes, the blue green algae are biochemically and structurally similar to the chloroplasts of many eukaryotes. As symbionts, blue green algae are often found in the cells of amoebas, flagellated protozoa, green algae, and diatoms (Droop 1963; Geitler 1959). These observations practically establish that the chloroplasts in both the higher and lower forms of eukaryotes were derived from blue green algae through symbiosis. In addition, symbiosis between blue green algae and other organisms is continuously taking place.

The blue green algae have been in existence for about 2.7 billion years (Schopf 1967), while the red algae have been in existence probably less than 650 million. On this basis, some investigators have questioned whether the origin of the chloro-plasts in the red algae is blue green algae. In my opinion, for the occurrence of symbiosis, it is the cell structure and function that are the important factors, not the history of survival of the organisms. Evidently, the simple cellular structure and the photosynthetic property of the blue green algae are biological features that favor such a process. This explains the presence of blue green algae as symbionts in many living organisms.

Autotrophic algae are found as symbionts in more than 150 genera of inverte-brates (Smith, Church, and McCarthy 1969), in addition to higher plants and other algae. The number of such associations observed in living organisms sug-gests that they arose with relative ease.

Based on the similarities and differences between the biochemistry of the pro-karyotic blue green algae on the one hand and the chloroplasts of the different algal divisions on the other, the evolution of chloroplasts in the prokaryotes and eukaryotes is postulated (Raven 1970). From cholorophyll a and probably cer-tain carotenoids, three biochemically distinct derivatives arise: (1) the living blue green algae, in which phycobilins evolved; (2) the "green prokaryotes" in which chlorophyll b evolved; and (3) the "yellow prokaryotes," in which chloro-phyll c evolved. Each of these groups had entered into symbiotic associations with primitive eukaryotic cells probably more than once. Derivatives 2 and 3 no longer exist as free-living organisms. The chloroplasts of the Rhodophyta (red algae) and Cryptophyceae (motile brown green algae, sometimes included in the Pyrrhophyta) were derived from symbionts of group 1; those of Chlorophyta (green algae) and Euglenophyta (euglenoids) from group 2; and those of Phaeophyta (brown algae), Chrysophyta (yellow green algae), Xanthophyceae, and Pyrrhophyta (dinoflagel-lates, see chap. 4) from group 3.

It may be said that there is concrete evidence that in the cells of eukaryotes the mitochondria were derived from bacteria and the chloroplasts from blue green algae by symbiosis in the remote past of evolution. In view of the fact that the virus genome can be incorporated into the bacterial genome, as will be discussed next, why did the genomes of these organelles not integrate into the genomes of the hosts? It may be because there is a great difference in the chromosome structure between eukaryotes and prokaryotes. Also there is a great amount of DNA in the organelles for the eukaryotic nuclei to accommodate. These are merely speculations and we have no good answer at present.

GENOME INTEGRATION

The intimate association of viral DNA with bacterial, plant, or animal cell DNA has provocative implications. Such integration can be called symbiosis in a broad sense; the mechanisms and the consequences of gene integration not only concern the origin of viruses and the relation of some animal viruses to their host cells but are of profound implication in the evolution of genes. The integration of viral genomes with bacterial or animal cell genomes is different from the symbiosis of mitochondria and plastids; that is, the viral genome is incorporated into the host genome, and this process can be observed experimentally.

We can imagine that in the primeval soup there were numerous virus- or bacteriumlike primitive organisms. As coupling or polymerization of amino acids or peptides is assumed in the evolution of macromolecules, genome integration would result in an increase of the size of DNA molecules and might be accounted an essential process of evolution in primitive organisms.

Lysogeny

When a sensitive bacterial strain is infected with bacteriophages, two alternative responses may result. One is the lysis of the bacterium cells and release of the newly formed infectious progeny bacteriophages after a latent period. The process takes place by first the vegetative production of the bacteriophage DNA and then the production of proteins needed for transmission of DNA to other cells. In some cells they survive after the infection, and the bacteriophage genome is incorporated into the bacterial genome. This process of infection is called lysogeny. Bacteria-carrying noninfectious bacteriophages are said to be lysogenized. In lysogenized bacteria, bacteriophage DNA persists in a nonvegetative form and replicates synchronously with the cell genome. There is now good genetic evidence that the viral genome constitutes an integral part of the host genome.

The process of integration is believed to be achieved by crossing-over between the host chromosome and the viral chromosome (Campbell 1962), a hypothesis well confirmed by various chemical and biological techniques. Prior to integration, the viral chromosome forms a circle and attaches to a specific region of the host chromosome. Both chromosomes then break and rejoin in such a way that the ends of the viral chromosome join to the broken ends of the bacterial chromosome instead of to each other. As a result, the bacteriophage DNA is inserted linearly into the bacterial chromosome at a specific locus (fig. 3.4). If the formation of the circle by end joining and its opening by recombination occur at different places on the circular structure, then the order of prophage markers is a cyclic permutation of the vegetative order. Detachment of the prophage occurs by an inverse recombinational event, and the circle is then opened at the correct point to regenerate the linear bacteriophage order.

The Campbell model provides an accurate description of the foregoing type of integration. The main features of the model—circularization, recombination, and linear insertion—are supported by a large body of indirect and direct evidence (Singer 1968).

Oncogenesis

Oncogenesis is the production or causation of tumor. Oncogenic viruses are divided into two major classes: those having DNA as a genome (DNA viruses) and those containing RNA (RNA viruses). Infection with oncogenic DNA viruses results in either productive infection, in which the virus replicates and the cell dies; or abortive infection, in which virus replication is blocked out, but under certain conditions the abortively infected cells may be transformed.

Polyoma (Py) and simian virus 40 (SV40) are tumor viruses (Green 1970). Polyoma virus was isolated from mouse cell cultures inoculated with organ extracts of leukemic mice and of mouse mammary cancers. The virus induces tumors of different histological types in mice, hamsters, and other rodents, but it never produces leukemia. The SV40 virus replicates extensively in normal and established lines of African green monkey kidney cells, and replicates to a lesser extent in human cells (Kit, Dubbs, and Somers 1971). Both Py and SV40 are small double-stranded DNA viruses (Kit, Dubbs, and Somers 1971). They contain 12% DNA and 88% protein, and possess a single circular duplex DNA molecule that is both infectious and oncogenic. The SV40 DNA has a molecular weight of 2.3×10^6 daltons to 2.5×10^6 daltons, and Py DNA has a molecular weight of 2.9×10^6 daltons to 3.4×10^6 daltons. No homology was detected between the DNAs of Py and SV40. The SV40 and Py genomes consist of about 4,500 nucleotide base pairs that contain sufficient information for about 1,500 amino acids.

Cells transformed by Py and SV40 (Benjamin 1966) are found to synthesize virus-specific RNA and thus must contain persisting viral genes. Studies by Sambrook and co-workers (1968) suggest that SV40 DNA is integrated into cellular DNA. They showed the presence of viral DNA sequences in isolated chromosomes of the cells, using DNA hybridization techniques. They also found that the DNA sequences are linked to high molecular cellular DNA by covalent bonds. Further evidence of integration has been provided by the demonstration of Py and SV40 transplantation antigens. Animals immunized with cells transformed by the virus, or with the virus itself, rejected the tumor cell transplants.

Working with mouse leukemia, a number of investigators have found that some viral genes have been incorporated into the chromosomes of certain inbred strains of mice (Rowe 1973). Using the nucleotide hybridization methods, Spiegelman (1973) found the presence of the viral genes in the genomes of a number of mammalian species. Although the issue on the viral genes as an etiological factor

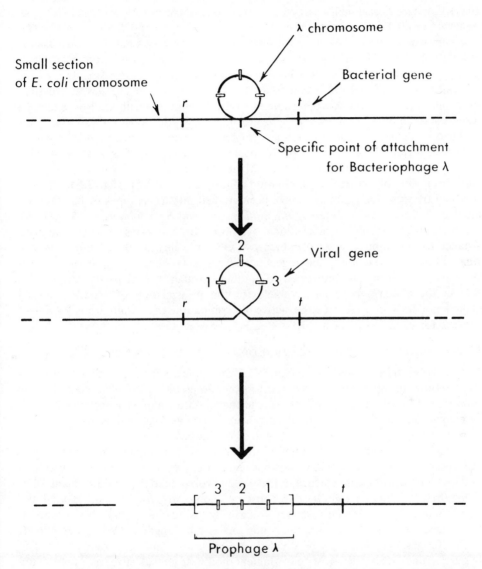

Fig. 3.4. Insertion of the chromosome of bacteriophage into the *E. coli* chromosome by crossing-over (Campbell hypothesis). (Reprinted, by permission, from James D. Watson, *Molecular Biology of the Gene*, 2d ed. © 1970 by J. D. Watson; W. A. Benjamin, Inc., Menlo Park, Calif.)

of leukemia is still highly controversial, their wide presence in the genomes of many species is of great interest from the evolutionary point of view.

Transduction

Bacteriophages occasionally carry genes from one bacterium to another. This phenomenon, called transduction, was discovered by Zinder and Lederberg (1952). It occurs when a virus particle is formed that carries a very small portion (usually less than 1% to 2%) of its host chromosome. When this virus particle (a transducing bacteriophage) reaches a host bacterium, the chromosome fragment of the original host is injected into the new host cell. This fragment engages in crossing-over with the new host chromosome, thereby integrating with it. For example, when bacteriophage particles P1 were grown on a strain of *E. coli* able to grow on lactose, a small number of the bacteriophages carried the gene (lac^+) involved in lactose metabolism. Addition of the bacteriophages to an *E. coli* strain unable to use lactose (lac^-) transformed a small number of the lac^- bacteria to the lac^+ form by means of genetic recombination (Davis et al. 1967) (fig. 3.5).

Many transducible characters have been studied in various types of bacteria, including *D. pneumoniae, Haemophilus influenzae,* and *S. typhimurium. Salmonella* has many traits, including biochemical synthesis, fermentations of sugars, and resistance to antiobiotics, most of which can evidently be transduced one unit at a time. Thus, transduction is phylogenetically a special form of symbiosis, incorporating genes from one individual to another through a third party. The passive role of the bacteriophage as a vehicle for the transmission of genetic material should not be confused with the active determination of certain traits by latent viruses, for example, lysogeny.

HETEROKARYOSIS AND THE EVOLUTION OF SEXUAL REPRODUCTION

In line with symbiosis and the evolution of chloroplasts, it is appropriate to introduce another group of organisms and to examine its biological characteristics and especially its method of reproduction: the fungi. As an important group of primitive organisms in phylogeny, the occurrence of both asexual and sexual reproduction is of special interest from the point of evolution of sex.

Fungi, being devoid of chlorophyll, are unable to manufacture their own food; they obtain energy by breaking down organic matter in dead plant and animal bodies. Classified for convenience in the old division Thallophyta of the plant kingdom, their origin and affinities are obscure. Some mycologists still defend the theory that fungi originated from green algae by losing chlorophyll; but a modern view places them in neither the plant nor the animal kingdom (Whittaker 1969).

The fungus life cycle may be divided into an asexual and a sexual phase. The hyphae (fungi with a microscopic filament body structure) give rise to asexual spores that germinate and again produce hyphae. Thus the asexual cycle may repeat itself indefinitely under certain environmental conditions. The same hyphae may give rise to sex cells, which unite and initiate the sexual phase (fig. 3.6). In addition to these two regular types of reproduction, Hansen and Smith (1932) observed cells with more than a single nucleus in a common cytoplasm, in the species Fungi Imperfecti, resulting perhaps from some reproductive process of neither of the above types. Such cells are referred to as heterokaryon and the property as heterokaryosis.

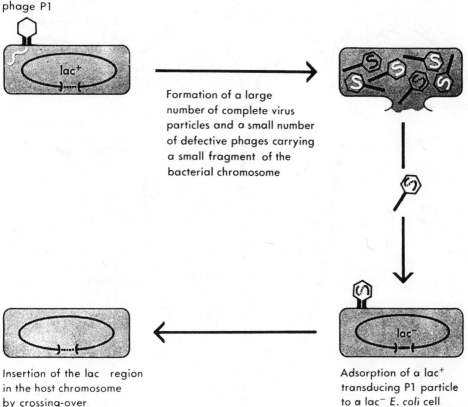

Infection of E. coli by phage P1

Formation of a large number of complete virus particles and a small number of defective phages carrying a small fragment of the bacterial chromosome

Insertion of the lac region in the host chromosome by crossing-over

Adsorption of a lac$^+$ transducing P1 particle to a lac$^-$ E. coli cell.

Fig. 3.5. Passive transfer of genetic material from one bacterium to another by means of carrier bacteriophage particles (transduction). (Reprinted, by permission, from James D. Watson, *Molecular Biology of the Gene*, 2d ed. © 1970 by J. D. Watson; W. A. Benjamin, Inc., Menlo Park, Calif.)

Heterokaryotic associations are not confined to combinations within species. Bisexual heterokaryons of the fungi *N. crassa* have occurred (Lindegren 1934). They can also be experimentally produced (Dodge 1942; Pontecorvo 1946). An interesting experiment on heterokaryosis was carried out by Beadle and Coonrat (1944), who induced lethal mutations in *Neurospora* by ultraviolet light. Because of the mutation, homokaryotic segregates are inviable, but cells with different nuclei (one with the mutation and the other without) sustain each other given a stabilized or balanced heterokaryon, an obligate symbiosis comparable to balanced lethal heterozygotes in *Drosophila*. This is a form of biological intracytoplasmic "coexistence," different from the above described cases of chloroplasts and mitochondria with nuclei.

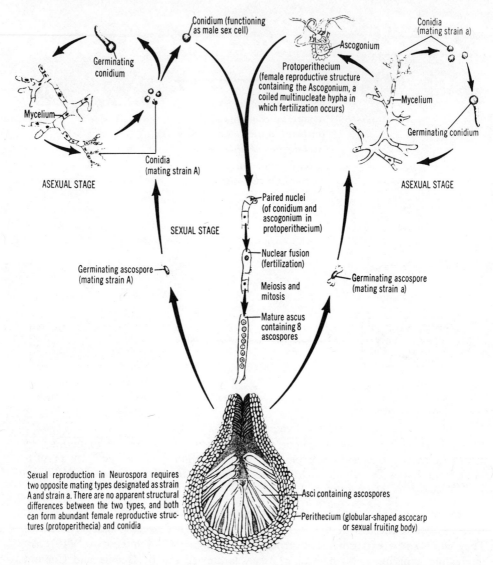

Fig. 3.6. Life cycle of the pink bread mold, *N. crassa.* (Reprinted, by permission, from Nason, 1965.)

Heterokaryotic nuclei can be separated and again assume asexual or sexual reproduction under favorable environmental conditions, and both contribute equally to the life of the cells. It should also be noted that the structure of the ascus (spore sac) (fig. 3.6) of *Neurospora* is such that it can hold a few spores at a certain stage of the sexual cycle. Such a structure appears as a primitive form of the ovary in higher organisms. It is therefore suggested that heterokaryosis was an early phylogenetic process toward the development of sexuality.

Darlington (1939) points out that in the evolution of sexual reproduction, nuclear fusion and meiosis must have occurred simultaneously, since one without the other would be not only useless but detrimental. He claims meiosis to be a consequence of nuclear fusion since, without such chromosome reduction, nuclear materials would accumulate continuously and become harmful to the cells. Nuclear fusion and meiosis could arise as a result of a single evolutionary change (Beadle and Coonradt 1944) (fig. 3.7), a scheme of sexual reproduction that could be applied in principle to any primitive organisms, such as Myxomycetes (slime mold), some algae, and numerous protozoa, particularly those with multinucleated cells.

Interspecific heterokaryons tend to be unstable but can be maintained under selective pressure. Recent studies on heterokaryon formation among different strains of *N. sitophila* showed that a gene controlling heterokaryon stability and stable heterokaryons can be produced from certain strains (Mishra 1971). This result is an indication that genetic compatibility in heterokaryosis may involve immune systems as in higher organisms, and may also provide a reason why heterokaryosis takes place only with difficulty in higher forms and is less productive than sexual recombination for potentially adaptive variations.

Heterokaryosis can be produced experimentally by fusing two somatic cells of the same individual or two different individuals from same or different species. This is a further evidence of the occurrence of symbiosis in nature. In biology, cell fusion has became a rather useful technique in studying somatic cell genetics and normal and abnormal cell growth in vitro; its discussion is here omitted.

PHYLOGENIC CLASSIFICATION AND EVOLUTION FROM PROKARYOTES TO EUKARYOTES

The recognition of the characteristics in both structure, function, and possible origins of the cytoplasmic organelles poses another dimension for phylogenic consideration. The traditional classification of all the organisms into two kingdoms (fig. 3.8) has been found unsatisfactory for many reasons; the most obvious one is the impossibility of clear division of the unicellular organims into plants or animals.

A number of investigators suggested a third kingdom for the lower organisms, and proposed the term *Protoctista* or *Protista* for them. Copeland (1956) further separated the lower organisms into two kingdoms: Monera for organisms without nuclei, such as bacteria and blue green algae, and Protoctista for the lower nuclear organisms, such as the protozoa, the red and brown algae, and the fungi.

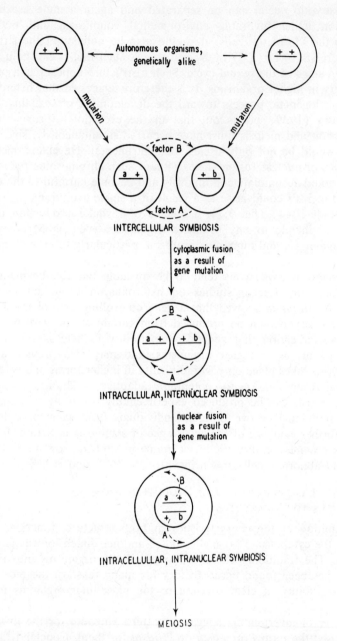

Fig. 3.7. Postulated steps in the evolution of sexual reproduction. (Reprinted by permission, from G. W. Beadle and V. L. Coonradt, *Genetics* 29(1944):291–308.)

Lately, Whittaker (1969) proposed the five-kingdom classification by separating the fungi out of the Protoctista (fig. 3.9): (1) Monera for prokaryotic cells without nucleic membrane (mitochondria, and so on), with the predominant mode of life being absorption, (2) Protista for eukaryotic unicellular organisms, (3) Fungi for multinucleate organisms with eukaryotic nuclei without plastids and photosynthetic pigments, (4) Plantae for the higher plants, and (5) Animalia for the higher animals. It can be seen that Whittaker's system is based on both morphology and function of the cells.

We should remember that phylogenic classifications are merely a convenience for study; there may always be instances where evolutionary changes escape the

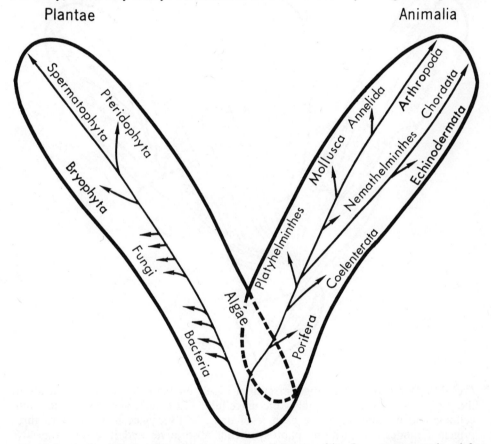

Fig. 3.8. A simplified evolutionary scheme of the two-kingdom system as it might have appeared early in the century. The plant kingdom comprised four divisions: Thallophyta (algae, bacteria, fungi), Bryophyta, Pteridophyta, and Spermatophyta. Only major animal phyla are indicated. (Reprinted, by permission, from R. H. Whittaker, *Science* 163(1969):150–60. © 1969 by the American Association for the Advancement of Science.)

rules of classification. However, it is important to understand the genetic relationships between organisms and their possible origins and deviations, although placing them in one system or another is a very minor problem. It may not be too much of a generalization to say that the basic units of the higher organisms of either the plant or animal kingdom were derived by symbiosis of prokaryotes. Symbiosis is one of the processes by which a higher level of cellular function is attained.

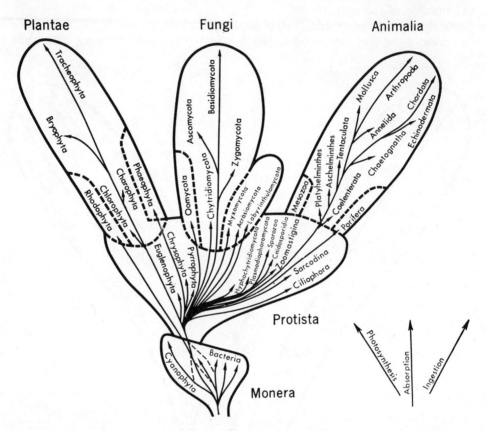

Fig. 3.9. Whittaker's five-kingdom system based on three levels of organization: the prokaryotic (Monera), eukaryotic unicellular (Protista), and eukaryotic multicellular and multinucleate. On each level there is divergence in relation to three principal modes of nutrition: photosynthetic, absorptive, and ingestive. Ingestive nutrition is lacking in the Monera, and the three modes are continuous along numerous evolutionary lines in the Protista; but on the multicellular-multinucleate level the nutritive modes lead to the widely different kinds of organization that characterize the three higher kingdoms: Plantae, Fungi, and Animalia. (Reprinted, by permission, from R. H. Whittaker, *Science* 163(1969):150–60. © 1969 by the American Association for the Advancement of Science.)

Margulis (1968) has made a serious effort to assess the role of symbiosis in evolution. She explored the geochemical and paleontological records and data of cytology, microbiology, and biochemistry in order to construct a comprehensive picture of the evolution of eukaryotic cells or organisms. In the context of symbiotic theory, she postulated that eukaryotic cells originated between 0.5 and 1.0 billion years ago (perhaps a billion years after the evolution of O_2-forming blue green algae) when the presence of atmospheric oxygen allowed the evolution of aerobic bacteria. The ancestral eukaryote or eukaryotes represented the fusion of an aerobic bacterium and another prokaryotic cell. The subsequent addition of a

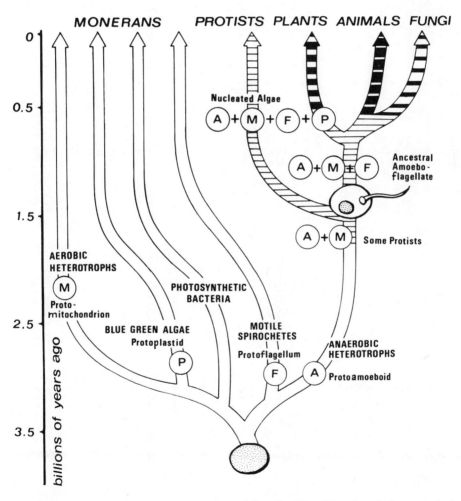

Fig. 3.10. The symbiotic theory of the origin of cells. (Reprinted, by permission, from Margulis, 1974.)

TABLE 3.2 Summary of the Five Kingdoms

Kingdom	Examples of Organisms	Genetic Organization	Approximate Time of Diversification and Documented First Appearance (millions of years ago)	Major Traits that Environmental Selection Pressures Acted on to Produce	Major Significant Selective Factor in the Environment
Monera	All prokaryotes: bacteria, blue green algae, mycelial bacteria, gliding bacteria, and so on	Prokaryote chromonema merozygotes only; sex unidirectional	Early-middle Precambrian (3000–1000); Fig Tree microfossils (> 3000); Bulawayan stromatolites (> 3000)	Ultraviolet photoprotection, photosynthesis, motility, and aerobiosis	Solar radiation, increasing atmospheric oxygen concentration, depletion of nutrients
Protista	All eukaryotic algae: green, yellow green, red and brown, and golden yellow; all protozoa; flagellated fungi; slime molds; slime net molds	Eukaryote chromosomes, ploidy levels, meiotic sexual systems vary, gametes and zygotes	Late Precambrian, early Paleozoic (1500–500): possibly Bitter Springs microflora (≈ 900); Ediacaran animals (750)	Mendelian genetic systems, mitosis and meiosis: obligate recombination each generation; phagocytosis, pinocytosis, intracellular motility	Depletion of nutrients
Animalia	Metazoa: all animals developing from blastulas	Diploids, meiosis precedes gametogenesis	Phanerozoic (600 on); Ediacaran fauna (750)	Tissue development for heterotrophic specializations; ingestive nutrition	Transitions from aquatic to terrestrial and aerial environments
Plantae	Metaphyta: all green plants developing from embryos	Alternation between haplo- and diplophases	Phanerozoic (600 on); rhyniophytes, Downtonian, Wales, Czechoslovakia, New York State (405)	Tissue development for autotrophic specializations; photosynthetic nutrition	Transitions from aquatic to terrestrial environments
Fungi	Amastigomycota: conjugation fungi, sac fungi (molds), club fungi (mushrooms), yeasts	Haploid and dikaryotic; zygote formation followed by meiosis and haploid spore formation	Phanerozoic (600 on); Rhynie chert, *Paleomyces asteroxyli* (400)	Advanced mycelial development; absorptive nutrition	Transitions from aquatic to terrestrial environments; nature of nutrient source, nature of host

SOURCE: Reprinted, by permission, from Margulis, 1974.

spirochetelike organism provided the motile form, subsequently active in the differentiation of centrioles (a minute cell organoid that is the focus of an aster in mitosis [see chap. 4 under dinoflagellates] and the mitotic apparatus). A summary of her symbiotic theory of the origin of cells is given in figure 3.10, and of the selection factors and selection pressure involved at the early stage of evolution in table 3.2. Cohen (1970) suggested the possibility that the formation of higher plants represents the infection of eukaryotic algae by a blue green alga.

SUMMARY

The evidence is strong in support of a general theme that, as a device for self-perpetuation, symbiosis preceded sexual reproduction. Since it allowed an organism to take genetic materials from another far different from itself phylogenetically, symbiosis was a device that speeded up evolution even faster than gene duplication. Once organisms evolved to higher levels, however, such a device could not operate easily because of physical and biological barriers; thus sexual reproduction became a necessary consequence of evolution.

Four

the evolution of chromosomes

The replication and transmission of chromosomes from one cell generation to another are usually accurate and precise, but now and then errors or accidents may occur. These may result in abnormal types of transmission or chromosome rearrangements. Such changes usually have ill effects on the organisms and are lost immediately, but occasionally some prove beneficial and become fixed in the population. These changes can build up and result in characteristic differences between species in chromosome number, size, and many other morphological features important for speciation.

However, the evolution of chromosome organization from the level of prokaryotes to that of eukaryotes is least understood. It was a major step in organic evolution and probably took place over a long period of time. Recent studies have revealed evidence bearing on such organizational changes, and there is also much information on the chemical structure of chromosomes that has evolutionary implications. Therefore, before discussing chromosome rearrangements, we will first examine the chromosome organization and structure in a number of species.

EVOLUTION OF CHROMOSOME STRUCTURE AND ORGANIZATION

Fundamental differences exist between prokaryotes and eukaryotes in chromosome structure and organization. In their search for an organism with intermediate structures on which to construct a lineage, investigators have so far found only the dinoflagellate, an alga with cellular and nuclear structures similar to those of eukaryotes in some respects and prokaryotes in others.

Prokaryotes

The genome of prokaryotes consists of a single linear structure of the DNA molecule, often joined at the ends to form a ring. The DNA content for viruses and bacteria are shown in table 4.1. The chromosome of the ϕX174 virus contains 2.6×10^{-6} pg with about 5,500 nucleotides, which corresponds to about 5 genes. *Escherichia coli* contains 4×10^{-3} pg, with about 4,500,000 nucleotide pairs, corresponding to 4,500 genes. There is no nuclear membrane in the prokaryotes. The DNA sequences are unique, that is, each gene is represented only once. Without proteins attached, the DNA is referred to as "naked."

TABLE 4.1 DNA per Haploid Genome

Organism		DNA (picograms)	Length (μ)
Viruses	ØX 174	2.6 $\times 10^{-6}$	2
	Lambda	50 $\times 10^{-6}$	17
	T2	208 $\times 10^{-6}$	55
	Fowl pox	382 $\times 10^{-6}$	91
Bacteria	Mycoplasma	840 $\times 10^{-6}$	265
	E. coli	4,000 $\times 10^{-6}$	1,200
Fungi	Yeast	22,000 $\times 10^{-6}$	
Protozoa	Aslasia longa	1.5	
	Trypanosoma evansi	0.2	
	Plasmodium berghei	0.06	
Sponges	Tube sponge	0.06	
Coelenterate	Cassiopeia	0.33	
Echinoderms	Sea urchin (Lytechinus)	0.9	
Mollusks	Snail (Tectarius)	0.7	
Crustacea	Crab (Plagusia)	1.5	
Insects	Drosophila	0.2	
	Chironomus	0.2	
Chordates	Amphioxus	0.6	
Fish	Lamprey (Petromyzon)	2.5	
	Carp	1.7	
	Shad	1.0	
Amphibians	Toad, Bufo bufo	7.3	
	Amphiuma	84.0	
Reptiles	Snapping turtle	2.5	
Birds	Domestic fowl	1.2	
Mammals	Man	3.2	
Plants	Euglena gracilis	3	
	Aquilegia sp.	0.6	
	Gladiolus sp.	3	
	Lilium longiflorum	53	
	Tradescantia	58	

SOURCE: Reprinted, by permission, from H. Ris and D. F. Kubai, *Ann. Rev. Genet.* 4(1970):263-94.

The DNA in bacteria is double stranded, forming the helical structure first deduced by Watson and Crick. The nuclei are composed entirely of bundles of fine DNA fibers with diameters of 30 Å to 60 Å, arranged as zigzag chains loosely held together at the turning points with non-DNA materials.

The whole genome behaves as a single replication unit with a DNA synthesis rate of about 30μ/min (Ris and Kubai 1970), which is at least fifteen times greater than that of eukaryotes. The nuclear division of prokaryotes (such as bacteria), is invariably direct and does not involve visible changes in the nuclear organization as in higher organisms. Before division the nucleus is compact and stands out sharply in the cytoplasm. After division is complete, each resulting nucleus soon assumes a compact shape in preparation for the next division.

Eukaryotes

In contrast to that of the prokaryotes, the genome of eukaryotes consists of two or more linear structures of DNA molecules associated with basic proteins, and contains many times more DNA, a considerable fraction of which generally consists of redundant sequences. The rate of DNA synthesis is about 0.5μ/min to 2μ/min. There are numerous initiation points for the synthesis within each DNA sequence. Ris (1961) suggested restricting the term *chromosome* to the structure of eukaryotes since it represents more than a length of DNA; he suggested *genophore* as a general term for a structure containing genetic materials of either DNA alone or DNA plus associated proteins. The DNA content and the basic proteins of the chromosomes will be further discussed in the next section.

In eukaryote chromosomes (fig. 4.1) there are specialized regions believed to be associated with activation of the genes or nuclear division and to serve as markers for chromosome studies, although few of their biological functions and mechanisms relating to evolution are clearly understood.

Chromomeres. Chromomeres are denser and thicker beadlike chromatins linearly arranged along the chromosomes of the prophase in mitosis and meiosis. The size and sequence of chromomeres form a pattern that is constant for a particular stage of a given cell type but need not be the same in different cells within an organism. Both chromomeres and interchromomeric regions contain DNA (Pelling 1966). In addition, both regions consist of the same kind of DNA histone (DNH) fibers (Ris 1961). They are tightly coiled in the chromomeres but straight and parallel to the chromosome axis in the interchromomeric regions.

It has been suggested, but not generally accepted, that chromomeres are the units of replication, transcription, mutation, and recombination. This view is based on the observation of discontinuous DNA synthesis (see Mulder, van Duijn, and Gloor 1968) and the concept of independent replication of chromomere DNA (Keyl 1965). The evidence that localized mutations in specific bands, puffs, and loops are considered structural manifestations of gene activity is interpreted to mean that chromomeres function as units.

Centromere or Kinetochore. The centromere represents the region (or regions) of each chromosome with which the spindle fibers become associated during mitosis and meiosis. A localized centromere representing a permanently localized region occurs in most species. Its position on the chromosome is used as a reference point for identifying different types of chromosomes and chromosome rearrangements. Metacentric chromosomes are those whose centromeres are localized in a roughly median position; submetacentric chromosomes are those whose cen-

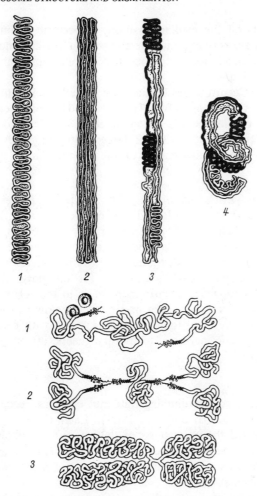

Fig. 4.1. The "folded fiber model" of chromosome structure. According to this model the chromatid consists of one (or a few) elementary fibrils (diameter 200 Å to 250 Å): *Top*, (*1*), transverse folding; (*2*), longitudinal folding; (*3*), combined transverse and longitudinal folding; and (*4*), quarternary coiling superimposed on a chromatid of the folding type 3. *Bottom*, organization and replication of the chromosome as interpreted by the folding fiber concept. Each interphase chromatin fiber consists of a single DNA molecule held in a regular secondary helix by its protein coat (*1*). The replication of the fiber proceeds sequentially from an end toward the middle (*2*). After replication, the daughter fibers fold up to form the "condensed" metaphase chromosome (*3*); this folding is postulated to be accomplished by contractile protein molecules in the sheath of the fibers. (Reprinted, by permission, from Rieger, Michaelis, and Green, 1968.)

tromeres are slightly off the median position; acrocentric chromosomes are those whose centrometers are very close to one end so that one chromosome arm is small and the other very much longer; telocentric chromosomes are those with centromeres at the terminals.

In plants, there are centromeres that are not visually well defined in the chromosome. They are referred to as diffuse (nonlocalized) centromeres. Therefore, the attachment of the spindle fibers is not confined to a strictly localized segment of the chromosome (Rieger, Michaelis, and Green 1968); in a few plant species, the spindles attach over the entire length of the chromosome (see Ris and Kubai 1970).

Heterochromatin and Euchromatin. In most interphase nuclei, chromatin exists in two major states: highly condensed (chromocenters) or dispersed. Heits in 1928 (see Ris and Kubai 1970) demonstrated that chromocenters represent specific chromosome regions, unraveled in telophase and remaining condensed throughout the cell cycle. He introduced the term *heterochromatin* to denote these chromosome regions, and *euchromatin* for the parts that underwent the normal uncoiling in telophase. Euchromatin refers to the regions of chromosomes that show the normal cycle of coiling and normal staining properties. A number of unique properties have been found to be associated with heterochromatin (Ris and Kubai 1970): genetic inertness, temporary genetic inactivity, and a high content of redundant DNA.

Dinoflagellates

The distinctive nature of the dinoflagellate nucleus was recognized as early as 1885 by Butschli (see Kubai and Ris 1969). Later Chatton (1920, see Kubai and Ris 1969) stressed the uniqueness of the nuclear morphology and cell division of dinoflagellates, and suggested the terms *dinokaryon* and *dinomitosis* to describe their characteristics. Recently, the unusual features of nuclear division have begun to be understood.

The cellular structure of dinoflagellates is similar to that of the eukaryotes, with typical membrane-bound organelles, such as mitochondria, chloroplasts, and nuclei (Leadbeater and Dodge 1966). But the nuclear structure is unlike that of eukaryotes, bearing a closer resemblance to that of the prokaryotes in its absence of the chromosome-coiling cycle. The chromosomes maintain the same morphology throughout the cell cycle, and are distinctly visible as rod-shaped bodies when seen through a phase microscope (fig. 4.2). This is in sharp contrast to the chromosomes of eukaryotes, which remain uncoiled during the interphase and are recognizable as individual bodies only during cell division.

In prokaryotes, the replication and division of the genophore does not involve a complex apparatus (as in the eukaryotes). In the dinoflagellates, the DNA is distributed in different fibrous bodies. We may expect that there are many mitotic spindle formations during division to assure equal distribution of the DNA fibrils, but under a light microscope, typical mitotic spindles are not found. Leadbeater and Dodge (1967) found extranuclear microtubules within the cytoplasmic invaginations. It was found that in dividing cells most chromosomes become V-shaped, and the apexes of the V are in contact with the membrane surrounding the cytoplasmic channels. It has been suggested that these channels are involved in the separation of the daughter cells (Kubai and Ris 1969), the phenomenon bearing some similarities to that of chromosome separation during nuclear division in eukaryotic cells.

The complexity of the nuclear and cytoplasmic structures in the eukaryotic cells suggests that evolution of chromosome structure and organization from prokaryotes to eukaryotes is not a simple process. Cell division in eukaryotes involves a concerted action of the whole cell. As discussed in chapter 3, the chloroplasts of dinoflagellates are considered to be derived from yellow prokaryotes. The pres-

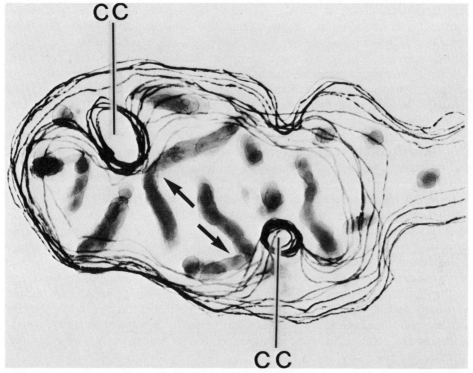

Fig. 4.2. Dinoflagellate chromosomes in the early division nucleus. The chromosomes are longer and thinner in comparison with chromosomes in the nondividing nucleus. The apexes of V-shaped chromosomes (*arrows*) are intimately associated with membrane surrounding cylindrical channels (cc), which may be involved in the separation of daughter chromosomes. (Reprinted, by permission, from D. F. Kubai and H. Ris, *J. Cell Biol.* 40(1969):508–28.)

ence of chloroplasts, mitochondria, and some crude form of nuclear membrane in the dinoflagellates is an indication that they are evolving toward eukaryotes. We are fortunate to have discovered such an intermediate. It is possible that both more and less advanced forms of nuclear evolution than dinoflagellates may be discovered in the future, so that the gap of knowledge on evolutional changes from prokaryotes to eukaryotes can be filled.

CHEMICAL VARIATIONS IN THE CHROMOSOMES OF DIFFERENT SPECIES

DNA Content

The DNA content of nuclei in various organisms is shown in table 4.1. Apparently viruses and bacteria contain the least DNA; unicellular organisms, such as algae and primitive metazoans, contain more than microorganisms, but still much less than the higher organisms. It appears that in the lower organic forms an increase in the amount of DNA is related to the degree of evolutionary advancement. This is not so in the higher plants and animals; for instance, the DNA content of *Gladiolus* is about five times that of *Aquilegia*. And, in vertebrates, the *Amphiuma* has about twenty-five times more DNA in its nuclei than does man. Evidently, in higher organisms, the relative amount of DNA in the nucleus is not an indication of advanced evolution; the arrangement, not the amount, of DNA becomes the important factor for the functioning of the cell and the organism as a whole.

In eukaryotes, DNA is subdivided into several chromosomes that aid in handling these unwieldy fibers. Each chromosome contains many times the DNA of the prokaryotic genome. If the DNA in the longest human chromosome (5.0μ at metaphase) is in a single piece, it is about 16 cm long (Ris and Kubai 1970). How is this enormous length of DNA packed into a chromatid?

The question here is whether the DNA of a chromosome exists as one or as many molecules. Autoradiographic and chemical evidence suggests the existence of a large number of simultaneously replicating units. These subunits are joined by alkali-labile linkers (Lett, Klucis, and Sun 1970). It is generally assumed that the redundant or repetitive DNA sequences occur either tandemly along the length of the chromosome or in a laterally parallel sequence (or stranded) (see various proposed models, fig. 4.1). Large-scale tandem repetitions have been found in the chromosomes of Diptera as repetitive band patterns. Thomas et al. (1970) demonstrated the actual existence of these repetitive sequences in the DNA of trout, salmon, and calf.

Histone

Histones are the basic protein associated with DNA. The DNA is coiled or folded in the DNH fibers (fig. 4.3). It is suggested that the nonhelical carboxy-terminal end of the histone molecule with a cluster of basic amino acids binds to DNA (Ris and Kubai 1970).

So far it is thought that histone performs two functions: (1) to pack a large amount of DNA into a manageable unit and (2) to regulate DNA transcription.

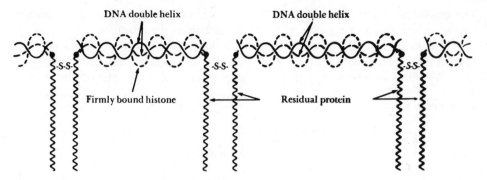

Fig. 4.3. A proposed model for structure of chromosomal fibers. In this structure, units of residual protein are assumed to alternate with units of DNA. Each unit of DNA would be attached at both ends to residual protein except the end DNA unit. The number of peptide chains in residual protein units is not specified. According to Dounce, whether the bulk of the residual protein is structural or regulatory remains to be determined. Disulfide bonds represented by -S-S-. (Reprinted, by permission, from A. L. Dounce, *Am. Scientist* 59(1971):74–83, journal of Sigma Xi, the Scientific Research Society of North America, Inc.)

The presence of histone in eukaryotes, not prokaryotes, is perhaps a most important step of evolution. It is possible that a chromosome at the eukaryotic level of organization had to be attained before any form of cellular differentiation could be reached.

The molecular weight of histone ranges from 11,000 to 21,000. Since the total weight ratio of histone to DNA is approximately 1 to 1.2 for different tissues, it can be calculated that a given amount of DNA with a molecular weight of about 10^6 (corresponding roughly to one cistron) combines with about fifty histone molecules (1,000,000/21,000).

Five major fractions of histone in approximately equal amounts have been observed in a wide variety of animals and plants, and in all chromosomes irrespective of tissue, phase of cell cycle, and stage in development of the organisms. Up to the present time, no one has been able to show any variation in histone content in chromosome sections differing in activity, such as between puffed and non-puffed regions or between heterochromatin and euchromatin (Comings 1967; Gorovsky and Woodard 1967). Practically no differences were found in the amino acid sequence of the homologous histones between different tissues or different vertebrate species (Ris and Kubai 1970). This indicates the conservative nature of the histone in ontogeny and phylogeny.

DNA Activation

In eukaryotes, only a small fraction of DNA can act as a template for transcription, and 90% of the chromatin must be considered inactive (Paul and Gilmour 1968; Smith, Church, and McCarthy 1969). The enlarged lampbrush chromosomes in the amphibian nucleus are believed to be an indication of gene activation.

The amphibian nucleus consists of a linear series of condensed chromatin masses known as chromomeres; a pair of loops project from the chromomere (fig. 4.4). These loops consist of a single DNA double helix that is almost completely un-raveled and covered with more or less compact ribonucleoprotein fibers and gran-ules. Amazing details within these loops were revealed by Miller and Beatty (1969). Granules, approximately 100 Å in diameter, are regularly spaced on the 50-Å DNA loop axis, perhaps representing the RNA polymerases. Projecting from

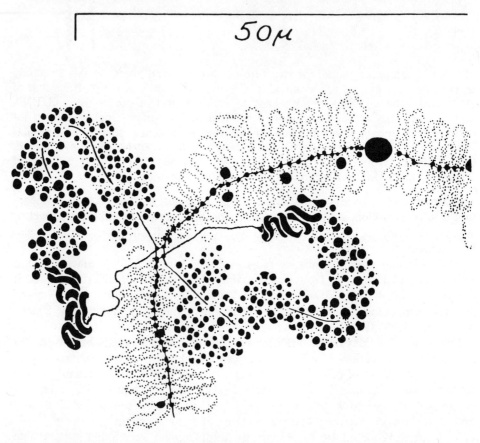

Fig. 4.4. The left end of chromosome XII from an oocyte of the newt *Triturus cistatus*. During the lampbrush phase, a large number of "normal" loop pairs, in addition to the "giant granular loops," are shown. All loops are asymmetrical (one end is thinner than the other). In the giant granular loops, the thin insertion con-sists of a fine thread leading to a dense, contorted region. Each loop possesses a delicate axis, whose diameter is less than 100 Å. The loops arise from chromomeres lying in the chromosome axis. All loops contain RNA, and the loop axis must contain DNA. (Reprinted, by permission, from J. G. Gall and H. G. Callan, *Proc. Nat. Acad. Sci. U.S.* 48(1962):562–70.)

each granule is a thread of RNA integrated with protein. Apparently the newly synthesized RNA immediately combines with a nonhistone protein and remains on the DNA axis for some time before being transported to the cytoplasm. It is generally thought that the lampbrush chromosome loops in the amphibian represent sites of active RNA synthesis (Gall and Callan 1962).

Although little is known about the molecular organization of chromosome puffs (fig. 4.5), they are thought to be sites of gene inactivation. It appears that the DNA of the puffed region unravels and becomes associated with ribonucleoprotein, which is compressed into fibers and granules of various sizes (Ris and Kubai 1970) as in the lampbrush loops. As previously mentioned, histochemical studies have demonstrated that the DNA-histone ratio is the same in the puffs as in the inactive bands (Gorovsky and Woodard 1967).

The fact that such a large portion of the DNA or of the genome is inactive leads naturally to the question: Does this portion of DNA presently serve any useful purpose at all, or did it perhaps only function earlier in evolution? No completely satisfactory answer can yet be given. It seems logical to assume, however, that through the history of the evolution of any species many genes useful at one period may have become useless later, even though carried along for some time by the organisms. This assumption would explain the large variations in the DNA content among the higher species. The amount of DNA represents the amount of genetic information. But, after reaching a certain level, evolutionary advancement depends on the specificity of the genetic information and the mechanisms used to control it. In direct relevance to the latter statement will be our discussion of ontogeny and phylogeny in the next chapter.

SOME EVOLUTIONARY FEATURES OF CROSSING-OVER

Meiosis, a type of cell division, is basic to maintaining an identical chromosome number between parent and offspring in sexually reproducing organisms. In the absence of meiosis, fertilization as an integral part of the sexual cycle would lead to a geometric increase in chromosome number. Chromosomal variations and/or rearrangements arise mostly during the course of meiosis. It is amazing that in the majority of sexually reproducing organisms, the sequences of pairing, chiasma formation, and segregation occur with complete regularity. The close similarity existing among organisms having wide phylogenetic differences (White 1945) suggests that the evolution of meiosis, possibly from mitosis, is very old, possibly a parallel of the evolution of sexual reproduction.

One of the novel mechanisms in meiosis is crossing-over, which permits relative freedom of gene exchange between homologous chromosomes. This mechanism provides some flexibility for limited freedom of gene segregation, and is well utilized by organisms in evolution. The theoretical importance of linkage and crossing-over in relation to natural selection will be discussed in chapter 6.

Crossing-over between homologous chromosomes is rather universal among species, but there are some rare exceptions. For example, no crossing-over occurs

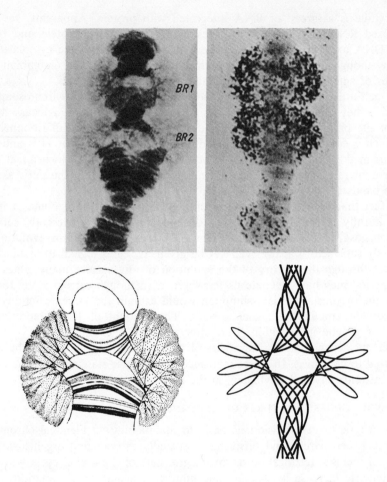

Fig. 4.5. Top left, two large chromosome puffs (Balbiani rings [BR1 and BR2]) in the giant salivary chromosome IV of *Chironomus tentans* (Diptera). Puffs contain large amounts of proteins and RNA; DNA is certainly present but difficult to identify. A puff originates from a defined band on the chromosome; it swells up and becomes diffuse. During puff, each of the many thousand homologous chromomeres in a chromosome band unfolds into a long looplike strand up to 10μ in length, and each probably consists of a single DNA double helix. *Top right,* incorporation of tritiated-uridine as evidence of RNA activity. (Reprinted, by permission, from Beermann, 1967.) *Bottom left,* the puffed portion of a chromosome of *Chironomus. Bottom right,* the interpretation of the chromosomal fibers corresponding to the strands of DNA in the region of the puff. (Reprinted, by permission, from Rieger, Michaelis, and Green, 1968.)

in male *Drosophila* and in female silkworms. In male *Drosophila* the autosomes pair, but do not pass through the later stages of prophase and diakinesis. At the stage corresponding to diakinesis, the two homologues lie parallel to one another. No chiasmata have been observed. White assumed that this type of meiosis is found in most or all of the higher Diptera. Chiasma frequencies and crossing-over are higher in female than in male house mice (Crew and Koller 1932; Slizynski 1960). Similarly, chiasma frequencies occur at a higher rate in female than in male cells of *Dendrocoelum* (Pastor and Callan 1952), *Fritillaria*, and *Lilium* (Fogwill 1958).

It has been suggested that reduced crossing-over in the male cells may result from limitations imposed by cell size (Brat 1966). Haldane (1922), however, suggested that where sex differences exist, there is less crossing-over in hetero-gametic sex. From an evolutionary standpoint, Darlington (1959) claimed that the complete suppression of crossing-over in males stabilizes the gene combinations in one sex and facilitates the establishment of favorable new combinations. In an extremely outbred and short-lived species, such as *Drosophila*, the absence of crossing-over in males is a desirable feature, allowing the desirable genes to be transmitted together. However, the significance of anomalous meiosis in evolution remains far from clear.

There is no crossing-over between the heterogametic sex chromosomes, such as X and Y. This is expected evolutionally, since crossing-over between X and Y would eliminate sex differences. Moses, Counce, and Paulson (1975) observed partial synapsis of human X and Y chromosomes with an electron microscope and suggested that the distal part of the short arm of the X chromosome is ho-mologous with the short arm of the Y chromosome. This observation seems to indicate that there was a common origin for the X and Y chromosomes; however, their homology must have been lost during evolution to the extent that it is not sufficient for crossing-over to occur.

TRANSLOCATION

A structural change in a chromosome causing a portion of one chromosome to be transferred to another position on the same or another chromosome by a process other than normal crossing-over is called translocation. It is referred to as simple translocation when only one chromosome segment is transferred. When two are exchanged, it becomes a reciprocal translocation. The exchanged segments need not be of equal length. Simple translocations rarely occur, but reciprocal trans-locations are known in both natural and domestic populations.

Reciprocal translocation homozygotes may show little difference cytologically from the normal type. Only when the exchanged segments are of an unequal length can differences be observed. In heterozygotes, however, translocation can be detected cytologically. When they are reciprocal, two new chromosomes are constituted, which may vary in size depending on the points of breakage.

The cytology of translocation is a complicated subject in itself. For our purpose, we need discuss only a simple case: the symmetrical and reciprocal translocation that frequently occurs. Three zygotic types can be present in the population: (1) standard, nontranslocation homozygotes, (2) translocation homozygotes, and (3) translocation heterozygotes (fig. 4.6).

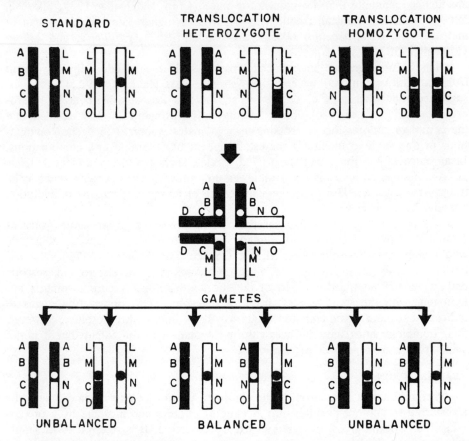

Fig. 4.6. Gametic production by a reciprocal translocation heterozygote. *Top*, standard, translocation homozygote and translocation heterozygote. *Middle*, characteristic crosslike configuration produced by the heterozygote at pachytene phase in meiosis. *Bottom*, resulting balanced and imbalanced gametes.

Translocation homozygotes do not show cytological peculiarities during meiosis, and transmission of chromosomes from one generation to another is usually the same as in the standard types. In fact, each chromosome of a species may have acquired translocated segments in various regions and lengths during evolution through different phylogenetic levels. As soon as the translocation is fixed in the population, it cannot be cytologically recognized unless comparison can be made with a closely related species or population in which homologous chromosomes can be identified, such as in *Drosophila*.

However, in translocation heterozygotes, complications arise in attaining close pairing of homologous parts at meiosis. In this case, the chromosomes ordinarily assume a characteristic crosslike configuration at pachytene that may open up at a later stage as a four-membered ring or become transformed into a figure-eight arrangement. At the completion of meiosis, one-third of the gametes have the complete chromosome complements (standard and balanced translocation types), and two-thirds of them carry the imbalanced translocations (one segment is duplicated and another is deleted) (fig. 4.6).

It should be pointed out that reciprocal translocations not only change the spatial relationships of genes, but may also result in the occurrence of various types of mutations. It must be borne in mind that the point of chromosome breakage can occur within genes, as well as between them. If there is breakage between the genes, new genes may be formed at the joining point, in addition to the occurrence of spatial changes. The new gene types may be that of a deletion, or a combination of a partial nucleotide sequence of one gene with that of another, so that various mutational consequences may result.

Reciprocal translocations have occurred in a number of species and races of *Datura*. Bergner (1943) found many different types of interchanged chromosomes in six *Datura* species. There is indirect evidence that reciprocal translocations may have also occurred in the higher animals. Painter (see Snyder and David 1957) found that the amount of chromatin material in the rat and mouse was almost identical, but that the distribution of chromatin among the chromosomes differed greatly between these two closely related species. Thus they speculated that the divergence of the mouse and rat from a common ancestor probably occurred through a number of translocations. Snyder and David (1957) claimed that with mutation added to translocation, a distinct species was formed. As possible evidence of its occurrences in the ancestors of these species, many translocations have been found in different chromosomes of the laboratory mouse. Paralogous genes on different chromosomes, as discussed in chapter 2, must have resulted from translocation.

INVERSION

Inversion is a chromosomal structural change characterized by the reversal (with reference to an arbitrarily assigned standard arrangement of the linkage group) of a chromosome segment and the gene sequence it contained. Inversion results from chromosomal breakage and reunion such that the segment between the two breaking points is rejoined following a 180° end-to-end rotation. Inversions are classified into different types according to the locations and number of inversions within a chromosome.

1. *Simple inversion.* In simple inversion, only one segment in a chromosome is inverted. Simple inversion can be further divided into (*a*) paracentric inversion, in which the inverted segment occurs entirely in one of the chromosome arms,

that is, on one side of the centromere and (*b*) pericentric inversion, in which the inverted segment includes the centromere itself, involving a portion of both chromosome arms.

2. *Complex inversion.* In complex inversion, more than one segment of the same chromosome is inverted. This is further classified into (*a*) independent inversions, in which inverted segments are separated by noninverted segments; (*b*) tandem inversions, in which inverted segments are adjacent; and (*c*) inclusion inversion, in which inversion occurs within a longer inverted segment. There are other complex inversions, but it is unnecessary to list them all here. Two types are illustrated in figure 4.7.

Complications in meiosis may arise in the inversion heterozygote. Depending on the length of the inverted segment, a loop may be formed in the meiotic pairing of normal and inverted chromosomes, since the inverted segment affixes itself to its homologous partner by the pairing of homologous loci. This loop formation is generally used as a criterion for detecting inversion. However, when the inverted segment is so short that a loop cannot be formed, or when the inverted segment is too long, chromosomes may pair without formation of a loop.

Except in organisms with no crossing-over, meiosis of individuals heterozygous for inversion may lead to inviable offspring when crossing-over occurs within the inverted region: pericentric-inversion heterozygotes produce duplication-deficient gamete chromatids; paracentric-inversion heterozygotes produce acentric and dicentric chromatids. But crossing-over in the inverted regions is rare, especially when the inverted segments are small. This property has been utilized by organisms to transmit genes favored by natural selection in a single unit form. What genetic effects can be caused by inversion? One result may be that the genes at each end of the inverted segment attain a new neighborhood and that, due to the alteration in their spatial relationship, their function may also change. This phenomenon, called position effect, was discovered by Sturtevant approximately fifty years ago. Inversion may also cause additional genetic changes. During the process of inversion, if breakages should occur within a gene, gene mutations, as discussed under translocation, will result. Thus, inversion may cause both structural and functional changes of genes.

The presence of giant chromosomes in the salivary glands of *Drosophila* has facilitated cytological studies. Much information on inversions in different species of *Drosophila* has now become available, certain aspects of which illustrate major points concerning problems of chromosomal inversion and evolution.

Pericentric inversions are rare in *Drosophila,* whereas paracentric inversion is a common feature (Stone 1962). Several meiotic specializations exist that tend to reduce the frequency of inviable offspring of individuals heterozygous for inversions in *Drosophila.* First, crossing-over occurs only in meiosis in the female; second, the four products of meiosis are so oriented that a normal nucleus usually

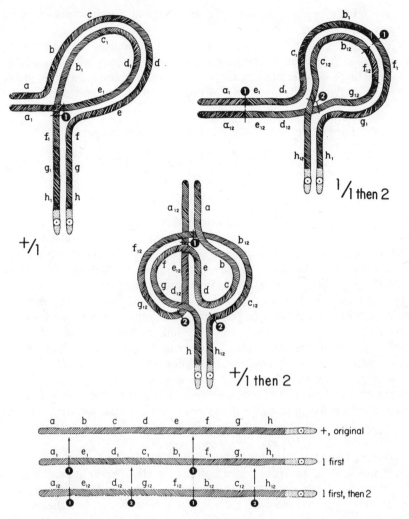

Fig. 4.7. A sequence of two inversions added to a chromosome where the breakage points, and therefore inversion, overlap. The sequence of events illustrated here is $+ \to 1 \to 1, 2$ in the same chromosome. The possible heterozygous synaptic configurations are shown in cases where all possible sequences still exist in the species. The homozygous combinations are obvious from the diagrams of changed sequences. The locations of a few gene loci are indicated to make the sequences in the synaptic configurations clear. The determinate complexity of the banded pattern of the polytene chromosomes and the pairing of homologues allows one to determine the simple change in sequence easily. (Reprinted, by permission, from Stone, 1962.)

becomes the female pronucleus, with the acentric and dicentric nuclei being excluded from the zygote.

Using overlapping inversions, Sturtevant and Dobzhansky (1936) were able to reconstruct the phylogenetic relationships among species of *Drosophila*. Assume that the three gene sequences for a homologous chromosome, as illustrated in the bottom of figure 4.7, are found in three populations, and one for each as below:

Population	Gene Order in Chromosome
1	a b c d e f g h
2	a e d c b f g h
2	a e d c b f g h
3	a e d g f b c h

We can see that there is a direct relationship between populations 1 and 2, and populations 2 and 3. The chromosome of population 2 can be produced by inversion of the chromosome of population 1 with breakage points at a-b and e-f, or the chromosome of population 1 by inversion of the chromosome of population 2 with breakage points at a-e and b-f. In a similar manner, the relationship between populations 2 and 3 may be explained. But no direct relationship exists between populations 1 and 3. The phylogenetic relationship may be $1 \rightarrow 2 \rightarrow 3$, $3 \rightarrow 2 \rightarrow 1$, or $1 \leftrightarrow 2 \leftrightarrow 3$. Thus a possible linkage between the three populations, although not an ancestral relationship, can be constructed. The validity of this method was subsequently confirmed in different subgroups of *Drosophila* and in the discovery of several hypothetical gene orders.

Since the classical work of Sturtevant and Dobzhansky, several major studies have been made on the phylogeny of the genus *Drosophila*; all are based on the information on inversions (Patterson and Stone 1952; Stone, Guest, and Wilson 1960; Wasserman 1962, 1963).

The phylogenic relationships between different species in *D. melanica* are illustrated in figure 4.8 for inversions in chromosome III. The existence of a hypothetical population is proposed. This population with the C inversion survived for a short time and then became extinct, the reason being that the probability for a simultaneous occurrence of two inversions, such as C and D, in *D. melanura* and *D. eronotus*, and C and B inversions in *D. nigromelanica* is extremely small. Therefore, it is very likely that in such a hypothetical population, the C inversion would have occurred first, and B and D later, at separate times. The original author (Stalker 1966) has assumed *D. micromelanica* to be the standard sequence.

Chromosomal inversions have received a most thorough study in the genus *Drosophila* and have been observed to be relatively frequent. For example, the average inversion number of each species in *Drosophila* groups (Stalker 1966) is as follows: *D. melanica*, 23.5; *D. virilis*, 13.0; *D. repleta*, 3.1; and *D. cardini*, 9.6. Within the group *D. melanica*, the distribution of intraspecific inversions among six species is shown in table 4.2. The number of inversions is higher in the second and X chromosomes for species of this group (Stalker 1966).

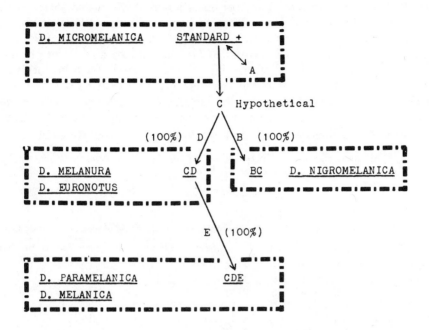

Fig. 4.8. The phylogeny based on inversion in chromosome III in the *D. melanica* group. The letters inside the boxes represent the heterozygous or intraspecific inversions; letters between boxes represent homozygous or interspecific inversions. For example, in *D. micromelanica* the third chromosome may show either the standard (+) arrangement or the inverted (A) arrangement. The two species *D. melanura* and *D. euronotus* have a single common sequence CD, which differs from the standard of *D. micromelanica* by the homozygous inversions C and D. The percentages indicate the proportion of the chromosome that could be homologized in any particular interspecific step. (Reprinted, by permission, from H. D. Stalker, *Genetics* 53(1966):327–42.)

Groups differ in frequency of inversions, as well as in chromosomes with the highest number of inversions. Why there should be such a large difference in the number of inversions among species, as well as among chromosomes (as in the *D. melanica* group), is not yet thoroughly understood. It has been suggested (Novitski 1961; Bernstein and Goldschmidt 1961) that chromosomes already heterozygous for inversions are especially liable to breakage or failure to heal following breakage, as was demonstrated by x-irradiation. Whether such instances would occur in natural conditions is not known (Stalker 1966).

In the *D. melanica,* the frequency of inversion heterozygotes within each species as shown above is large. However, no two species with heterozygotes having the same inversions were found (Stalker 1966). Only a very small number of hetero-zygotes for the same inversion in any two species were discovered in the *D. repleta* (Wasserman 1960) and *D. virilis* (Stone, Guest, and Wilson 1960) groups. But this is not the case in the *D. paulistorum* group, where a large number were found in different incipient species (Kastritsis 1967). This may be a transient phenome-non. Whether some of the heterozygotes may later drop out through loss or fixation when complete speciation is established remains uncertain (Kastritsis 1966). If the same heterozygotes persist throughout the species, it would indicate that they con-tribute fitness in general to most of the species. Stalker suggested that the intra-specific heterozygote inversions arose mostly following species formation.

TABLE 4.2 Distribution of Intraspecific Inversions among the Chromosomes
        of the *D. melanica* Group

Species	Total Inversions Known in Various Chromosomes					No. Population Samples Structurally Heterozygous in Various Number of Nonhomologous Chromosomes					
	XL-XR	II	III	IV	Total	None	One	Two	Three	Four	Total
D. micromelanica	1–2	9	1	1	14			2	4	1	7
D. nigromelanica	0–0	3	0	1	4	3	13	6			22
D. melanura	1–1	4	0	0	6			3			3
D. euronotus	0–1	17	0	0	18	1	20	5			26
D. paramelanica	4–5	7	0	0	16		3	10			13
D. melanica	1–3	16	0	2	22	6	20	9	1		36
Total	19	56	1	4	80	10	56	35	5	1	107
Percentage	24	70	1	5	100	9	52	33	5	1	100

NOTE: In the left half of the table the figures indicate the total known inversions in each chromosome for each species. In the right half the figures indicate the numbers of population samples showing no inversion variability, inversions in one chromosome, inversions in two nonhomologous chromosomes, and so on. In the case of *D. melanica*, all population samples are derived from laboratory stocks, in many instances a single stock representing a population. In the other species, almost all samples are based on studies of various numbers of wild-caught individuals. XL and XR refer to the left and right arm of the X chromosome, re-spectively.
SOURCE: Reprinted, by permission, from H. D. Stalker, *Genetics* 53 (1966): 327-42.

On the basis of 650 species in the genus *Drosophila,* the number of inversions fixed in the evolution of the genus has been estimated at approximately 6,100 to 36,500. This was a very conservative estimate; no additional allowance was made for greater cytological differences between or within species groups or subgenera. With the larger number of species now being proposed for the genus, the number of paracentric inversion differences fixed in its evolution probably lies between 22,000 and 56,000, while perhaps 18,000 to 28,000 are now heterozygous in many populations around the world (Stone, Guest, and Wilson 1960). We must remember that the same gene mutations may recur, whereas the same inversions may not (or only in very rare instances). In comparison with gene mutation, inversions are relatively infrequent. In spite of the rarity of such events and the fact that most of them are eliminated by natural selection, we find that a large number of inversions have still been fixed in the single genus *Drosophila*. This could be true in many other genera as well. We can imagine, then, that in organic evolution —allowing time—a most improbable event could establish itself. Chromosomal mutation may be useful for survival of a population, as well as for speciation.

Chromosome Number Variations

As White (1945) pointed out, the mechanism for mitosis in some organisms is best suited to accomplish the separation of a moderate number of chromosomes, and the spindles may be unable to deal efficiently with extremely high or low numbers. The biological mechanisms set the restrictions. The chief advantage of sexual reproduction lies in achieving a free distribution of genes within a breeding unit. For this purpose, it is desirable to have large numbers of chromosomes so that a freer combination of genes can be accomplished. On the other hand, for joint transmission of genes with interactive effect, restriction of free recombination is advantageous. For these reasons, it appears that an optimum number may be best for evolution.

According to White, chromosome evolution seems to obey what we might call the principle of homologous change, that is, the same type of chromosomal change occurs at different points in a phyletic line. He claimed that chromosomes were "organized bodies whose sequence of centromere, hetero- and euchromatic regions determines the mechanical relations of the chromosome to the spindle, and the amount and position of crossing-over." In order to establish an arrangement, it must not upset the genetic balance and must remain compatible with normal meiosis. The whole architecture of the cell itself imposes certain limitations on the types of chromosomal rearrangement.

Apparently, the classical view that chromosome numbers have no significant evolutionary importance has not been accepted by many modern investigators. Chromosomal number variations arise through fusion (fig. 4.9), duplication (polyploidy), or other structural rearrangements. Whenever there is a change in chromosome number, a corresponding change may occur not only in the number of genes, but also in the dosage balance between them.

Chromosome Number Reduction in Plants

Reduction in chromosome number has occurred in both the animal and the plant kingdom. It can take place through different processes of chromosomal rearrangement, followed by the loss of chromosome segments or a whole chromosome. In most cases, the zygotes fail to survive or else they develop into sterile adults due to removal of a large number of genes. The most frequent mechanism of reduction is through acrocentric fusion (White 1945), that is, two acrocentric chromosomes joining together at the centromere to form a metacentric chromosome, known as acrocentric fusion.

Acrocentric fusion may result in the deletion of a small region around the centromere in one of the chromosomes. As previously stated, the chromatin next to the centromere is usually heterochromatic and may be inactive. If so, such a deletion might have no significant biological effects. Chromosome fusion allows more chance for the favored genes present in the two original chromosomes to be transmitted together as one unit. It involves a change in the spatial relationship between genes, with an effect that may be similar to that in translocation. It also automatically causes an increase in the size of single chromosomes. Different types of chromosomal fusion or reduction have been described by White (1945) and Stebbins (1966).

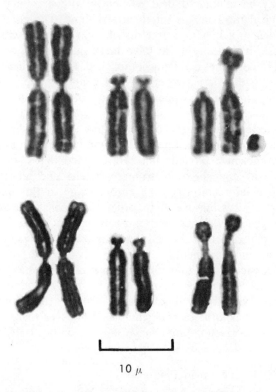

10 μ

Fig. 4.9. Karyotypes of the Indian deer, muntjac (*Muntiacus muntjak*), illustrating a type of chromosome fusion. This species usually has the low diploid number of 6 for the female (*bottom*) and 7 for the male (*top*). The X chromosome is the upper portion of pair 3, and is fused to the centromeric end of the acrocentric autosome. The Y chromosome is the tiny metacentric chromosome and remains separate (*top*). (Reprinted, by permission, from D. H. Wurster and K. Benirschke, *Science* 168(1970):1364–66. © 1970 by the American Association for the Advancement of Science.)

In plants the most primitive species of *Crepis* have the basic monoploid number of x = 6, while in other lines the number is reduced to 5, 4, and 3. Similar reductions in chromosome number have been observed in other genera of the tribe Cichoriaceae, in tribes of Compositae, and in many other families of the flowering plants (Stebbins 1966). According to Stebbins, the longest chromosome reduction series in plants occurs in the genus *Haplopappus,* with the highest number of x = 9 (as in the primitive species *H. eximus* Hall) through each individual number down to x = 2, such as in *H. gracilis* Gray (Jackson 1964).

Chromosome Number Reduction in Animals

The haploid chromosome numbers of most animal species range from 6 to 20. Outside this range, *Ascaris* has the lowest haploid number of 1, while the highest number recorded is 112 in the geometrid moth *Phigalia pedaria* (Renart 1933; see White 1945).

There is a great range of variation of chromosome numbers in many groups below the rank of class or subclass. Certain numbers occur more often than others. Cytogeneticists refer to these as type numbers. Thus, in groups of exoptergote insects, the type number (haploid) is 12 in Acrididae and 5 in *Drosophila* (fig. 4.10). In these groups, a single mode exists for the distribution of chromosome numbers.

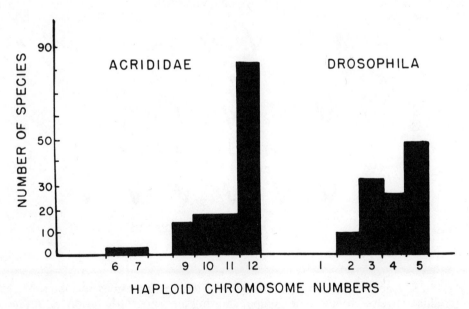

Fig. 4.10. Haploid chromosome number distribution in the group of Acrididae and *Drosophila*. After White 1945.

White believes that acrocentric fusion is a major factor in causing variation. This may be illustrated by the distribution of chromosome number and type in the grasshopper group (Acrididae), where the number of chromosome arms can easily be determined and remains rather constant (fig. 4.11). It can be seen in this figure that the reduction from *a* to *f* occurs through centric fusion. In the *Machaercocera mexicana* and *Philocleon anomalus* the X chromosome fuses with an autosome.

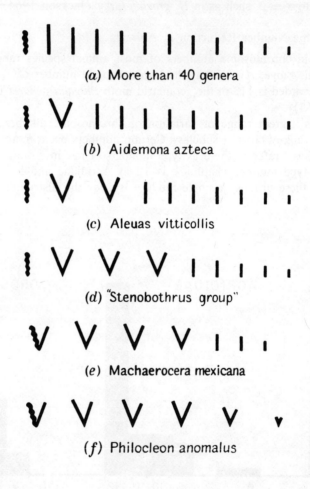

(*a*) More than 40 genera

(*b*) Aidemona azteca

(*c*) Aleuas vitticollis

(*d*) "Stenobothrus group"

(*e*) Machaerocera mexicana

(*f*) Philocleon anomalus

Fig. 4.11. Haploid chromosome number distribution in various members of the Acrididae (twelve-chromosome group), showing the acrocentric fusion of chromosomes in different members of the group. X chromosomes in each case are on the left, represented with a wavy line. (Reprinted, by permission, from White, 1945.)

In a few members of the twelve-chromosome group, a genuine reduction in the number of arms has occurred. In *Miramella* and *Zubowskya,* the smallest chromosome seems to have been lost through translocation (Makino 1936; Corey 1938), with its centromere having also disappeared. In *Niitakacris rosaceanum,* the medium-sized pairs of chromosomes have been eliminated (Helevig 1942); in Catantopinae, only nine acrocentric chromosomes are still present, with at least three pairs of autosomes no longer in existence. Chromosome numbers for various animal species are listed by Makino (1951).

The main heterochromatic regions of grasshopper autosomes are close to the centromere and are sometimes present in the distal ends. According to White (1945), the heterochromatic regions in the grasshopper have undergone repeated duplications and deletions in the course of evolution. The existence of heterochromatic regions next to the centromeres facilitates centric fusion, since a small deficiency or duplication of the heterochromatic material—generally considered to be biologically inert—will not affect the viability of individuals. It has been suggested (Darlington 1939) that groups in which no observable whole-arm interchanges occur may have no heterochromatic regions around the centromeres. White hesitated to assume that the frequency of centric fusion depends on the length of inert regions. However, he was certain that in many sections of the Acrididae, where large segments of heterochromatin surround the centromeres, centric fusions always take place.

Variations in the number of metacentric chromosomes of feral mice (*Mus musculus*) in alpine areas of Switzerland are shown in figure 4.12 as an illustration of the evolution of chromosome number being in process. In *M. musculus,* $2n = 40$, as found in the laboratory mouse, is taken as the standard number. All chromosomes are of the telocentric type. The metacentric chromosomes discovered in alpine populations have been found to result from acrocentric fusion (Gropp et al. 1972). Gropp et al. claimed that mice in the Valle di Poschiavo of southeastern Switzerland having a reduced chromosome number should be considered a separate species, since reduced fertility has been observed in the hybrids of these crossed with laboratory mice. The authors pointed out that the results may represent multiple and independent de novo chromosome fusion in the course of speciation.

It is generally believed that impairment of fertility usually occurs in heterozygotes with chromosomal mutation because of possible difficulties in meiosis. The very presence of differing chromosome numbers in various populations indicates that a selective advantage favors them in particular local environments, and that such an advantage in physiological fitness overweighs the disadvantages of hybrid infertility. Once homozygotes for fused chromosomes or any other type of rearrangement are produced, speciation will accelerate by both reproductive isolation and selection. This process, in turn, may lead to sympatric evolution.

In the past ten or fifteen years, chromosomal rearrangements of different types have been observed in many species. Investigators have attempted to establish phylogenetic relations between very similar species, based on these criteria. On the assumption that an increase in the submetacentric chromosome number results from a fusion of acrocentric chromosomes, Egozcue (1969) diagrammed the phylogenetic order in the Callithricidae, a family of New World monkeys (fig. 4.13). The amount of increase in metacentric chromosomes coincides with the amount of reduction in acrocentric chromosomes. (For a more complete discussion of chromosome rearrangements and evolution in other species, see the symposium papers in Benirschke [1969].)

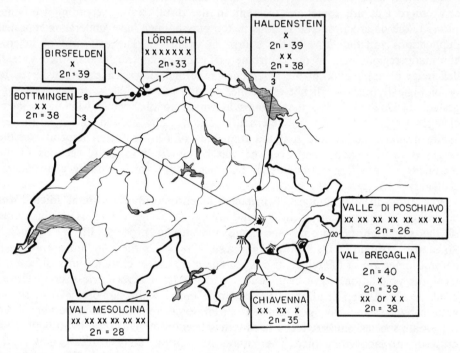

Fig. 4.12. Geographical map of Switzerland, showing mouse populations with different numbers of chromosomes resulting from acrocentric fusion. (Reprinted, by permission, from A. Gropp et al., *Chromosoma* 39(1972):265–88.)

Distributions of chromosome numbers are plotted in figure 4.14 for four groups of vertebrates: fish, amphibians, reptiles, and mammals. In the fish, including the Elasmobranchii and the Teleostomi, large gaps are present in the distribution, apparently resulting from polyploidy. The chromosome number for the bulk of the species is within the range of eighteen to twenty-six, with a major mode at twenty-six. One large gap is present in the distribution for the Amphibia; unlike the situation regarding the fish, extremely large numbers are not found. The majority of the species have chromosome numbers in the lower range of eleven to fourteen, with a major mode at twelve. From Amphibia to Reptilia to Mammalia, the mean chromosome numbers tend to increase, and large breaks in distributions are lacking. But sharp peaks in distribution, which may represent the type numbers for certain groups, are present. In the distribution for mammals, there are three breaks: nine, thirteen, and thirty-two, and a major peak at twenty-four. The chromosome numbers of some of the popular mammals, such as the mouse and the human, are twenty and twenty-three, respectively, which brings them into the middle range of distribution and very close to the major mode.

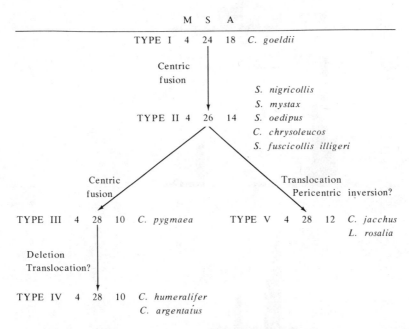

Fig. 4.13. Chromosomal evolution in the Callithricidae (primates). The chromosome complement is composed of three types: metacentrics (M), submetacentrics (S), and acrocentrics (A). Callithricidae includes *Callimico, Saguinus,* and *Leontideus.* (Reprinted, by permission, from Egozcue, 1969.)

Polyploidy in Animals

A cell or an organism possessing more than two haploid sets of chromosomes is referred to as polyploid. The occurrence of three haploid sets is called triploidy; four haploid sets, tetraploidy; and so on. Polyploids that consist of unlike chromosome sets are referred to as allopolyploid, occurring usually in species hybridization. Those that consist of two homologous chromosome sets are called autopolyploid, occurring sometimes in somatic cell mutations. Diploid cells with one or more extra chromosomes are referred to as polysomic. Cells with one extra chromosome (2n + 1) are called trisomic; with two extra chromosomes of the same pair (2n + 2), tetrasomic; with two extra chromosomes of different pairs (2n + 1 + 1), double trisomic; and so forth.

According to Muller (1925), the lack of polyploid species in animals results from the low probability that a union will take place between an unreduced egg and an unreduced sperm; similarly, the survival of a zygote produced through the

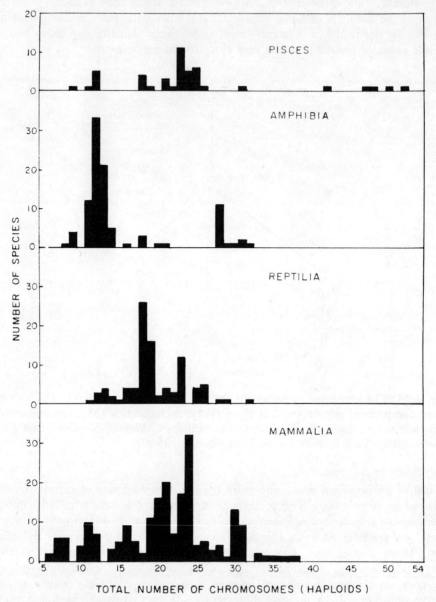

Fig. 4.14. Distribution of haploid chromosome numbers in the four groups of vertebrates: Pisces, Amphibia, Reptilia, and Mammalia. Data from Makino 1951.

union of a gamete from a tetraploid individual with one from a diploid individual is very improbable. Muller's second argument was that the survival of tetraploidy in the unbalanced heterogametic sex, such as XXXY in most vertebrates and ZZZW in some insects, would be difficult. Astaurov (1969) claimed that an imbalanced sex-determining mechanism was surmountable. Experimentally, he produced tetraploidal silkworms and found that the segregation ratio of female (ZZZW) to male (ZZZZ) equaled the normal 1:1 ratio.

Polyploidization may occur as the result of various modes of animal reproduction, the most common being bisexual reproduction, parthenogenesis in its various forms, hermaphroditism, and asexual or vegetative reproduction (Astaurov 1969). For example, polyploidy is rather widespread in earthworms, and Mudal (1951) suggests that bisexual polyploids have arisen indirectly from parthenogenesis in this group. The following chromosome complements exist in the bisexually reproducing shrimps: diploid genus *Lepidurus* (2n = 6), tetraploid (*Branchinecta* and *Chirocephala* genera, 4n = 16), and hexaploid (*Streptocephalus torvicornia,* 6n = 18). There are several species of polyploid insects, such as earwigs, stick insects, mantises, crickets, and mole-crickets. Astaurov (1969) suggested the possibility that they might also have arisen from parthenogenesis.

Although true polyploidy is rare in vertebrates, some evidence of it has recently been found. In *Coregonus,* 2n = 36 and 4n = 72 were reported by Kapka (see Astaurov 1969). In the fish family Cyprinidae, Ohno et al. (1967) found that the two barb species, *Barbus tetrazona* and *Barbus fasciatus,* have diploid chromosome numbers of 2n = 50 and 2n = 52, while the carp (*Cyprinus carpio*) and the goldfish (*Carassius auratus*) have a diploid number of approximately 104. The authors then suggested that goldfish and carp are tetraploid species in relation to other members of the family Cyprinidae. The wide range of chromosome numbers in fish, as previously discussed, may be due to polyploidy.

Polyploidy in Plants

Polyploidy is considered a significant factor in plant evolution since it seems to occur more commonly in plants than in animals. In *Crepis,* polyploidy complexes exhibit a complete series of intergradations and recombinations of the characteristics found in the diploids (Babcock and Stebbins 1938). The North American complex of *Crepis* exhibits a pattern of variations that remains within the range of those belonging to the diploid species. These can be explained as arising from lack of sexual reproduction. However, there are sexually reproducing species, none of which have transgressed the limits of variation in the diploid species. According to Stebbins (1966), the failure of the polyploids to evolve new characteristics results from a specific handicap, the presence of many duplicated gene loci.

However, other species of *Crepis* exist that show the adaptive significance of polyploidy. The species complex of *Crepis runcinata,* which inhabit the Rocky Mountain region, consist entirely of sexual diploids. This species is distributed along stream margins and other moist areas, either continuously or scattered, with

various groups showing little difference from each other. Polyploid species such as
C. acuminata, C. modosensis, and *C. occidentalis,* with marked drought resistance,
inhabit open mountain slopes. The difference in ecological adaptation seems to be
correlated with polyploidy in the dry mountain species, and diploidy in the moist
meadows. The genus *Betula* occupies a similar situation. Polyploid species inhabit
the dry areas, whereas diploid species of the genus *Alnus* are confined to lowlands
and stream sides (Stebbins 1966).

In flowering plants, the percentage of polyploid species appears to increase in
moving from warm to colder climates. The figures given below represent the findings of different investigators as summarized by Morton (1966):

Locality	Percent Polyploidy
Northern Sahara	38
Great Britain	53
Iceland	66
Greenland	71
Arctic Peary Land	86

The success of polyploidy is believed to result from the presence of a greater
variety of genes, which enable the plants to survive under unfavorable conditions.
Ordinarily hybrids from two diploid species are sterile. In some cases, polyploidy
derived from crosses between two or more dissimilar species restores fertility and
presents genetic characteristics of the parental species. Thus a reproducible genetic
system with a wide gene spectrum can be established. The genetic condition is
comparable to the heterosis that results from the cross between different inbred
strains, in which fitness is increased by both allelic and nonallelic interactions. In
each case, the organism is more vigorous and more easily adaptable to a new or
more hazardous environment.

Stebbins (1966) speculated that separate cycles of polyploidy occurred during
the evolution of plants. The basic chromosome number for the American *Crepis*
complex is $x = 11$. Species were derived from a European group with $x = 4$, and
another with $x = 7$. In the tribe Pomoideae of the family Rosaceae, the basic
number $x = 17$ probably resulted from amphiploidy (referring to the case where
somatic cells contain the diploid chromosome complement of both parental
species) by hybridization between ancestral species with $x = 9$ and $x = 8$ as their
basic numbers. This event occurred a long time ago, perhaps in the Tertiary or
Cretaceous period. Stebbins cites many other examples: in the spore-bearing plants
the lowest basic numbers are $x = 9$, 10, or 11, as in the genera *Selaginella* and
Isoetes; and the highest number, $2n = 400$ for *Tmesipteris*. Several cycles of polyploidy must have taken place in them. The most extensive development of polyploidy is in the family Ophioglossaceae, which comprises the most primitive ferns.
The lowest chromosome number is $x = 45$ in *Botrychium*. One of the species,
Ophioglossum reticulatum, has $2n = 1,260$ in two of its races, the highest number
in plants ever known.

Chromosome numbers in apocarpous dicotyledons are given in table 4.3. No
particular number can be referred to as the most primitive, but in general families

consisting of woody plants have higher basic numbers than do those that are predominantly herbaceous. The numbers x = 12, 13, 14, and 19 are possibly secondary polyploid derivatives (Stebbins 1966).

TABLE 4.3 Chromosome Numbers in Apocarpous Dicotyledons

Family	Basic Monoploid Number
Annonaceae* (six genera)	7, 8, 9
Calycanthaceae*	11
Cercidiphyllaceae*	19
Crossosomataceae*	12
Degeneriaceae*	12
Dilleniaceae* (three genera)	8, 13
Eupomatiaceae*	10
Illiciaceae*	14
Lauraceae* (six genera)	12
Leguminosae* (many genera)	6, 7, 8, 9, 10, 11, 12, 13, 14
Magnoliaceae* (six genera)	19
Nymphaeaceae	8, 10, 12, 14, 17
Paeoniaceae	5
Ranunculaceae	7, 8
Rosaceae	7, 8, 9
Saxifragaceae (s. str.)	7, 8, 9, 10, 11, 12, 13, 15
Schisandraceae*	14
Trochodendraceae*	19
Winteraceae*	13, 43

SOURCE: Reprinted, by permission, from G. L. Stebbins, *Science*, 152 (1966): 1463-69.
© 1966 by the American Association for the Advancement of Science.
*Woody.

CHROMOSOME SIZE VARIATIONS

In several groups of higher plants, a correlation apparently exists between chromosome size and ecological adaptation. Avdulov (1931; see Stebbins 1966) pointed
out that tribes and genera of plants that grow in tropical zones, or plants in tem-
perate regions that grow only during the warm season, have small and medium-
sized chromosomes, while those growing in cold regions have large chromosomes.
This agrees with observations of chromosome size in the family Leguminosae and
the order Liliales (Stebbins 1966). The biological and adaptive significance is
seen as being related to the mitotic rhythm. Van't Hof and Sparrow (1963) compared six genera of angiosperms with medium to large-sized chromosomes, and
found that nuclear volume is positively correlated with the length of the mitotic
cycle.

Chromosome size and nuclear volume are associated with the distribution of
heterochromatin. Avdulov (1931, see Stebbins 1966) discovered that in different

species of grass, heterochromatin is scattered throughout the nuclei of those species having large chromosomes, whereas in the species with small chromosomes the heterochromatin aggregated in regions of the centromeres. This is also the case in *Crepis.* As a general rule, large chromosomes (or large DNA contents) are associated with relatively high amounts of heterochromatin; plants with small chromosomes have proportionately less. The genetic significance of heterochromatin is not yet fully understood. It is generally agreed that when chromatin becomes heterochromatic, it ceases its activity in DNA replication, although the same region may have been active in earlier developmental stages of an organism (Brown 1966). In *Drosophila,* when chromosome translocation occurs, genes in the translocated region express themselves irregularly if the translocated chromosome segment was connected with a heterochromatic region.

The overall evidence leads us to conclude that large chromosomes in both plants and animals indicate an increase in heterochromatin and in gene number. But some genes are not active, or possibly a large number of duplications have occurred, therefore the total number of different gene-controlled metabolic processes may not necessarily increase.

Stebbins (1966) offers a theory on the evolution of heterochromatization as a process of adaptation, believing that it slows down gene activity. He bases this theory on Stern's thesis (1963) that the action of similar gene units or groups is coordinated through a system of macro-operons with one gene in a series activating the next. Thus in a particular biological system affected by a series of gene units, one finds that the larger their number, the slower the rate of action in the system; thus development in particular stages slows down. Conversely, euchromatization speeds up the rate of gene action. According to Stebbins, natural selection has adjusted cellular metabolism in order to complete certain rates of development in one of two ways: (1) heterochromatization or (2) reduction of gene units to fit the growth patterns of various species in different environments.

X-Chromosome Inactivation

In mammals, the X-linked genes are present in a single dose in males (hemizygotes) and in a double dose in females. Apparently evolutionary changes have been made to compensate for the dosage differences between males and females. One of the X chromosomes in females became inactive in the somatic cells early in development (although not in birds). Such compensation was called X-chromosome inactivation (Lyon 1961). The concept of gene dosage compensation was introduced by Muller (1932). He proposed that the presence of a certain gene cancels the effects of different doses of other given genes. The mechanism of inaction still remains obscure, but the hypothesis is supported by various types of evidence. First, inactivated X chromosomes remain heterochromatic. This was noted by Barr and Bertram (1949) before the hypothesis was proposed. They observed a distinctly stained body in the nuclei of neurons of the female cat, but not in the males. Now it is referred to as the Barr body and has been observed in the females of different types of mammalian cells. Mosaics were found for genetic markers of X-chromosome linked genes, such as glucose-6-phosphate dehydrogenase deficiency, in heteroxygous women. This is because some of the X chromosomes carrying the normal gene are inactivated. In the mouse (*M. musculus*) the XO females (one X chromosome only) are found to be fertile and physically fit (Welshons and Russell 1959), suggesting no inactivation because of the single X chromosome.

Ohno (1969) hypothesized that in mammals the X chromosomes are evolutionarily conservative. According to estimated amount of chromatin or size, the X chromosome is about 5% of the chromatin of the haploid autosomes through practically all the mammalian species. Many genes on the X chromosomes are found to be homologous between different species. Thus it is assumed that since the reptilian stage, despite the extensive speciation in the mammals, the original X chromosome in mammals has been preserved in toto without much change. However, the Y chromosome in mammals has been miniaturized and it is still questionable whether the Y chromosomes carry any genes with a function other than male sex determination or male characterization.

CHROMOSOME VARIATIONS IN HUMAN POPULATIONS

Chromosomal variations occur in our own species (table 4.4). The data have been taken from four different populations obtained through a cytological analysis of leukocytes in newborn infants (Lubs and Ruddle 1970; Ratcliffe et al. 1970). Although the sex chromosomes represent only two of the total forty-six, the incidence of abnormalities is much higher than that of an individual autosome. The incidence of chromosomal abnormalities is associated with the age of the mother: the higher the age, the greater the incidence of abnormalities. Except for some autosomal trisomies showing Down's syndrome at birth, most babies with chromosomal abnormalities appear normal, although some may develop physiological and mental abnormalities later.

TABLE 4.4 Percentage of Chromosomal Variations in Human Populations

Geographic Location	No. of People Examined	XYY	XXY	XXX	XO	Translocation	Aneuploidy (Trisomy)	Others
Ontario	1,066	0.37	0.09			0	0.09	0
New Haven, Conn.	4,500	0.07	0.09	0.07	0.02	0.14	0.12	
Boston	1,931	0	0.21			0.16	0	0.26
Edinburgh	1,931	0.14	0.08			0.17	0.14	0.03

SOURCE: New Haven study, Lubs and Ruddle 1970; others, Ratcliffe et al. 1970.
NOTE: The New Haven study includes both males and females. All the others include males only.

SUMMARY

We assume that the evolution of chromosome structure in eukaryotes originated from the genophore of prokaryotes, and that the process of meiosis evolved from mitosis. The intermediate evolutionary stages in each process are poorly understood if not actually unknown. Currently the best available evidence for establishing links comes from dinoflagellates, which have a cellular structure comparable to that of the eukaryotes and a nuclear structure similar to that of the prokaroytes. Different types of intermediate divisional processes may remain undiscovered in still other lower organisms.

In the process of nuclear division, variations occur in the crossing-over frequencies between sexes in different species, but this fact is insufficient to explain the evolutionary process from mitosis to meiosis. At present all we can be certain of is that the evolution of chromosomes and meiosis in the eukaryotes is very old, and that possibly it developed parallel to that of sexual reproduction. It is interesting to note that both the basic structures of chromosomes and their replication remain generally invariable throughout all species in the eukaryotes.

It has been pointed out by cytogeneticists (White 1964; Darlington 1966; Stebbins 1966; Carson 1970) that the importance of chromosomal variation in evolution has been underestimated. I share this view. Chromosomal rearrangements play certain specific roles in evolution that gene mutations cannot.

The essence of speciation lies in isolation. The geographical isolation of one group from another of the same species is an important factor in speciation. When two groups are separated, evolutionary factors operate independently in each, a point further discussed in chapter 7. Eventually sufficient genetic differences caused by either gene mutation or chromosomal rearrangements will be established so that interbreeding between individuals of different groups becomes impossible. Species so formed are called allopatric, and it is generally agreed that geographic isolation principally influences allopatric speciation.

What is still not entirely understood is the formation of sympatric species, that is, two or more of the same genus coexisting in the same geographic locality. Many investigators believe (Mayr 1963) that a population cannot evolve into different species if they inhabit the same locality, and that sympatric species were possibly allopatric but have been brought together in one way or another after achieving sexual isolation. The genetic reason behind this is that sexual differences in behavior and reproduction are influenced by many genes that carry additive and interactive effects. A single gene mutation alone is not sufficient to cause sexual isolation.

Chromosomal rearrangements, however, can be a more powerful genetic mechanism than gene mutation for achieving sexual isolation and establishing new populations. When a chromosomal mutation, say acentric fusion, occurs in a population, passing from the heterozygous to homozygous stage, as through a bottleneck, this represents the most difficult method for its establishment. We have mentioned that one possible assistance would be that the heterozygote is highly favored by natural selection in physiological fitness. Another circumstance would be a small population size, so that heterozygotes would have more chance to breed among themselves. Once a sufficient number of homozygotes is present in the population, they can interbreed. With the help of natural selection, a new chromosome type can be rapidly and firmly established. (Whatever other possible biological or populational mechanisms may exist, there must be various ways for a population to go through such a difficult evolutionary journey. Otherwise, no new chromosome rearrangement could be evolved.) These populations may then be on their way to the formation of a sympatric species.

If a new species is ever established from its parental species while inhabiting the same locality, abrupt and large genetic changes would have had to take place in order to assure sufficient reproductive isolation from the start. In this case chromosomal mutation is a plausible mechanism. Speciation will be fully discussed in chapter 7.

Gene interaction is an important property in evolution. Organisms have taken advantage of this by joining the favored and interactive genes to a linkage group, so that their chances of being transmitted as a unit are increased. When these genes occur in separate linkage groups, they can become linked by chromosomal fusion or translocation. Chromosomal inversion further reduces segregation. Perhaps in each species an optimum size and number of chromosomes exists (White 1945).

The significance of chromosomal structural rearrangement is known to many geneticists. There is, however, an additional feature in chromosomal change that I believe has evolutionary significance but that has received insufficient attention, that is, the elimination of outdated genes by chromosomal rearrangement. During the evolutionary development in each species, genes that were possibly active and useful in earlier periods become inactive or useless following genetic and natural environmental changes. For the benefit of the organism, some may be eliminated through gene deletion (as discussed in chap. 2) or through chromosomal deletion. The large number of heterochromatic regions present in each species does not necessarily imply that these regions consist of outdated genes. Some may be active in certain developmental stages of an organism (Brown 1966). It is likely, however, that many are evolutionary relics, merely carried over by organisms generation after generation because their elimination is difficult. Chromosomal deletion should prove the most efficient mechanism for elimination. Indeed, as in the case of acentric fusion, some deletions may always be involved in the centromere region. As this region is in most cases heterochromatic, this may be one reason why such deletions do not significantly affect the fitness of the organisms. It may be extending a hypothesis too far to say that the genes around the centromere are in fact outdated genes; more evidence is needed on this point.

Sexual reproduction cannot succeed without meiosis, and during meiosis various incidents involving the chromosomes can occur. Many facts concerning chromosomal replication and the crossing-over process remain unknown. As is true for gene mutations, chromosomal rearrangements are random events; they occur more rarely than gene mutations, but they do play a specific role in evolution.

Five

evolution of structure and function

For an organism, it is important that necessary genetic information or genes be available for survival and adaptation. But when sufficient genetic information has evolved through the processes previously mentioned, it is equally important to have a program by which the use of certain information at certain stages in the development of an organism can be arranged. Such selective use of genetic information may be called gene regulation, which in turn may depend on other genes or environmental agents.

Unfortunately, gene regulation underlying morphogenesis and differentiation is one of the least accessible areas in biological science. We are far from complete understanding, for instance, of the genetic mechanisms involved in embryonic development leading from one stage to the next. Nevertheless, recent advances in various fields of biology have shed light on many puzzling development phenomena so that we can begin to appreciate, in principle, the genetic basis for growth and development and the biological characteristics of the phyletic lines.

GENIC REGULATION OF DEVELOPMENT

How is genetic information released or utilized for growth, development, and other biological activties in the same organisms that are its carriers? The mechanisms involved in the complete process are definitely complex and may vary in the biological systems and species to be discussed. However, some models and hypotheses, developed from experimental evidence, will be useful in thinking about evolution of a biological system.

Genes are generally classified into two types according to function: structural and regulatory. Structural genes determine the primary amino acid sequence of the polypeptide chains that constitute the different protein molecules. In addition, these genes may also influence the secondary, tertiary, or quaternary structure of the proteins. The existence of structural genes was first demonstrated in microorganisms; in the past decade, a surprisingly large number have been identified in the higher organisms, including those that specify the polypeptide sequence of isoenzymes, serum proteins, hemoglobins, and hormonal substances. It is because of the later discoveries that the term *structural gene* has acquired more precise meaning. Regulatory genes are concerned with control of structural gene expression.

Genetic studies in bacteria have provided useful evidence for unraveling various intricacies in the complex system of gene transcription. *Escherichia coli* normally possesses very low levels of the enzymes necessary to metabolize unfamiliar substances. Only when the organism is exposed to them is the synthesis of the enzymes required for their metabolism stimulated. This process is called enzyme induction, and the produced enzymes are known as inducible enzymes. The control of inducible enzyme synthesis is highly specific: lactose induces only the synthesis of lactose-metabolizing enzymes, maltose only maltose-metabolizing enzymes, and so on. It was discovered that certain mutations caused an uncontrolled synthesis of specific proteins irrespective of the presence or absence of the inducing agents to which the wild type was sensitive (table 5.1). These mutations often simultaneously affect the synthesis of several proteins. They are invariably mapped at a gene distinct from those that govern the structure of these enzymes. The mutant genes do not modify the molecular properties of the proteins. Such control processes for protein synthesis in both wild and mutant types of *E. coli* have led Jacob and Monod (1961) to formulate their well-known operon model of gene regulation.

TABLE 5.1 The Effect of Mutation of the Regulator Gene i

Genotypes	Noninduced			Induced		
	β-galactosidase	Galactoside permease	Galactoside transacetylase	β-galactosidase	Galactoside permease	Galactoside transacetylase
1. $i^+z^+y^+$	<0.1	<1	<1	100	100	100
2. $i^-z^+y^+$	120	120	120	120	120	120
3. $i^+z^-y^+/Fi^-z^+y^+(*)$	2	2	2	200	250	250
4. $i^-z^+y^+/Fi^+z^+y^-(*)$	2	2	2	250	120	120
5. $i^-z^-y^+/Fi^-z^+y^+$	250	250	250	200	250	250
6. $\Delta_{izy}/Fi^-z^+y^+$	200	200	200	200	200	200

SOURCE: Reprinted, by permission, from F. Jacob and J. Monod, *Cold Spring Harbor Symp. Quant. Biol* 26 (1961): 193-211. © 1961.

NOTES: i^+ = the wild-type allele; i^- = the mutant type. Line 1 and 2 represent cells in the haploid condition and 3 to 6 in the diploid condition. Δ_{izy} represents deletion. Note the mutant genes for i are invariably recessive to the wild type.

*The high levels of the structural gene productions in noninduced cultures of i^+/i^- are due to the presence of a small fraction of homozygous i^-/i^- recombinants.

This model describes the mechanisms involved in gene expression. Structural genes produce messenger RNA (mRNA) complementary to the DNA of the gene. When the messenger synthesis begins at a certain point of the DNA strands, it elongates by adding nucleotides. The growing messenger is firmly bonded to the template DNA by the DNA-dependent RNA polymerase. After its synthesis is completed, mRNA goes to the cytoplasm where it attaches to the ribosome.

Through the aid of transfer RNA, the protein synthesis takes place. The process of producing mRNA from DNA is called gene transcription, and that of producing the polypeptide from mRNA is called translation. The synthesis of mRNA is a sequential and oriented process controlled by a gene called an operator. A single operator may control the transcription of several structural genes. Such a group of genes constitutes an operon. The operator is affected by a regulator gene, so that a regulator gene produces a cytoplasmic repressor that tends to associate with a specific operator. This interaction blocks the initiation of transcription for the structural genes of the operon.

Since Jacob and Monod presented their hypothesis on gene regulation, numerous studies have produced much more descriptive information on the process. The present concept of gene regulation in the *lac* operon of *E. coli* is illustrated in figure 5.1 (Reznikoff 1972). In this model, the *lac* operon consists of three structural genes (z,y,a) arranged in linear fashion, with three regulatory genes (i,p,o) controlling their transcription. The i gene codes for the synthesis of a repressor protein that binds to the operator (o). Thus it prevents the RNA polymerase that binds to the promotor (p) from transcribing the structural genes. The promotor is the target site for catabolite repression (the inhibition of enzyme induction by glucose or related substances in *E. coli*), with a catabolite gene activator protein (CAP)[1] (Riggs, Reiness, and Zubay 1971) and adenosine $3':5'$-cyclic phosphate (cAMP) playing the active part in catabolite repression. When the inducer, a galactoside, combines with the repressor, the repressor dissociates from the operator and the polymerase begins its transcription. Most cells of *E. coli* mutants in i or o genes produce beta-galactosidase in the absence of an inducer. This occurs in i mutants because the repressor formed does not bind to the operator. In the o mutants the operator is changed so that the repressor will not recognize it. It is assumed (Ohno 1972) that, in order to exist as a self-sustaining regulation system, a regulator gene such as the *lac i* in *E. coli* cannot be under regulation itself from other genes. The repressor, as a regulator gene product, must be a protein in order to bind the inducer and the DNA of the operator site.

In bacteria, there are two contrasting systems controlling the transcription of the structural genes. One is the induction of synthesis of an enzyme by the presence of its substrate. The substrate is called an inducer. In the absence of an appropriate inducer, the repressor is allowed to inhibit the transcription of the structural gene. Such a system is referred to as being under positive control. The production of the enzyme, galactosidase, by the presence of galactoside is an

---

1. Catabolite gene activator protein (CAP or CGA) is a regulatory protein or a DNA-binding activator. It is a dimer with subunits of 22,000 daltons, and it has a substantial affinity for DNA. This affinity is strengthened by cAMP and strongly inhibited by guanine $3':5'$-cyclic phosphate (cGMP). The presence of both cAMP and CAP is required for the expression of the *lac* operon. Thus they are a part of the controlling system. The expression is inhibited by cGMP. The inference is that CAP protein activates the *lac* by binding the DNA under the influence of cAMP.

example. In the case of enzymes that catalyze biosynthesis of essential cell constituents, a reverse situation has been observed. That is, the synthesis of an enzyme, instead of being induced by its substrate, is repressed by the presence of the
products of the reaction that it catalyzes. Examples for this system are the protein
production of *trpA* and *trpB* genes (controlling tryptophan synthesis) of *E. coli*.
They are repressed by the presence of tryptophan in the growth medium. Similarly,
if histidine is added to the culture medium for *Salmonella*, the enzymes for the
synthesis of histidine are not produced by the organisms. In the absence of histidine, they are synthesized in amounts necessary to meet the demands of cells.
This system is referred to as being under negative control. That is, in the absence
of the gene production, the repressor does not function, permitting synthesis of
the necessary enzymes.

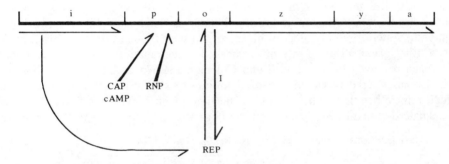

Fig. 5.1. Genes controlling lactose utilization (lac region) of the *E. coli* chromosome. This region consists of three regulator genes, *i*, *p*, and *o*, and three structural
genes, *z*, *y*, *a*. The lac *i* codes for the repressor, (REP); *o* represents the operator
and *p* the promotor gene. The lac *z* codes for β-galactosidase, *y* for β-galactoside
permease, and *a* for thiogalactoside transacetylase. RNA polymerase (RNP) binds
to and initiates mRNA synthesis at the promotor, *p*. The binding of RNP to *p* is
dependent on the presence of the catabolite gene activator protein (CAP) and
cyclic adenosine 3′:5′-monophosphate (cAMP). Transcription continues until the
RNP reaches the terminator of the DNA; the REP prevents transcriptions of the
lac operon by binding to the operator, *o*. The inducer (I) prevents the REP from
binding to *o*, and allows transcription to occur.

   The two differing systems of gene regulation are both adaptive and of evolutionary significance. The positive control system is for the production of enzymes
that metabolize nutrients to supply energy to the organisms. Since it would be
extremely inefficient for the cells to synthesize the enzymes when no substrates are
present to react, a positive control system has evolved so that only in the presence
of a substrate can the genes for enzyme production be activated. The negative
control system appears to function in the production of cell building blocks, that
is, amino acids. As long as the environment supplies these, there is no need for
the genetic system to produce them. Only when the need arises and there is no

outside supply will the organisms put their own genetic machinery into operation. This function requires the negative control system.

To what extent the operon theory can explain gene regulations in organisms other than bacteria is yet to be ascertained. There is some preliminary evidence that seems to suggest the existence of a positive control of gene transcription. For example, the simple deprivation of a carbon source results in the formation of a variety of enzymes for the transport and metabolism of sugars in *Neurospora* (Metzenberg 1972). Paigen (1970) reported that in mice one of the postulated regulatory mutants had negative control analogous to the repressor system of bacteria. Thus he proposed that, in eukaryotes, the regulatory control of enzyme synthesis is positive. For example, the gene *Lv* in mice affects delta-aminolevulinate dehydratase synthesis (Russell and Coleman 1963). But in the case of catalase, there is a gene affecting the rate of degradation rather than synthesis (table 5.2). Note that in the mouse system the regulating substances are gene products of the animal itself, not of the environment. This is perhaps a highly complex self-regulating system present solely in higher organisms. Evidence concerning a number of other genes affecting rate of synthesis and degradation of different enzymes have been reviewed by Kandutsch and Coleman (1967) and Paigen (1970).

It was discovered that some people continue to synthesize fetal hemoglobin at a high rate rather than cutting it off at the time of birth. The condition, inherited as a single dominant gene located on the chromosome very close to the structural

TABLE 5.2 The Effects of Genes on Catalase Realization in Mice

	DBA / 2	C57BL / 6	C57BL / An, He, Ha
Genotype			
Presumed catalase structural gene	Cs^a/Cs^a	Cs^g/Cs^g	Cs^g/Cs^g
Degradation gene	Ce/Ce	Ce/Ce	ce/ce
Liver			
Rate of synthesis	1.0	1.0	1.0
Rate of degradation	2.0	2.0	1.0
No. of molecules per cell	0.5	0.5	1.0
Activity per molecule	2.0	1.0	1.0
Observed activity	1.0	0.5	1.0
Kidney			
Rate of synthesis	1.0	1.0	1.0
Rate of degradation	1.0	1.0	1.0
No. of molecules per cell	1.0	1.0	1.0
Activity per molecule	2.0	1.0	1.0
Observed activity	2.0	1.0	1.0

All values are expressed as relative to the same quantity in the C57BL / An, He, Ha phenotype.

SOURCE: Reprinted, by permission, from Paigen, 1970.
NOTE: *Cs* represents the structural gene and *Ce* the gene for degradation.

genes for beta and delta hemoglobins, appears to be a defect in the operon that contains these genes (Sutton 1960). The defect is perhaps in the regulatory gene or in the controlling system of the structural genes.

A most interesting case in the evolution of gene regulation occurs in the production of different beta hemoglobins in sheep and goats. It was found that there are two forms of beta hemoglobin in sheep, beta A and beta C. Beta C is a neonatal hemoglobin present at birth; in adults it is replaced by beta A. When an adult sheep is anemic, however, beta C hemoglobin reappears in the blood. The goat possesses the same switching mechanism and, in addition, it has multiple forms of beta hemoglobin. It has been discovered that erythroprotein, a hormonal substance that stimulates red blood cell production, is involved in the regulating mechanism for this process. When the sheep is anemic, its erythroprotein level rises. The effect of this protein on the production of beta C hemoglobin occurs at the gene transcription; this result has been confirmed by cell cultures.[2]

Humans do not have a gene homologous to that for beta C hemoglobin in sheep and goats, and therefore do not possess this regulating mechanism. It appears that goats are among those who can inhabit a wide range of altitudes, and the mechanism may be intended to aid in their adaptation to higher altitude conditions.

The discovery of cAMP as an intermediary for the hormone effects on gene expression provided a step forward in the understanding of gene regulation mechanisms in higher organisms. Except for the cyclic phosphate diester, cAMP is identical with AMP of the DNA molecule (fig. 5.2). Since the discovery of its role in the regulation of sugar metabolism in the liver by insulin and epinephrine (Sutherland, Øye, and Butcher 1965), the function of cAMP in mediating the action of many hormones has been tested in many biological systems. It has now been generally recognized that cAMP plays the role of an intermediary in the action of practically all peptide hormones.

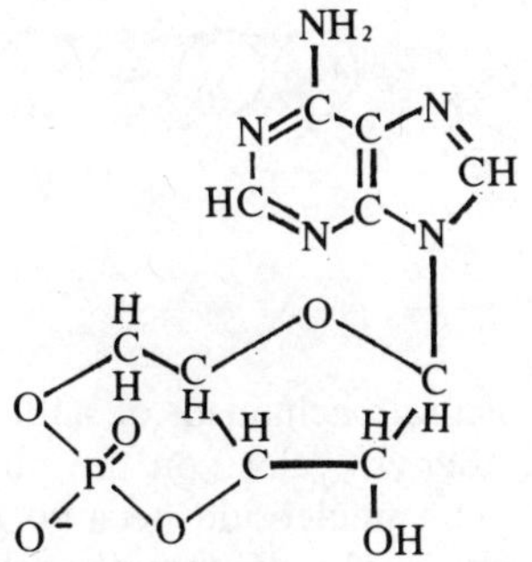

Fig. 5.2. Chemical structure of cyclic adenosine 3':5'-monophosphate (cAMP).

2. W. F. Anderson 1974: personal communication.

In his proposed model for the mechanisms involved in gene regulation, Sutherland (1972) concluded that the hormones are the first messengers, acting on the
enzyme adenyl cyclase, which is present in the cell membrane not only of mammalian tissues but of a variety of other phyla as well. He hypothesized that adenyl
cyclase consists of two subunits: one of which is species or tissue specific and
contains the receptor site for the hormone; the other a catalyst in the production
of cAMP from ATP. As the second messenger, cAMP acts on a protein receptor
that combines with the repressor or inducer for the transcription or translation of
the structural genes (Emmer et al. 1970). (It may be reasonable to assume that
the adenyl cyclase system existed earlier than the genes for some specific hormones.) On this hypothesis, a specific model for the action of adrenocorticotrophic
hormone (ACTH) on the synthesis of adrenal cortical hormones was constructed
(fig. 5.3).

The rapidity and diversity of the response to the steroid hormones is remarkable. Investigators have long been searching for the basis of such reactions. Some
preliminary evidence suggests that the primary effects take place on the gene

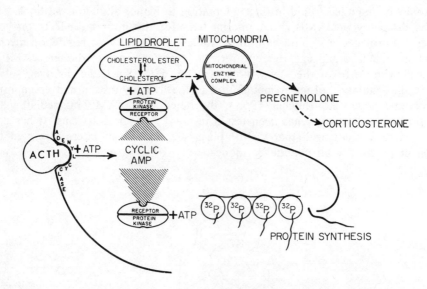

Fig. 5.3. Hypothetical model of the mechanism of action of ACTH in the adrenal
cortical cell: ACTH activated adenyl cyclase on the plasma membrane, catalyzing
the formation of cyclic AMP. The nucleotide became bound to the receptor and
activated protein kinase, which catalyzed the phosphorylation of a ribosomal
moiety, thereby modulating the translation of stable mRNA/mRNAs. This resulted
in the induction of the regulator-protein, which, by an unknown mechanism, facilitated the translocation of cholesterol from the lipid droplet to the mitochondrion.
(Reprinted, by permission, from L. D. Garren et al., *Recent Progr. Hormone Res.*
27(1971):433–78.)

translation level (Mueller et al. 1972). According to these authors, the general mechanisms are similar to those of the receptor model of cAMP operation for the peptide hormones. The difference in operation between the peptide and steroid hormones is that the steroid hormones can enter cells, and there they themselves, or their metabolites (or derivatives), participate in gene transcription or translation. The peptide hormones cannot cross the cell membranes; they operate through the cyclase and cAMP system. Both classes appear to operate on the receptor principle in gene regulation, but much more supporting evidence for such a hypothesis is still needed.

At the present stage, we can visualize the hormones as modulating the expression of multiple genes by controlling the availability or state of specific proteins essential in the nucleus for the transcription or translation of the structural genes. Some hormones are the products of specific genes; thus their production is also subject to regulation by inducers or repressors. In this way one gene affects the expression of another.

In general, a gene cannot be activated without an inducer. From what sources do inducers originate? An inducer can be a small molecule, such as a nutrient material from the outside environment, or it can exist within an organism as a metabolite, hormone, or other cellular or humoral substance. But the ultimate source of many inducers is the natural environment. We observe, then, the importance of this environmental effect on gene regulation; changes in environment feed directly or indirectly into the gene regulation system, with the necessary consequences on physiological fitness and adaptation.

However, the recognition of hormonal substances as inducers can be considered as one major advance in the evolution of higher organisms. Such control or regulation of gene action has made the development and maintenance of life relatively independent of the environment. For example, adrenal cortical hormone secretion is dependent on ACTH from the pituitary gland. But the adrenal cortical hormone in turn affects ACTH output, so that a high level of the adrenal cortical hormone inhibits ACTH production, and a low level increases it. Thus, ACTH and adrenal corticoid are maintained at optimum levels for the required regulation of gene action. This type of physiological self-regulation, termed homeostasis, is well known in vertebrates. With the recent advance in the understanding of the role of hormones in gene action, the importance of this feedback control for the transcription or translation of genes emerges clearly.

DIFFERENTIATION AND MORPHOGENESIS

The least understood of the areas under scrutiny here is the mechanism by which gene action shapes the development of the organism, that is, the effects of genes on differentiation and morphogenesis, which constitute the basis for taxonomic classifications. The term *differentiation* describes the complexity of changes involved in the progressive diversification and specialization of cell structure and function, whereas *morphogenesis* includes changes in form resulting not only from an aggregation of molecules, but also from differential cell death, differential growth, and the displacement of cells and cell groups (DeHaan and Ebert 1964; Ebert 1965; Waddington 1966). Therefore, differentiation may be regarded as an outward sign of selective gene action and a reflection of changes in biochemical processes, following the release of information encoded in one-dimensional form by the genes (Allen 1965). Through folding, aggregation, and redistribution of cells, morphogenesis is responsible for further shaping of a structure from a one-dimensional to a three-dimensional form.

Two genetic hypotheses have been proposed for differentiation and morphogenesis (Ebert and Kaighn 1966): differential release of information and differential replication. The former concept has already won general acceptance, agreeing as it does with the operon model of Jacob and Monod. One supporting piece of evidence involves the heterochromatic regions of the chromosomes, which are considered to be a sign of genes in the inactive state, although they may have been active and nonheterochromatic in the earlier stages of organism development. After genes are inactivated they remain irreversibly "silent." X-chromosome inactivation, first described by Lyon (the so-called Lyon hypothesis, discussed in chap. 4) is an example. Inactivation takes place in early embryogenesis. Supporting evidence for the theory of differential replication lies in the fact that during development of the salivary gland nuclei in *Drosophila*, euchromatic regions of the chromosomes replicate, whereas heterochromatic regions around the centromeres do not (Rudkin and Schultz 1961; Schultz 1965). Furthermore, in *Ascaris* the heterochromatic regions are lost at early mitosis in the somatic cells, but are retained in the germ lines.

Selective gene action has actually been demonstrated in studies on cell surface antigens. Paramecia possess sets of genes for cell surface antigens detected serologically, and only one of these genes is found to be active at a given time of their life cycle (Sonneborn 1948). It has been shown that in mice, thymocytes and lymphocytes have their specific antigens that are not present in other tissue cells (Boyse 1971). Each of these antigens is determined by a gene, most of which have been located on the linkage map. In a multicellular organism, each somatic cell possesses an identical set of genes. The phenotypic difference between them, such as in the antigens, must be the result of differential gene action.

The protein hormones that are coded by specific genes and produced by different endocrine cells provide further evidence of differential gene action.

The various cell types present in higher organisms do not occur in random fashion throughout the body. Neither form nor structure is arbitrary; it arises epigenetically through cell movement. The results of Mintz (1967) in fusing mouse eggs provide an interesting illustration of this. She fused eggs of two different genotypes, one pigmented and the other albino, from the two-cell stage to the morula stage. These eggs were transplanted to female mice at a corresponding state of pregnancy, and the mothers were delivered of litters with mosaic coat colors. The standard coat color pattern contained a total of seventeen bands; three on the head, six on the body, and eight on the tail. Mintz proposed that the bands were clones of melanoblasts, each derived from a single cell, each of which originated from some of the neural crest cells and migrated toward the body wall at eight to twelve days of embryonic life. With the standard pattern of seventeen clones for each side, there are a total of thirty-four primordial melanoblasts. The question as to how this particular number of cells was selected for melanoblast differentiation and then dispatched to different parts of the body cannot be answered at present.

ONTOGENY AND PHYLOGENY

Ontogeny concerns the history of development of an individual organism and phylogeny the history of development of a race, species, genus, and so on. The development of a higher animal through stages resembling those of simple organisms at different levels of evolution was noticed as early as 1821 by Meckel. An evolutionary doctrine was formulated in the belief that each individual in the course of development passes through a series of stages corresponding to ancestral types. Ernest Haeckel (1834–1919) proposed that "the ontogeny is a brief and rapid recapitulation of the phylogeny."

Organisms evolve from simple to complex forms through many different stages. Unicellular organisms (fig. 5.4, *a*) become multicellular through the production of a cohesive property such that individual cells form into a cluster. Through arrangement of the cells, either a hollow shell (blastula) or a solid mass (morula) can be formed. The blastulas of calcareous sponges illustrate this stage (fig. 5.4, *b*).

Further evolution from the blastula stage is accomplished by invagination of a layer of cells to form a double cup, or by migration of cells from the surface into the hollow space forming a simple gastraea. When cells enter the cavity, the organism flattens so that the cells can maintain contact with the substratum (cells in the inner shell of the gastraea) in order to solve their nutritional, respiratory, and excretory problems. Such organisms were referred to as planuloid by Willmer (1970) (fig. 5.4, *c*). He further pointed out that the flattening of a spherical organism (that has already developed the anterior-posterior axis) initiates the dorsoventral sides and bilateral symmetry (fig. 5.4, *c*). It is believed that, in vertebrate phylogeny, passing through the planuloid stage is an important step.

From the planuloid stage, an organism may evolve into something resembling the present Acoela (fig. 5.4, *d*), a small delicate marine animal, free-swimming among the rocks and seaweed along the shore. The Acoela are more complex than planulae. They have an epidermis that consists of a mixture of ciliated, goblet (containing mucin and bulged out like a goblet), epitheliomuscular, and neuroepithelial cells. Organisms further advanced toward the vertebrates may be represented by the rhabdocoeloid (fig. 5.4, *e*) or nemerteoid (fig. 5.4, *f*). Of existing organisms, planarians are an example of the rhabdocoeloid organisms. Planarians have a more complete neuromuscular system than Acoela and a well-developed epidermis. They have a simple alimentary system with mouth and gut, but no anus, and much of the digestion of food is intracellular. They have developed a urogenital system. In some existing rhabdocoeles, the developing ova are entirely surrounded by close-fitting follicle cells and the sperm develops in fluid-filled sacs.

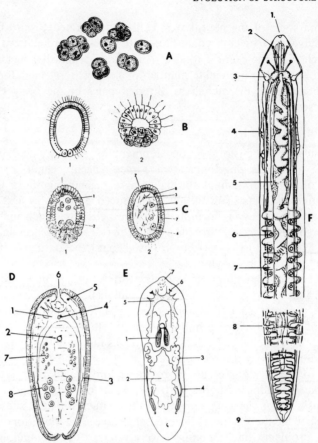

Fig. 5.4. Structures of organisms from lower to higher forms. (*a*), Greenslime. The pleurococcus consists of a single cell, a unit of protoplasm with a definite cell wall. The food intake and by-products discharge through the wall. When a cell divides in two, the daughter cells may be separated or cling together. After Gruenberg 1929. (*b1* and *b2*), Typical blastulas of calcareous sponges: note that some of the cells have flagella. (*c1* and *c2*), Horizonal sections through planuloid organisms, such as the sponge: (*c1*), Simple planuloid, the cavity filled with dehisced cells derived from two classes of epitheliocytes; (*c2*), More advanced planuloid: (*1*), premechanocyte; (*2*), preamoebocyte; (*3*), mechanocyte; (*4*), mesohyl (middle substance, analogus to mesoderm); (*5*), collagen fiber; (*6*), amoebocyte; (*7*), epitheliomuscular cell; (*8*), nerve cell. (*d*), Acoeloid organism: (*1*), nervous system; (*2*), pharynx; (*3*), muscle; (*4*), statocyst; (*5*), eye; (*6*), frontal organ; (*7*), male germ cells; (*8*), female germ cells. (*e*), Rhabdocoeloid organism: (*1*), pharynx; (*2*), intestine; (*3*), gonad; (*4*), nephridia; (*5*), brain and nerve cords; (*6*), eyespot; (*7*), frontal organ. (*f*), Nemerteoid organism: (*1*), frontal organ; (*2*), eyecup with nerve; (*3*), cephalic organ; (*4*), nephridium with nephridiopore; (*5*), pharynx; (*6*), gonocoel; (*7*), intestinal diverticulum; (*8*), gonopore; (*9*), anus. (Reprinted, by permission, from Willmer, 1970.)

The nemertines have most of the essential organs and tissues of vertebrates. Details of the phylogeny sketched above are given by Hyman (1951) and Willmer (1970).

A comparison of various lower organisms with developmental stages of a higher organism, such as the mouse, is illustrated in figure 5.5. Higher organisms, such as vertebrates, start as fertilized ova, very much like unicellular organisms. The ovum develops through a two-cell phase, then four-cell, and so on, to the morula and gastrula stages, which bear a close similarity to the blastulas of calcareous sponges. The most widespread among embryo types of different species was the blastula, largely because of its ability to maintain stable and maximal contact with the outside environment for acquisition of nutrients. Later the ectoderm, mesoderm, and endoderm were represented by separate layers of cells and the embryo resembled hydra or coral. At a more advanced stage the embryo corresponded to fishlike organisms. An evolutionary doctrine was formulated in the belief that each individual in the course of development passed through a series of stages corresponding to ancestral types. Such progressive changes were aimed at achieving a more independent or self-sufficient system.

In general, higher organisms possess more genetic information than the lower forms. All and the same genes are present in each somatic cell from single fertilized ovum to the well-differentiated types. But at a given point of time through the whole course of development, only a limited amount of genetic information is used or only a limited number of genes are active. It may be said that in the higher organisms, such as the vertebrates, only the genes that correspond to those of the calcareous sponges are active at the morula and gastrula stages; at the stage of development of embryonic germ layers, only the genes corresponding to those of hydra and coral are active. Such selective gene activation has evolved through a long course of natural selection.

There are some hazy areas in gene activation. For instance, the presence of heterochromatin in certain regions is assumed to represent inactivation of genes involved in those regions. But it is not known whether the genes were active at previous stages of development in an organism or whether they were active in the early ancestors from which the present species or phyletic lines derived. It is very possible that both may be true. In any event, we assume that organic evolution includes not only the presence or absence of genes but also their activation and inactivation at appropriate times of development. The concept of evolution of genes is widened, including not only the growth and alternation of genetic information, but also the selective usage of it. Genetic evolution represents a revision of the whole genetic program.

EVOLUTION OF CELL STRUCTURES

Basic Cell Types

Two basic types of cells characterize primitive organisms: the flagellates and the phagocytes. The former are connected with the activity of the flagella or cilia (elongated fibers forming a cylinder that serve primarily as locomotor appendages).

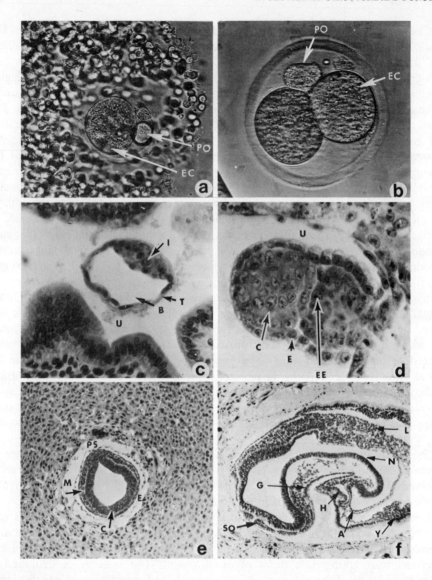

Fig. 5.5. Stages of development in mouse embryos: (*a*), fertilized egg from ovi-duct twenty hours after copulation; (*b*), two-cell egg, twenty-four hours after copu-lation; (*c*), blastocyst, 3½ days after copulation; (*d*), longitudinal section of five-day embryo; (*e*), cross section of seven-day embryo; (*f*), cross section of eight-day embryo. A, amnion; B, blastocoel; C, ectoderm; E, endoderm; EC, egg at one cell stage, two cell stage; Y, yoke sac; EE, extraembryonic ectoderm; G, foregut; H, heart; I, inner cell mass; L, allantois; M, mesoderm; N, head neural fold; PO, polar body; T, trophoblast; SO, somite; PS, primitive streak; U, uterine lumen. (Plates *c*–*f* from L. C. Stevens, The Jackson Laboratory.)

The latter concern the pinocytotic activity (digesting and absorbing) of amoebo-
cytes (Willmer 1970).

Naegleria gruberi is a unicellular organism, normally amoeboid, living in the
soil and feeding on bacteria. When put in water, its form changes (fig. 5.6), becoming polarized and flagellated. *Naegleria* cells possess all the essential biochemical activities and structures found in the cells of higher organisms. According to
Willmer, the *Naegleria* cell probably has a nucleus, a nuclear membrane, flagella,
mitochondria, cytoplasmic tubules, an endoplasmic reticulum, ribosomes, and a
cell membrane, with a DNA reproductive system well established. In other protozoa, efficiently formed mitotic and meiotic divisions, as well as differentiated sex
cells, can be observed. Since the protozoan stages, the evolution of cells has been
essentially toward a cooperative system of many cells having special structures and
functions.

Naegleria cells are sensitive to changes in osmotic pressure and ionic balance
that involve either maintenance of their amoeboid form or a change to the flagellate form. If glucose or glutamate are added to the medium, the flagellate form
returns to the amoeboid form. When the ionic strength of Ca^{++}, Na^+, K^+ falls below
0.05M in the medium, *Naegleria* of amoeboid form change rapidly to flagellate
forms.

The general activity of *Naegleria* in the amoeboid phase is quite different from
that in the flagellate phase, in which organisms are strongly polarized. The flagellate form is adopted at a low ionic concentration of the external environment,
while the amoeboid form prefers a high ionic concentration. Some time ago Holtfreter (1947) observed similar behavior in the epithelial cells of amphibian gas-
trulas. The cells had cilia beating at one end and amoeboid processes at the other.
Apparently, genes for flagella formation are also present in the *Naegleria* cells.
The appearance of either one phase or the other depends on the activation of
certain genes and the repression of others. The agents initiating gene regulation
are environmental, such as the ion concentrations or special chemical substances
in the medium. The shift from one phase to another is an evolved adaptive property of the cells. The property of ciliation in the cells of higher organisms may be
traceable far back to their unicellular ancestors.

A remarkable feature of *Naegleria* is its response to the introduction of steroid
hormones into the medium. Deoxycorticosterone and progesterone in a concentration of $10^{-4}M$ force the organisms into the amoeboid form. At lower concentrations
they may force the organisms into the flagellate form (Willmer 1970). (Cholesterol, with some similarity in basic chemical structure to the hormone, yet known
to be without hormonal activity, has no effect.) These hormones are synthesized
only by the sophisticated higher organisms. Once they were thought to be only
effective in vertebrates; later it was found that they were active in insects; and
now in almost the lowest cellular organism. Steroid hormones are known to act
on the transcription or the translation level of DNA. Apparently such reactive
mechanisms are present in unicellular organisms, although they do not produce
hormonal substances.

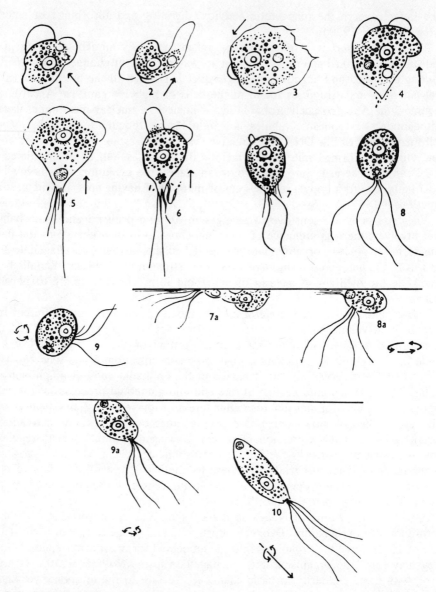

Fig. 5.6. Change of form of *Naegleria gruberi* when the organism is placed in dis-
tilled water. Arrows indicate the direction of motion: *(1–3),* amoeboid form; *(4),*
polarized form; *(5 and 6),* filiform pseudopodia present at posterior pole; *(7–10),*
development of active flagella; *(7a–9a),* the way events would appear if seen from
side when the amoeba leaves the surface of the coverglass; in *(4–10)* the contractile vacuole, in its different phases, is situated at the posterior pole. (Reprinted, by
permission, from Willmer, 1970.)

The skin of the nemertine (belonging to the suborder of Nemertea[3]) consists of ciliated cells and two varieties of secretory cells with basophilic and acidophilic properties derived from cell types corresponding to the flagellated and amoeboid cells (Willmer 1970). Except for sensory perception, the three types of cells carry out the function of skin. In the pharynx and the upper part of the alimentary canal of the nemertine, the cells can also be classified into three types similar to those in the skin, but they have specialized functions. In the pharynx, some cells produce acid, some absorb iodine, and others may perform functions of the pituitary and salivary glands of mammals. Others probably represent the origin of the chloride cells of fish gills, and the acid substance secreting oxyntic cells of the mammalian stomach. In the cephalic organ of a *Lineus* (a flatworm belonging to the phylum Platyhelminthes, order Turbellaria), the epithelial cells of the tubular structures could be developed from the ciliated cells, and the secretory and vesicular cells from the mucoid types. Similarly, in the intestine there are two main types of cells, some ciliated and some that secrete mucoproteins. In the anterior region of the intestine there are cells with granules that have the same staining properties as the cells in the skin. In each tissue, cells perform different functions, but may have not yet evolved far enough to completely lose their original identity.

Endocrine Cells

Endocrine cells are highly specialized in function and play a most important role in ontogeny and adaptation. The endocrine tissues in vertebrates are characteristically epithelial or glandular, and are generally composed of different cell types. Each produces specific hormones. For example, the adrenal glands secrete a number of corticosteroids and adrenalin; the thyroid secretes iodinated tyrosine compounds and thyrocalcitonin; the pituitary gland produces various trophic and neurohypophyseal hormones; and the islets of the pancreas produce glucagon and insulin.

A more detailed view of the thyroid gland shows that it is composed of two types of cells: follicular and parafollicular. The former produce iodinated thyroid hormones; the latter produce the recently discovered thyrocalcitonin, a hormone that maintains the calcium and phosphate balance. The thyrocalcitonin cells have been identified as derivatives of ultimobranchial bodies derived from the fifth set of branchial pouches, later incorporated into the thyroid gland. The thyroid gland is derived by invaginating an entodermal pocket in the midplane at the level of the first set of pharyngeal pouches (see Arey 1946, pp. 211–15). In fish, the ultimobranchial bodies play the same role as the thyroid in mammals.

---

3. Nemertea is a subdivision of worms, mostly marine, with ciliated skin, a retractile proboscis, and simple generative organs. They are ribbon-shaped animals, more or less cylindrical in section. There are no exterior appendages of any kind, and their colors are often bright and varied. They were formerly arranged among Platyhelminthes in the order Turbellaria.

The thyroid gland in vertebrates is believed to have evolved in the pharyngeal region of the ancestors of the chordates, a major phylum of the animal kingdom including vertebrates and simply built marine vertebrate relatives. The thyroid is the equivalent of the endostyle (a pair of parallel longitudinal folds projecting into the pharyngeal cavity and bounding a furrow lined with glandular ciliated cells). During the embryonic stage, the thyroid develops from the epithelium of the pharyngeal floor, a region known for its uptake of iodine in urochordates (simple marine animals) and more advanced chordates (Barrington 1965). Iodine is bound in a region running along the pharyngeal groove known as the endostyle in the larva of the lamprey; triiodothyronine (a thyroid hormone more potent than thyroxine) and thyroxine, the most common thyroid hormones in vertebrates, are also found there.

Barrington and Franchi (1956) observed that in the ammocoete larva of lamprey, the binding of iodine occurs in a special group of ciliated cells. Depletion of the secretion of the iodinated compounds in these cells by treatment with antithyroid chemicals, such as thiouracil, is a response that also occurs in the thyroid glands of mammals. The cellular pattern of the pharynx of *Lineus* bears a great similarity to that in the endostyle of ammocoetes, the whole pharynx region picking up iodine more rapidly than the rest of the worm.

Barrington and Thorpe (1963) found that the enteropneust (*Saccoglossus horsti*) takes up iodine and stores it in the form of monoiodotyrosine. In higher organisms, the two most potent thyroid hormones are thyroxine (T_4) and triiodothyronine (T_3). The former is produced by coupling two molecules of diiodotyrosine, and the latter by coupling one molecule of monoiodotyrosine and one molecule of diiodotyrosine. Both monoiodotyrosine and diiodotyrosine have some weak

Monoiodotyrosine

Diiodotyrosine

Triiodothyronine (T_3)

Thyroxine (T_4)

hormonal activity. The presence of monoiodotyrosine in the enteropneusts indicates that these organisms have advanced to the stage of being capable of binding iodine, as do the nemertines, and of producing hormones of a simple and weak form; they have not yet developed the genetic apparatus or possessed genes to produce the enzymes required for further iodination and synthesis of the more potent thyroid hormones.

Cells metabolizing iodine are also distributed in organisms of the plant kingdom. The iodine picked up by freshwater algae and common land plants is negligible, but marine algae accumulate large amounts. In plants, iodine remains as iodine or iodotyrosine, but in the halophytic plants it is converted to the thyronine form. This influence of NaCl concentration is similar to its influence on the pharynx of *Lineus,* in which the rate of iodine uptake and conversion to thyronine is greater when the NaCl concentration is increased in seawater (Willmer 1970).

The original globin gene is assumed to have had a small number of nucleotides. The evolution began by increasing the size of the gene with lengthening the polynucleotide chain. Then, by repeated gene duplication followed by mutation (nucleotide base substitution), a number of genes performing specific functions in the same biological system were produced. In tracing the evolution of any biological system, this is perhaps a good model to be kept in mind. In regard to the evolution of the thyroid gland and its function, it is logical that some simple form of a gene for iodine metabolism in a crude fashion may have been present in the cells of early organisms. Evolution from this primitive gene to the more sophisticated forms present in mammals is through a similar process of duplication and mutation as assumed for the hemoglobin gene evolution.

The evolution of a feedback mechanism between the thyroid and pituitary glands was postulated by Willmer (1970), who suggested that since the epithelia of both the dorsal and ventral surfaces of the anterior regions of nemertines pick up iodine, an evolutionary change of activity in one region could lead to a modification of function in another. Anatomically, the ventral region could be envisaged as specializing its function in iodine binding and production of thyroid hormones in the form of thyrosine and triiodothyronine in the manner of vertebrates. Meanwhile, the dorsal region, retaining its speciality in iodine metabolism, was developing a control mechanism of iodinated compound production leading to the manufacture of products that stimulate the rate of iodination of cells in the ventral region. Such evolutionary development suggests the origin of the pituitary-thyroid relationship in the vertebrates. The advance from lower to higher forms of specialization and regulation in the biological systems of organisms needs to be more clearly established.

It is assumed that in certain regions of primitive organisms there are groups of cells whose evolution into individual organs resembles the differentiation of the embryonic cells of vertebrates into individual organs. Indeed, the morphological relationships and physiological control of many organs can be traced to a common embryonic tissue region. The evolution of the pituitary gland, referred to as the master gland of the endocrine system, must represent an organization of many different types of cells.

EVOLUTION OF ORGANS (STRUCTURES)

Fossil remains provide little information regarding the evolution of soft tissues. The study of the development and variations of these tissues in living forms brings another dimension to evolution. Traditionally, the study of comparative anatomy has provided most valuable information; recent advances in physiologic and biochemical techniques have added much illumination and a deeper understanding of both structure and function. Since a discussion of all the body's biological systems is impossible in this limited space, the structure of the pituitary glands will serve as an example of the evolution of organs.

The pituitary gland secretes proteinaceous hormones that are the productions of structural genes. The chemical properties and the amino acid sequences for many of them have been determined. The structural and hormonal differences among species provide, therefore, an interesting picture of the simultaneous evolution of structure and function.

The pituitary gland, or hypophysis cerebri, is an endocrine organ located in the sella turcica (Turkish saddle), a concave area in the sphenoid bone, a location within the body and well protected and difficult to reach. The gland is encapsulated by the dura mater, a membrane covering the brain, and it is attached to the bottom of the diencephalon (the posterior division of the forebrain) by a thin infundibulum (stalk). The pituitary gland consists of an anterior lobe (pars distalis), an intermediate lobe (pars intermedia), and a posterior lobe (pars nervosa). A sagittal section of the human pituitary gland is illustrated in figure 5.7.

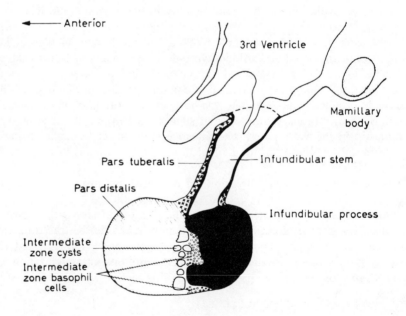

Fig. 5.7. Sagittal section through adult human pituitary. Note solid infundibular stem merging with median eminence, and "intermediate-zone" cysts and basophil cells, the latter lying in infundibular process. (Reprinted, by permission, from Sloper, 1966.)

The pituitary gland is of ectodermal origin; that is, it is derived from the outer layer of cells of the embryo. But its tissue comes from two different sources. The neurohypophysis originates from the infundibulum of the brain, the stalk of which permanently connects the posterior lobe with the hypothalamus, the ventral region of the diencephalon. The anterior and intermediate lobes are derived from Rathke's pouch, an outgrowth of the roof of the mouth. This pouch meets the infundibulum and separates from the buccal epithelium (the lining of the mouth). The intermediate lobe is derived from the wall of Rathke's pouch, and the anterior lobe develops by a thickening of the remaining part of the pouch. The pars tuberalis is an extension of the anterior lobe (fig. 5.7).

Within the subphylum of the vertebrates, pituitary glands are fairly constant in cellular structure, but there are gross morphological variations. The most apparent of these are the presence of a ventral lobe in elasmobranchs (a subclass of fishlike vertebrates including sharks, skates, and rays, having well-developed jaws), and no pars intermedia in birds and some mammals.

In the existing protochordate groups (amphioxus, tunicates), as well as in Crustacea (a class of the Arthropoda, comprising crabs, lobsters, shrimps, and so on), insects, and annelids (such as earthworms, leeches, various marine worms of the Annelida phylum), accumulating experimental evidence indicates that neurohypophysislike tissues or glands are widespread. The cells of these glands originate in the brain and their secretion is released by appropriate stimulation (Gorbman and Bern 1959). The sinus gland of crustaceans and the corpus cardiacum of insects are neurohypophysislike organs, for example. Furthermore, it has been found that there is similarity in the staining properties of the hagfish adenohypophyseal islets and the adjacent mucous epithelial cells due to the presence of glycoproteins in both. We can imagine that in the course of evolution, some mucous tissue next to the infundibulum evolved into adenohypophyseal cells. Such changes would necessitate the consolidation of a glandular structure and further genetic modifications for producing the hormonal substances.

A phylogenic relationship between a few species in the subphylum vertebrates is illustrated together with a sketch of the pituitary gland for each (fig. 5.8). The most distinguishing feature of the elasmobranch hypophysis is the ventral lobe, an elaborate organ not present in any other vertebrate. The homology of this structure is not yet ascertained. In the teleosts, the bony fish, the neurohypophysis is interdigitated with the adenohypophysis. The teleosts, like the cyclostomes and elasmobranchs, have no distinct posterior lobe. In primitive bony fish there are two diverging lines: the lobe-finned fish, represented by the lungfish, with a pituitary structure resembling that of Amphibia; and the ganoid fish, leading toward the structure in modern teleosts.

According to the ascending scale of vertebrate evolution, the Amphibia were the first to have a pars tuberalis. A separate blood supply and drainage are present in the pars nervosa, and a hypophyseal portal system exists between the median eminence and pars distalis. The pituitary gland of Reptilia is characterized by a

lack of the clear separation between the intermediate and the distal lobes discernible in that of Amphibia, and the pars intermedia is larger than that of any other
vertebrates. The avian pituitary gland is distinguished by the absence of an inter-

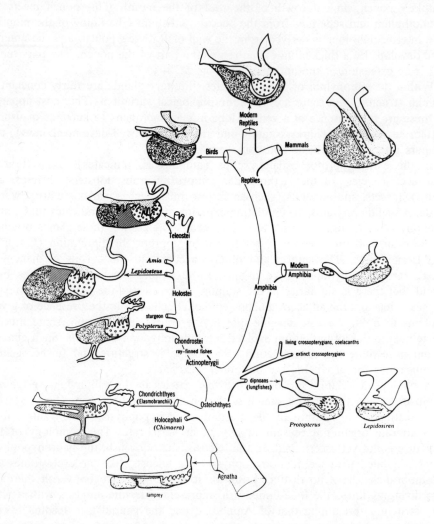

Fig. 5.8. Evolutionary relationships of the vertebrate groups (after Romer 1949),
and the structure of the pituitary gland in each group, as seen in a sagittal section; *sparse stippled*, neurohypophysis; *small circles*, pars intermedia; *small x's*,
pars tuberalis; *close stippled*, pars distalis; *oblique-lined*, differentiated proximal
zone in pars distalis; *vertical lined*, pars ventralis. (Reprinted, by permission, from
Gorbman and Bern, 1959.)

mediate lobe in the adult, although it is much like that of the reptiles in many other aspects. The pituitary glands of mammals are similar to those of reptiles and birds. In man, the intermediate lobe is well formed in early life, but gradually disappears. (For detailed information on the structure and development of the pituitary glands, see Gorbman and Bern [1959] and Sloper [1966].)

EVOLUTION OF HORMONES (FUNCTION)

The above brief discussion of variations in the gross morphology of the pituitary gland among species serves as an introduction to an examination of the evolutionary changes in its function, physiology, and behavior due to chemical changes in the hormones resulting from gene mutation. From these variations the evolutionary history of these species may be established. It will be seen also how the chemical changes of the hormones affect the physiology and behavior of the organism. There are nine known pituitary hormones, all peptides, the function of each of which is shown in table 5.3.

Neurohypophyseal Principles (Hormones of the Posterior Pituitary Lobe)

The two hormones of the pars nervosa, vasopressin and oxytocin, are octapeptides and consist of eight amino acids. The effect of vasopressin was first found in mammalian neurohypophyseal extracts by Oliver and Schafer in 1895 (Harris 1966). Other pharmacological activities were subsequently described as being antidiuretic and oxytocic, and as causing milk ejection. Oxytocin, originally chemically identified in neural lobe extracts from cattle and hogs, has since been so characterized in man, horse, sheep, and domestic fowl, its function being to cause contraction of the uterine muscles and milk ejection.

Three forms of vasopressin and two forms of oxytocin have now been identified in many species. Amino acid sequences for these hormones are as follows (Sawyer 1966, by permission of the author):

Lysine vasopressin

$$\text{Cys - Tyr - Phe - Glu(NH}_2\text{) - Asp(NH}_2\text{) - Cys - Pro - Lys - Gly(NH}_2\text{)}$$

Arginine vasopressin

$$\text{Cys - Tyr - Phe - Glu(NH}_2\text{) - Asp(NH}_2\text{) - Cys - Pro - Arg - Gly(NH}_2\text{)}$$

Arginine vasotocin

$$\text{Cys - Tyr - Ileu - Glu(NH}_2\text{) - Asp(NH}_2\text{) - Cys - Pro - Arg - Gly(NH}_2\text{)}$$

Oxytocin

$$\text{Cys - Tyr - Ileu - Glu(NH}_2\text{) - Asp(NH}_2\text{) - Cys - Pro - Leu - Gly(NH}_2\text{)}$$

Teleost oxytocinlike principle

$$\text{Cys - Tyr - Ileu   - Ser -   Asp(NH}_2\text{) - Cys - Pro - Ileu - Gly(NH}_2\text{)}$$

Both arginine vasopressin and lysine vasopressin are potent vasopressor peptides, the former having been found to be approximately six times as potent as the latter in inhibiting diuresis in dogs.

TABLE 5.3 Pituitary Gland Hormones

Hormones	Symbol	Sources
Thyrotrophin	TSH	Pars distalis
Follicle-stimulating hormone	FSH	Pars distalis
Luteinizing hormone	LH	Pars distalis
Adrenotrophin (adrenocorticotrophin)	ACTH	Pars distalis
Prolactin (mammotropin lactogenic hormone)	LTH	Pars distalis
Growth hormone (somatotropin)	GH	Pars distalis
Melanophore-stimulating hormone	MSH	Pars intermedia
Vasopressin		Pars nervosa
Oxytocin		Pars nervosa

The presence of the different forms in various species is shown in table 5.4. Arginine vasotocin appears to exist in the most primitive living vertebrates, such as the cyclostomes, and may have been the articulation of the ancestral neurohypophyseal principle. Arginine vasopressin could have arisen by the substitution of phenylalanine for isoleucine in position 3 of the amino acid sequence. The widespread occurrence of arginine vasopressin among known species suggests that lysine vasopressin could subsequently have evolved by the substitution of lysine for arginine in position 8 of the former. It is interesting to note that the frog (*Rana esculenta*) has a gene for mammalian oxytocin, apparently evolved from the fish oxytocin gene, and still has the same gene for vasotocin as the fish. Similarly, the domestic fowl has the gene for fish vasotocin as well as one for mammalian vasopressin (dimorphism), but in the domestic fowl oxytocin is of the mammalian type only. Recently it was discovered that both forms of vasopressin are present in laboratory mouse strains but, among five strains studied, lysine vasopressin was present in only one, and arginine vasopressin in the other four (Stewart 1971).

TABLE 5.4 Species Differences in Neurohypophyseal Hormones

Species	Lysine Vasopressin	Arginine Vasopressin	Arginine Vasotocin	Oxytocin	Teleost Oxytocinlike Principle
Domestic pig	+			+	
Domestic cattle		+		+	
Sheep		+		+	
Horse		+		+	
Man		+		+	
Mouse	+	+		+?	
Domestic fowl		+	+	+	
Frog (*Rana esculenta*)			+	+*	
Pollack (*Pollachius virens*)			+		+
Hake (*Urophycis tenuis*)			+		+
Bib cod (*Gadus luscus*)			+		+

SOURCE: Data from Sawyer 1966.
*Tentatively identified.

This intraspecies variation suggests that the interchange at position 8 from arginine to lysine may be quite frequent, and lysine vasopressin can apparently become adaptive in certain environments.

An evolutionary scheme for the structure of the neurohypophyseal hormones is proposed by Sawyer (1966) (fig. 5.9). The teleost oxytocinlike principle may have evolved from arginine vasotocin in two steps, the first of which may have been an exchange of isoleucine for arginine in position 8, the resultant peptide being the highly active 8-isoleucine oxytocin, with properties compatible with those found in the unknown oxytocinlike principle present in *Polypterus* (fish of the family of Polypteridae) pituitary glands. Sawyer (1966) speculated that 8-isoleucine oxytocin may have been a natural peptide present in early bony fish, the teleost oxytocinlike principle appearing later, perhaps after the *Polypterus* had separated from the ganoid teleosts.

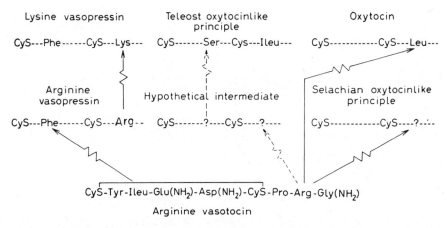

Fig. 5.9. Hypothetical scheme for the molecular evolution of active neurohypophyseal peptides. Locations of possible amino acid changes are suggested by the zigzag lines. It is assumed that arginine vasotocin was the ancestral peptide and that principles changed during vertebrate phylogeny by steps involving one amino acid substitution at a time. (Reprinted, by permission, from Sawyer, 1966.)

The above-proposed models on the evolution of neurohypophyseal hormones were based solely on amino acid variations in polypeptides. The total picture remains unclear unless one considers genetic determinants. For this, I propose a genetic model (fig. 5.10). The presence of both vasopressin and oxytocin together in the same organisms establishes, first, that there are two structural genes, one coding for vasopressin and the other for oxytocin. Dimorphisms may exist for each, such as has been shown in the mouse. Second, the presence of both the vasotocin and the oxytocinlike principle in relatively primitive species of fish would suggest the evolution, in mammals, of the vasopressin gene from the vasotocin gene, and the oxytocin gene from that of the oxytocinlike principle; or, alternatively, the vasopressin gene from that of the oxytocinlike principle and the oxytocin gene from the vasotocin gene. It should be recognized, however, that the two structural genes must arise by gene duplication in toto from a single ancestral gene at the beginning (followed by mutation of the nucleotide base substitution type) because of the similarity of the two hormones in both function and chemical structure. Present data do not indicate which is the ancestral form—perhaps neither is.

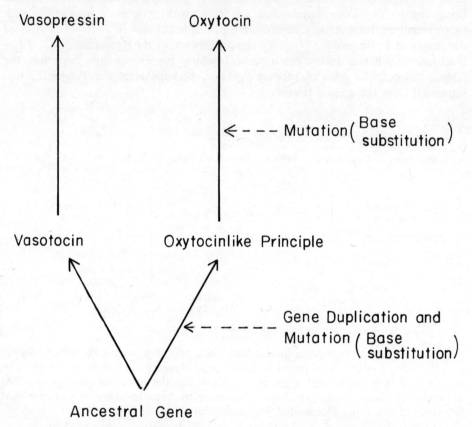

Fig. 5.10. Evolution of the genes for the neurohypophyseal hormones by duplication of the ancestral gene followed by mutation.

Intermediate Lobe Hormones

The intermediate lobe of the pituitary gland secretes hormones that stimulate the melanophore cells, causing a dispersion of pigment granules that darken skin color. In lower vertebrates such as fish, amphibians, and reptiles, these hormones exhibit a striking dermal pigmentary response to changes in the environment, such as temperature, humidity, and illumination, and are therefore of adaptive significance. The fact that they are generally produced in almost all chordates, from the tunicates to man, suggests other functions in addition to the known chromatic effects in the cold-blooded vertebrates.

The first intermediate lobe hormone studied was a peptide isolated from pig pituitary gland, containing eighteen amino acids and called β-MSH, with an amino acid sequence as shown in figure 5.11 (Harris and Roos 1959; Harris 1959; Geschwind, Li, and Barnafi, 1956, 1957). Later another form of N-acetyl tridecapeptide amide α-MSH was found (Harris 1959), consisting of thirteen amino acids. On a molar basis, the effect of β-MSH is half that of α-MSH. So far no variant has been reported for α-MSH, although variants apparently exist in β-MSH among species.

The active portion of MSH is known to be the heptapeptide from position 7 to, position 13 of the amino acid sequence, because this portion is constant for all known naturally occurring melanocyte-stimulating hormones. The amino acids on either side of this portion are species specific. The variations in these regions among species arose through mutation of the structural gene, and it is thought that the amino acids in these regions play a role in facilitating the attachment of the peptide hormone to its receptor site in the cells. Among the few species shown, man apparently has the beta form, with the largest number of amino acids.

It should be pointed out here that there are similarities between the melanocyte-stimulating hormones and ACTH in both chemical structure and function. With thirty-nine amino acids, ACTH possesses the active portion of the MSH sequence. It is known that when ACTH is applied clinically for illness involving adrenal cortical insufficiency, the patient's skin darkens. In view of these relationships between ACTH and MSH, and the common origin of the intermediate and anterior lobes of the pituitary gland, there must be an evolutionary connection between the genes that encode these hormones. One was possibly derived from the other by duplication, but we do not know yet which was the ancestral gene. It is more likely that the gene for MSH arose by partial duplication of the ACTH gene because the highly important function of the adrenal cortexes suggests that ACTH is essential and evolved earlier. We may be sure that the gene for β-MSH is derived from the α-MSH gene because of the wide presence of α-MSH in all species.

Anterior Pituitary Hormones

As the most important lobe of one of the most important organs in vertebrates, the anterior lobe of the pituitary gland secretes hormones vital for maintaining growth, reproduction and effecting behavior, and the fundamentals of fitness and adaptation. There are no other organs in the body of this small a size that have such a significant function. Studies on physiology and chemistry of these hormones have begun to reveal some of their evolutionary histories in the genes, as well as the effects of the evolutionary changes.

Adrenocortiocotrophic Hormones (ACTH). Adrenocorticotrophic hormones act on the adrenal cortex to stimulate hormonal production. The amino acid sequence

Corticotrophins
(all species)
Ser·Tyr·Ser·Met·Glu·His·Phe·Arg·Try·Gly·Lys·Pro·Val...Phe
1   2   3   4   5   6   7   8   9   10   11   12   13   39

MSH (all species)
CH CO·Ser·Tyr·Ser·Met·Glu·His·Phe·Arg·Try·Gly·Lys·Pro·Val·NH₂

α-MSH (pig)
Asp·Glu·Gly·Pro·Tyr·Lys·Met·Glu·His·Phe·Arg·Try·Gly·Ser·Pro·Pro·Lys·Asp

β-MSH (cow)
Asp·Ser·Gly·Pro·Tyr·Lys·Met·Glu·His·Phe·Arg·Try·Gly·Ser·Pro·Pro·Lys·Asp
1   2   3   4   5   6   7   8   9   10   11   12   13   14   15   16   17   18

β-MSH (horse)
Asp·Glu·Gly·Pro·Tyr·Lys·Met·Glu·His·Phe·Arg·Try·Gly·Ser·Pro·Arg·Lys·Asp

β-MSH (monkey)
Asp·Glu·Gly·Pro·Tyr·Arg·Met·Glu·His·Phe·Arg·Try·Gly·Ser·Pro·Pro·Lys·Asp

β-MSH (man)
Ala·Glu·Lys·Lys·Asp·Glu·Gly·Pro·Tyr·Arg·Met·Glu·His·Phe·Arg·Try·Gly·Ser·Pro·Pro·Lys·Asp
1   2   3   4   5   6   7   8   9   10   11   12   13   14   15   16   17   18   19   20   21   22

Fig. 5.11. Amino acid sequences of melanocyte-stimulating hormones of different species. (Reprinted, by permission, from I. Harris, 1966.)

analysis of ACTH showed that the twenty-four amino acids from the N terminal of the polypeptide chain are identical for cattle, sheep, and swine. In the remaining fifteen amino acids, the three species are identical in the six residues from the C terminal, and the variations between them occur only in the region of the remaining nine residues (Dayhoff 1969). Apparently, the biological activities of ACTH mainly depend on the twenty-four residues from the N terminal, since the chemical removal of the remaining fifteen amino acids shows little impairment of its function.

Growth Hormone. Among the pituitary hormones, the growth hormone (GH) contains the largest number of amino acids. Li (1959) found that cattle GH consists of 416 amino acids and partial enzymatic digestion of up to 25% of the amino acids does not impair its hormonal effects. A most distinct characteristic of GH is species specificity. The primates, for example, respond to their own GH but not to cattle GH, and fish GH extract is without effect on rats, but fish respond to cattle GH. Apparently GH of higher species is effective in lower species, but not in the reverse direction.

Some of the known differences between species in physical and chemical properties of GH are shown in table 5.5. One major difference is found at the end of the peptide chain: while cattle and sheep growth hormones have two N terminals, those of the whale, monkey, and man have one. There is no difference for the C terminal amino acid, but there are variations in the number of cystine and other amino acid residues. Very probably there are repeating sequences in such a long peptide chain, since hormonal activity is not lost after enzymatic digestion of up to 25%. The evolution of the gene coding GH possibly occurred by duplication followed by a series of mutations.

TABLE 5.5 Growth Hormone Differences in Chemical Structures between Species

	Ox	Sheep	Whale (Humpback)	Monkey (Macaque)	Man
Molecular weight	45,000	47,400	39,000	25,400	27,100
No. of cystine residue	4	5	3	4	2
N terminal residue(s)	phe ala]	phe ala]	phe	phe	phe
C terminal residues	-ala-phe-phe	-tyr-ala-phe	-leu-ala-phe	-ala-gly-phe	-tyr-leu-phe

SOURCE: Reprinted, by permission, from I. I. Geschwind, 1959.

Prolactin. The prolactin hormone, present in the pituitary gland of all vertebrates, has various effects in different species, such as osmoregulatory activity in teleosts, water-drive activity and larval growth activity in amphibians, crop stimulation in birds, and mammotropic and luteotrophic activities (acting on the occurrence of corpora lutea in the ovaries, known to be essential for maintaining pregnancy at the early stage) in mammals. No other hormone has so many widely different effects as have been claimed for prolactin. These variations suggest that prolactin may have undergone evolutionary changes to achieve added functions in the course of vertebrate phylogeny, providing a basis for physiological transitions from one state to another.

It has been found that, in efts, the land-living stage of this species of newt could be induced to return to water prematurely by injection of mammalian prolactin (Chadwick 1941; Grant and Pickford 1959). This hormonal effect is referred to as water-drive activity. The physiological mechanism involved for such behavioral changes must be highly complex, involving important integumentary changes. Experimental evidence discloses that water-drive activity is associated with the pituitary hormones in all vertebrates tested, with the possible exception of cyclostomes. The essential substance is believed to be prolactin, or prolactinlike principles, which suggests that an active site for water-drive in the prolactin of mammals is still retained.

Prolactin has growth-promoting effects practically throughout the vertebrate species, although the relation of body growth to fat production and carbohydrate metabolism requires further investigation. It has been found that prolactin causes fattening in birds, but this is taken as a physiological preparation for migration. The recent studies on growth effect in amphibians are of interest; prolactin possibly interacts with thyroid-stimulating hormone (TSH) and GH. It was suggested by Bern and Nicoll (1968) that in amphibians the growth-promoting effects of prolactin may have been replaced by other hormones during a later period of development. If this is the case, such sequential actions of the hormonal substances closely resemble the presence of individual forms of hemoglobin at different stages of development in mammals, as previously discussed.

Two striking effects of prolactin are stimulation of crop "milk" formation in pigeons and doves, and brood patch development in a number of species, indicating its fundamental role in ensuring the provision of nourishment for the young. The brood patch involves defeatheration in a region on the venter of some avian species, so that a highly vascularized and edematous skin region facilitates heat transfer from the parent bird to its eggs. In some species, prolactin interacts with estrogen to achieve this effect; the female bird develops the brood patch and incubates. In others, there is interaction of prolactin with androgen (male sex hormones); it was found that in male Wilson's phalaropes (*Steganopus tricolor*), the pituitary prolactin content was 3.5 times higher than that in the females, which neither develop brood patches nor incubate (Nicoll, Pfeiffer, and Revold 1967). There was some confusion on the crop effect of different fish pituitary glands. It has become clear that fish pituitary glands generally do not contain the hormone with the crop-stimulating biological activity characteristic of prolactins of the lungfish-tetrapod type. These findings disclose some landmarks in the evolution of the prolactin molecule. The hair-pulling in female rabbits prior to parturition apparently represents a similar type of behavior, although experimental proof for the effect of prolactin in this connection is not yet available.

Another major function of prolactin is its effect on mammary gland development, which is synergistic with sex hormones and adrenal cortical hormones in stimulating the development of the duct system of the mammary gland. Pituitary extractions from lower vertebrates produce no such response in mammals as shown from prolactin treatment. Thus Bern and Nicoll (1968) concluded that there is a difference between land-living vertebrates and their piscine predecessors with respect to the mammary-stimulating substances and their homologues in nonmammalian vertebrates. It would appear that in the vertebrate species the prolactin hormone has gone through adaptive evolutionary changes. Its mammotropic activity closely parallels pigeon crop secretion activity. The prolactin of the African lungfish (Protopterus) shows effective pigeon crop secretion activity, but this is not true of the cyclostomes, chondrichthyeans (a class comprising cartilaginous fish with well-developed jaws), and teleosts.

With all of the above evidence, an evolutionary scheme has been derived (fig. 5.12). Apparently some of the changes in prolactin that affect pigeon crop secretion and mammotropic activities took place before the target organ structures were evolved. On this ground, Geschwind (1967) claimed that hormones have not so much changed in the course of evolution, but rather that the sensitivity of the target organs has increased. Actually, both the hormone itself and the target organs are evolving; such joint changes are, in fact, characteristic of evolutionary development. Many interesting chemical and physiological relationships remain to be disclosed once the amino acid sequence of the molecule has been determined. Nevertheless, from what information we now have it is clear that prolactin is highly important in vertebrate evolution, essentially enhancing reproductive fitness. The preparation of birds for migration to breeding zones, analogous to the amphibian water-drive; the osmotic preadaption of euryhaline (the ability to live in water in a wide range of salinity) fish; the induction of brood behavior in birds; and the obvious effects of milk secretion in birds' crops and mammalian gland development are all elements contributing to Darwinian fitness.

Thyrotrophic Hormone (TSH), Follicle-stimulating Hormone (FSH), Luteinizing Hormone (LH). Thyrotrophic hormone is a substance secreted by the pituitary gland that stimulates thyroidal hormonal production. Both FSH and LH are referred to as gonadotrophic hormones, FSH causing the ripening of the ovarian follicles, and LH inducing ovulation and maintenance of the corpus luteum, which is essential at least in the earlier part of pregnancy. Because of the chemical difficulties involved in purifying small amounts of the material, characteristics and molecular structures have been established for LH and TSH in only a limited number of species (table 5.6).

Each of the above hormones is a glycoprotein containing two subunits, alpha and beta. The amino acid sequence for the alpha and beta subunits of TSH and LH have been worked out by Pierce (1971), although there are still some slight ambiguities in certain parts of these chains. Both TSH-α and LH-α consist of ninety-six amino acid residues with carbohydrate moieties at positions 56 and 82.

Except in some very minor points the two subunits are completely identical; TSH-β consists of 113 amino acids and LH-β 120. When properly aligned, their residues are about 50% identical. The differences in sequence between the alpha and either one of the beta subunits is much greater than between the two beta subunits.

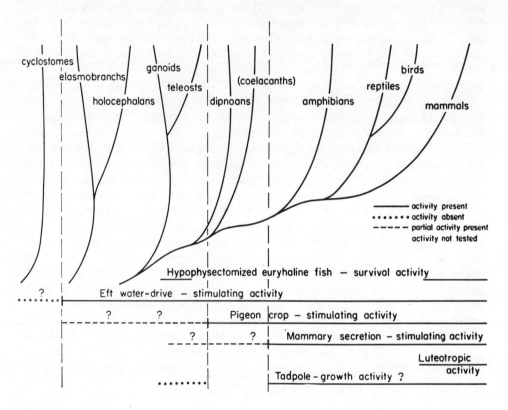

Fig. 5.12. Distribution of several of the activities of prolactin among the vertebrates. A scheme for the evolution of the hormone is proposed. (Reprinted, by permission, from H. A. Bern and C. S. Nicoll, *Recent Progr. Hormone Res.* 24 (1968):681–720.)

In testing for functional differences between the subunits of TSH and LH, the specific activity, as far as gonadotrophic effects are concerned, is determined more by the beta than by the alpha subunits. TSH-α + LH-β is about equally potent as the LH standard. Apparently TSH-α + LH-β is not as potent as LH-α + LH-β, suggesting that minor differences between TSH-α and LH-α in the amino acid sequences are responsible for the differences in activity. The alpha and beta subunits of TSH and LH are so different that no genetic relationship can exist between them. In contrast to the conservative nature of the alpha subunits, it is possible that the beta subunits were initially rather similar, and may, therefore, have been

TABLE 5.6 Gonadotrophic Activity of TSH-α plus LH-β

Potency Relative to NIH-LH-S1 Standard	
TSH-α	<0.02
LH-α	0.04
TSH-β	0.06
LH-β	0.08
LH-α + LH-β	2.6
TSH-α + LH-β	0.9
LH	2.5
TSH	0.02

SOURCE: Reprinted, by permission, from J. G. Pierce, *Endocrinology* 89(1971):1331–44.

derived from the same gene by duplication; perhaps it was ancient and went through a large number of subsequent mutations.

Parallel to the functional similarities between TSH and LH are the similarities of the cells that secrete them. Both TSH and LH cells in the pituitary gland are basophylic and can be differentiated only by cell morphology under the light microscope (fig. 5.13). The evolution of these cell types were perhaps parallel with the evolution of the genes coding the hormones. I suggest that there are two ancestor genes, one coding the alpha component and the other the beta component. These ancestor genes had more thyrotrophic activity than luteinizing activity. Later each one had a duplication, and resulted in the LH-α and LH-β, respectively. By further mutation, the ancestor genes became more specialized for TSH activity and the new genes for LH activity. It is very likely that the beta gene duplication occurred earlier than the alpha gene duplication. During the time the LH-β gene was established and before the LH-α gene was formed, both TSH-β and LH-β shared the same alpha component to form complete TSH and LH molecules. The evidence for this supposition is the much larger number of differences in the amino acid residues between the two beta components than between the two alpha components. Much more information on species differences in these components is needed to demonstrate the details of their evolutionary history. We have seen the hormonal changes through gene mutation and their effects on physiology and behavior, the basic elements of adaptation. It may not be too much of an exaggeration to say that in higher species the evolution of hormones plays a leading role in evolution. It is perhaps the basis for many associated morphological changes that take place subsequently.

PIGMENTATION

The presence of various colors in plants and animals is important to the fitness of the organism in respect to camouflage, reflection and absorption of light, and so on. We are generally aware of the role of pigmentation for successful adaption to a changing environment. Coat color in animals is important for concealment from predators as well as for epigamic and aposematic display. Animals are equipped with a number of different types of color genes normally concealed because of recessiveness. The switch from one color to another can be achieved when environments or circumstances demand.

In mammals, similar coat colors often appear in different species. They are transmitted in the same Mendelian manner. Such facts can best be explained by the supposition that the gene for a particular coat color feature of different species is derived from a common ancestral gene, and that the particular mechanism for the operation of the gene is still preserved. That is, the gene is still performing a similar function and producing similar effects as when it was carried by the ancestors of the present species. There is a homology for coat color genes among species (Searle 1968).

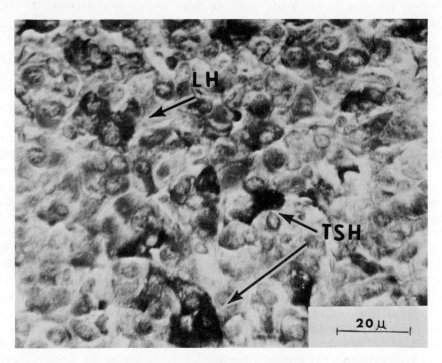

Fig. 5.13. Photomicrograph of a mouse pituitary section. Note the angular shaped and heavily stained cytoplasmic granules of TSH cells compared with the relatively light stained and round LH cells. Both cell types show purplish blue color by Halmi stain (×980).

Pigment formation, and the distribution, size, and shape of pigment granules have been more thoroughly studied in mice than in any other mammals, resulting in the discovery that more than fifty genes influence pigmentation. Pigment in the mouse is produced by melanocytes, a special type of cell derived from melanoblasts. Melanoblasts originate in the neural crest, from which they migrate to their final sites during the eighth to twelfth day of gestation (fig. 5.14). Melanocytes are present in the hair follicles of the dermis. They produce pigment granules classified as eumelanin and phaeomelanin. The former produces a black or brown, and the latter a yellow coat color. The number, size, shape, and clumping, plus the arrangement of the granules within the medullary cells and along the hair, are important for the color appearance (Russell 1948). Naturally occurring melanin is a polymer of indole-5,6-quinone that combines with protein to form melanin granules. The indole-5,6-quinone is derived from the multistep oxidation of tyrosine by the enzyme tyrosinase. It is not known, however, whether tyrosinase is required for each step (Wolfe and Coleman 1967).

The following list gives the alleles at the agouti locus in the mouse and their phenotypes:

Alleles	Phenotypes
A^y	Yellow, lethal when homozygous
A^{vy}	Viable yellow
A^{iy}	Intermediate yellow
A^w	Light-bellied agouti
A^i	Intermediate agouti
A^+	Gray-bellied agouti
A^s	Agouti-suppressor
a^{td}	Tanoid
a^t	Black and tan
a^x	Lethal nonagouti
a^m	Mottled agouti
a	Nonagouti
a^e	Extreme nonagouti

This list is arranged in descending order of dominance. (For a detailed description of the hair and pigment granules, see Wolfe and Coleman [1967] and Searle [1968].)

Agouti alleles present in different mammalian populations are shown in table 5.7. It is possible that more allelic forms than those listed actually exist, but they have not been observed. From the geographical distribution of polymorphisms in species (table 5.8), Searle found that among the total number of species half of them show a common pattern of ecological preferences. That is, the melanic form is most common in humid forest areas. He concluded that the pattern agrees with Gloger's rule (see Searle 1968) that dark color due to melanin pigmentation is more strongly developed in warm and humid environments.

The agouti gene is certainly of evolutionary antiquity, as agouti hairs are found in monotremes, marsupials, and almost all eutherian orders. Its survival may be due to other effects in addition to its influence on pigmentation. Pleiotropic effects, the effect of a single gene on several biological systems, is a rather common property for genes of higher organisms. The most well-known case is the A^y gene; heterozygotes are obese and have yellow coats, homozygous embryos die at the

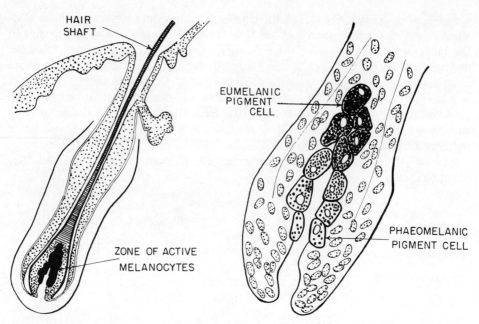

Fig. 5.14. Left, distribution of melanocytes in the bulb of the hair follicle and the hair shaft. *Right,* agouti hairbulb, with both eumelanin- and phaeomelanin-forming cells. (Reprinted, by permission, from Searle, 1968.)

TABLE 5.7 Agouti Alleles Appearing in Different Mammalian Species

Species	A^{vy}	A^w	A^+	a^w	a
House mouse	+	+	+	+	+
Norway rat		+			+
Black rat		+	+		+
Guinea pig		+	+		+
Rabbit		+		+	+
Cat			+		+
Dog	+	+		+	+
Fox (*Vulpes fulva*)	+				+
Pig			+		+
Horse*		+	+	+	+

SOURCE: Reprinted, by permission, from Searle, 1968.

NOTE: a^w = light-bellied nonagouti. See table 5.1 for the other gene symbols. The effect of a homologous gene may be slightly different on different species.

*Dr. William E. Castle had made a thorough study on coat color inheritance in the horse, and the coat color variation in this species is great. I believe that the dun is due to the presence of A^w, the bay to A^+, and the light black and solid black to a^w and a, respectively. According to my observation of Mongolian horse populations (semiwild), solid black horses are extremely rare.

blastocyst stage. Another allele, a^x, also causes lethality in homozygotes. The agouti locus is known to influence tumor susceptibility. In my laboratory, it was found that A^wa mice had a lower level of circulating leukocytes than aa mice with identical genetic backgrounds.

The above discussion of pigmentation, with special reference to the agouti locus, represents a particular condition limited to mammals. Variegation in pigment patterns exists generally among higher organisms, crossing classes, phyla, and even kingdoms. Although the wide variations in color type, size, and arrangement or combination of patterns in different species can hardly be generalized for gene homology, it would not be surprising if the basic mechanism for control of gene expression or gene regulation was the same or similar. This remains to be seen. But the widespread occurrence among species is an indication of the importance of color variegation in adaptation and selection. A few examples are given in which some genetic basis has been established.

McClintock (1967) analyzed the production of variegated patterns in the seeds of maize. These patterns are due to the presence of two different alleles; each produces its own pigmentation patterns independently of the other, and the two patterns overlap (fig. 5.15, a). In insects, a wide range of distribution patterns of black pigment on a yellow background is found in the elytra (hardened forewings of insects) of the ladybird beetle (*Harmonia axyridis*) (fig. 5.15, b). At first, breeding tests produced surprising and puzzling results (Tan 1946). Later it was found that all patterns could be traced to the action of a single chromosomal locus, with each allele responsible for one distinctive pattern of pigmentation.

In birds such as barred Plymouth Rock chicks, the dark and white band patterns in the feathers are due to the presence or absence of melanin pigment granules in the dark and light bands, respectively (fig. 5.15, c). This pattern is influenced by a sex linked gene, B. In birds, the male has two sex chromosomes, ZZ, and the female has one sex chromosome, ZO. When a female barred Plymouth Rock (B) is mated to a male Rhode Island Red (b/b), the F_1 hybrid females have red feathers (b) and the F_1 males have barred feathers (B/b). Ursprung (1966) proposed the following explanation for the genetic mechanism of the band formation. Melanocytes are always present in the skin of the birds, whether banded, all black, or even all white feathers are produced. Complete absence of pigment would very likely be due to a blockage of melanin synthesis or of differentiation of melanocytes. Alternatively, the white bands may be due to the exhaustion of substrate for melanin production, such as tyrosine, or dopa (dihydroxyphenylalanine) after a few cycles, so that no melanin can be formed until the substrate returns to a sufficiently high level. Such periodic exhaustion and accumulation of substrate results in the rhythmical production and release of melanin for barred patterns.

Six sex-linked loci for coat color variegation are known in mice, two in hamsters, and one in cats (fig. 5.15, d). For detailed descriptions of these genes, see Wolfe and Coleman (1967) and Searle (1968). Coat color variegation is very common in carnivores but has not been well studied. It can be assumed that some

TABLE 5.8 Percentage Distributions of Melanic Forms of Mammals in Relation to Habitat

Name	Species	Where Found	Maximum Frequency of Melanic Form	Peculiarity of Habitat	Other Information
Brush possum	Trichosurus vulpecula	Tasmania	100%	Dense forest with high rainfall	Black form not common anywhere in Australia
Black rat	Rattus rattus	Ports in northern Europe	High	Commensal with man	Black form regarded as subspecies *rattus*. Color inherited as dominant (E^d)
European hamster	Cricetus cricetus	Parts of Russia and Germany	~ 25%	Humid forest—steppe	Black coat inherited as dominant
Muskrat	Ondatra zibethica	Parts of United States, for example, Virginia, North Carolina	100%	?	Two melanic forms: black and tan (probably a^t) and uniform black (a)
Water vole	Arvicola amphibius	Scotland, East Anglia	100%	?	Scottish subspecies *A.a.reta* is dark or completely black, while *A.a.amphibius* has black populations in the fens
Red squirrel	Sciurus vulgaris	Eurasia, but not Britain	100%	Deciduous forest	Black or dark subspecies to south of range and in Japan, dimorphic subspecies north of these; probable intermediate inheritance
Gray squirrel	Sciurus carolinensis	United States, for example, Pennsylvania	40%	High forest, for example, beech, birch, maple	Usually dark agouti, but some jet-black. Latter with recessive inheritance (a?), former probably intermediate
Western pocket gopher	Thomomys townsendii	Idaho	10%	?	

Golden-mantled marmot	*Marmota flaviventris*	Wyoming	25%	?	Black form most common in canyons, and so on, of Teton range, Wyoming
Grasshopper mouse	*Onychomys leucogaster*	Utah	?	Mountainous areas	Dark phase recessive to pale
Multimammate mouse	*Praomys morio*	Kigezi, Uganda	~100%	Moist montane forest	Not nonagouti, but a reduction of the agouti band
White-throated pack rat	*Neotoma albigula*	Texas, New Mexico	~ 70%	Tertiary lava beds	Nonagouti
Rabbit	*Oryctolagus cuniculus*	Tasmania	35%	Cleared rain forest over 1,500 ft.	Probably caused by homozygosity for nonagouti *a*
Hare	*Lepus timidus*	Western Russia (Ukraine)	?	Headwaters of river Desna	Recessive inheritance
Red fox	*Vulpes vulpes*	Alaska, Canada, Russia	100%	Northern part of range	Silver fox is a type of melanic. Different semidominant genes involved in Alaska and other regions
Raccoon	*Procyon lotor*	United States	?	Woods	
Leopard	*Panthera pardus*	Southeast Asia, African mountains	Over 50%	Rain forest	Melanic form known as black panther; dominant gene seems responsible

SOURCE: Reprinted, by permission, from from Searle, 1968.

Fig. 5.15. (*a*), Variegated corn kernels. (Reprinted, by permission, from McClintock, 1967.) (*b*), Band pattern of a ladybird beetle; the actual colors are black and yellow in the peripheries of the bands. The genetic mechanisms affecting the color variegation in both the corn kernels and the beetle elytra are similar. (Reprinted, by permission, from Farb, 1962.) (*c*), Feather color patterns of the barred Plymouth Rock chick due to the presence of a sex-linked gene *B*. (Reprinted, by permission, from Lamoreux, 1941.) (*d*), Variegated coat color of a cat due to the presence of the sex-linked gene *To* (tortoiseshell) and *Ta* (tabby); the dark and gray patches represent black and yellow colors. (Reprinted, by permission, from Searle, 1968.)

variegations are due to sex-linked genes, as in the cat. From the effect of the *B* gene in birds, we can speculate that some genes controlling banding of feathers may be sex linked in different avian species. It is interesting to see that genes influencing aposematic display or sexual selection are distributed on the sex chromosomes in both birds and mammals. Such characteristics persisting in species provide highly attractive speculative material regarding gene homology, if not sufficiently documented for proof.

CONTINUOUS AND DISCONTINUOUS VARIATIONS

As one descends the taxonomic hierarchy, variations are always observed within each category: the various organisms are arranged into a hierarchy of discontinuous clusters. It has generally been maintained that the observed discontinuity in present living species is illusory, to the extent that it is a consequence of the extinction of intermediates that actually survived in the past. From this viewpoint, it could be assumed that almost imperceptible transitions would be observed from representatives of every species that has ever existed. The neo-Darwinian viewpoint is in favor of the assumption that an approximately continuous series would have existed, even though not at a single time. This means that evolution is taken in small steps, in what is called microevolution.

This concept is built on a statistical analysis of the permutation of a large number of gene combinations. If an entire array of genotypes were formed, and the resulting genotypes, each differing from another by one gene, were represented, they would show a distribution line close to one smooth curve. The fact that we have not seen such an orderly array of organisms is because many forms were lost through the process of natural selection, and those that are actually observed are only the survivors that happened to possess particular combinations of characteristics distributable to "adaptive peaks" (Wright 1932b). In such a system, closely related adaptive peaks represent a species, small ranges represent genera, and larger ranges represent families and higher groups. Discontinuity results from the impossibility of forming an entire series of genotypes.

An alternative view is that larger steps of evolution may have been achieved through some systemic mutation of chromosomal level or single gene mutation of strong effect (Goldschmidt 1940). Evolution can occur at macroscopic levels.

These two different schools of thought have been in sharp contrast. Apparently, microevolution has generally been favored. But conclusions have been formed on a more or less hypothetical basis, leaving the actual problem unsolved. In the past twenty or thirty years, there has been much genetic evidence brought to bear on this issue, so that it can now be discussed more fruitfully and on more concrete genetic terms.

Continuous Variations

In genetics, one defines discontinuous variation as discrete distribution, so that the total number of individuals in a population can be separated into two or more categories without overlapping, and one individual can be assigned one, and only one, of these categories. All polymorphic traits represent discontinuous variation. Variations not of the discontinuous type are therefore continuous, such as, in man, skull length and breadth, and stature. Continuous variation is hypothetical, based on the assumption that if the sample size were infinitely large, the distribution with respect to a specific trait would approach a smooth curve.

Now we come back to the question of continuous variation and microevolution. The best way of approaching the problem is to study genetic variations within species, the causes of these variations, and the relative magnitude of effect of genes. There are numerous selection and crossbreeding experiments in plants and animals, revealing the genetics of size, yields, and performance. These types of characters are all distributed continuously within a population of a species. The results practically all indicate that there are a large number of genes involved for each, and each gene produces small effect. The work of Thoday and his colleagues has demonstrated and located some of these genes on the chromosomes that affect the number of bristles in *Drosophila*. In my laboratory, two strains of mice, different in body size, have been established. The large strain (LG) is about twice the size of the small strain (SM) (fig. 5.16). The two strains have no overlapping for their body size distributions. It has been estimated that at least eleven individual loci are involved (Chai 1956). Each contributes an effect no more than 1 gm of body weight (Chai 1961). Large size differences are present in every animal species; sharp contrast is particularly noticeable between different breeds of dogs and horses. These variations are all due to a large number of genes, and each shows small effect.

These genes—referred to as multiple factors or polygenes (Mather 1943), or isoalleles—are present in domestic as well as in wild populations of any species. At first polygenes were thought to represent a different class of genes with different physical and chemical properties. Now it is generally agreed that polygenes are the same as single major genes that contribute large effects. At this point, it can be said that there are genetic units that can cause microevolution. But this conclusion cannot exclude certain other possibilities.

Discontinuous Variations

Discontinuous variations include variations that may not be of macroevolutionary scale. But macroevolutionary variations as shown between taxonomic groups are always at or near the minimum level of discontinuity. Discontinuous variations can be caused by either chromosomal mutation or gene mutation.

Chromosomal Mutation. The issue of macroevolution, strongly advocated by Goldschmidt (1940), is based on his observation of chromosomal mutation that caused large phenotypic changes. As major evidence, he discussed the genetic basis fully in *The Material Basis of Evolution*. His main thesis is what he called "systemic mutation" resulting from chromosome rearrangements. He claimed that when a segment of a chromosome changed its order or location, such as in inversions and translocations, all the genes in the rearranged segment acted differently than when they were in the original sequence and location. He called the genetic change "systemic mutation," distinguishing it from gene mutation. The concept of systemic mutation is based on the position effect hypothesis. His evidence came from studies on differences between species in chromosome rearrangements of insects that were found along with abrupt physiological and gross morphological differences.

Simpson (1944) did not agree with Goldschmidt; he assumed that the large differences between species resulted from an accumulation of small changes with a possibility of the presence of some unknown intermediate types of organisms, which had become extinct before they reached an abundant number. Yet he apparently had some hesitation, since he said:

This and innumerable other examples show beyond reasonable doubt that the horizontal discontinuity between species, genera, and at least the next higher categories can arise by a process that is continuous vertically and that new types on these taxonomic levels often arise gradually at rates and in ways that are comparable to some sorts of subspecific differentiation and have greater results only because they have had longer duration. Two serious questions remain: whether this really is the usual or universal pattern on these levels, and whether it also occurs or is normal for still higher taxonomic categories. The facts are that many species and genera, indeed the majority, do appear suddenly in the record, differing sharply and in many ways from any earlier group, and that this appearance of discontinuity becomes more common the higher the level, until it is virtually universal as regards orders and all higher steps in the taxonomic hierarchy.

The basic issue of macroevolution may hinge on the question of the effects of chromosome rearrangement of the genes involved. Goldschmidt drew the analogy that, in the case of inversion, its effects are like the reverse of a sentence or part of a sentence in a text. This would not only alter the meaning but in most cases, destroy it. In the organism, it would cause most serious consequences and could possibly have lethal effects. Many inversions found in the laboratory mouse, however, have had no recognizable physiological and gross morphological effects (Roderick and Hawes 1970).

Now, knowing the chemical structure of the gene and the codons being made of triplet nucleotide bases, we can see the problem more clearly. What actually can happen in chromosome rearrangements depends on the points of chromosome breakage. That is, the breakage probably can occur either between or within genes, although we do not have direct evidence. Whether breakage between genes is more frequent than breakage within genes is not known. The chromosome structure in higher organisms is not yet completely understood. If the breakage occurs within genes at both ends of the involved segment, as in the case of inversion, there would be a reconstitution of two genes. This would be equivalent to two gene mutations occurring simultaneously. Also it can occur within a DNA triplet, which can lead to the formation of nonsense sets of triplets.

Gene Mutation. The most convincing evidence of gene mutation causing macro-evolution as conceived by Goldschmidt is the phenomenon of *homoeosis*. That is, the mutant tetraptera in *Drosophila* produced wings like those of another order, Diptera, while the mutants "proboscipedia," bithorax, and tetrapters all produce phenotypic characters of other orders of insects. One of the most genetically well-defined cases is the bithorax mutation in *Drosophila*. In the bithorax mutants, transformation involves portions of the mesothoracic, metathoracic, and first abdominal segments, arrangements shown in figure 5.17.

The variations are caused by a series of adjacent alleles at the locus *bx* on the third chromosome, with a recombination frequency of 1/10,000 gametes; they are

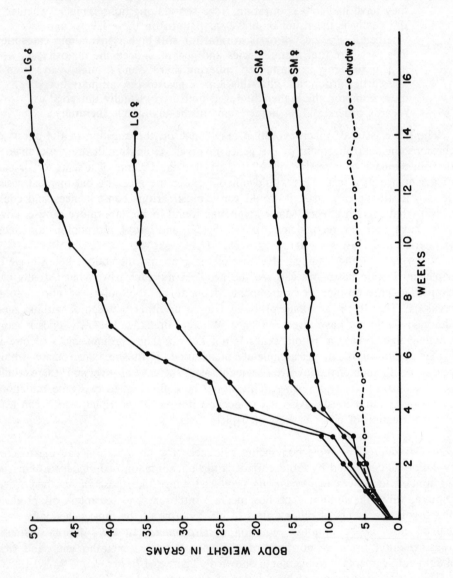

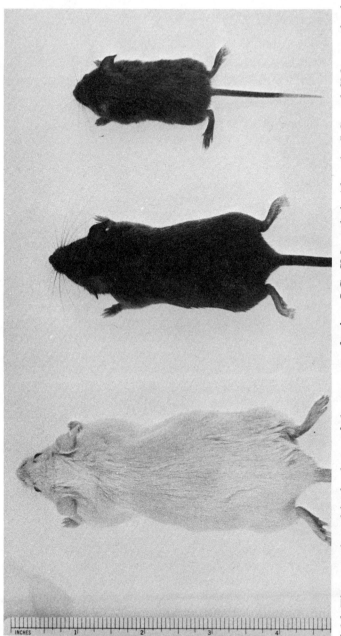

Fig. 5.16. The growth and body shape of three types of mice: LG, SM, and *dw/dw*; the LG and SM are selectively bred for large and small body size and then inbred. Both are genetically fixed. It was estimated that a minimum of eleven loci attributed to their difference. The *dw/dw* is a mutant, due to a single autosomal recessive gene, causing pituitary defects, with a definite lack of growth hormone production and thyrotropic hormone production in the pituitary. *Top,* growth curves. Notice that the LG mice had not only a greater rate but also a longer period of body growth than the SM mice. The *dw/dw* mice had normal growth up to about one week, and thereafter no growth throughout their whole life. *Bottom,* size and shape differences between these mice. The SM mouse, although small, has proportional body development. The *dw/dw* mouse is disproportional, with a marked longitudinal body reduction, thereby appearing stubby and fat.

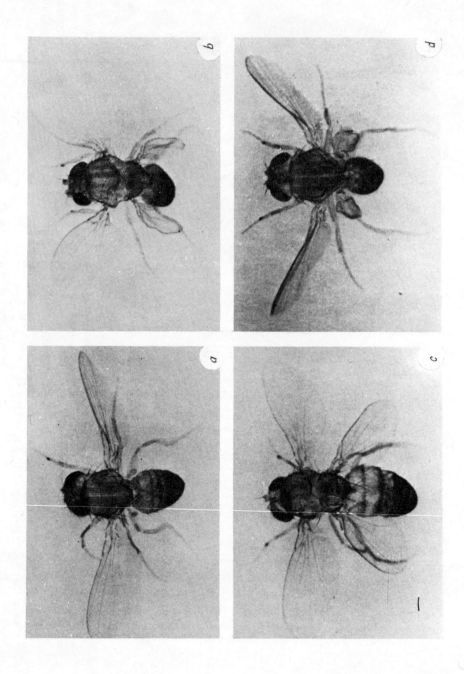

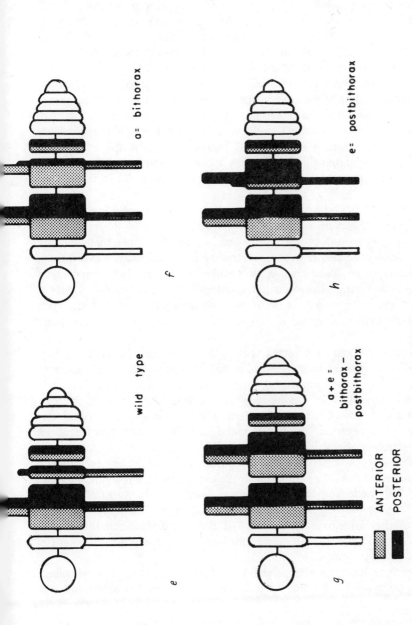

Fig. 5.17. (*a*), Wild-type male *Drosophila* (mesothoracic wings have been extended to show the dorsal metathoracic and first abdominal regions); (*b*), male with an extreme T1 transformation (genotype: bx^3/Ubx^{105}); (*c*), male showing a combination of extreme T1 and T2 transformations (genotype: $bx^3 \, pbx/Ubx^{105}$); (*d*), male with an extreme T2 transformation (genotype: pbx/Ubx^{105}). (*e-h*), Body segmentation plan of the *a* and *e* mutants compared with that found in the wild type and in the double mutant combination. Compare with figure 5.17*a*. (Reprinted, by permission, from Lewis, 1964.)

referred to as pseudoalleles.[4] We shall discuss only the *a* and *e* alleles here. The *a* allele causes the anterior portion of the metathorax to develop a structure closely resembling the anterior portion of the mesothorax, whereas the *e* allele causes the posterior portion of the metathorax to resemble the posterior portion of the meso-thorax. When *a* and *e* occur in the same fly, a complete mesothoracic modification of metathoracic segment is formed, resulting in a four-winged fly.

Lewis (1964) has discussed the effects of these genes in development and evolu-tion. It is assumed that some in this series caused the insects to evolve from multi-legged ancestors and, therefore, may have been the first to arise in evolution. Ap-parently a similar gene present in the Lepidoptera causes extra leglike appendages to arise in the larva and extra first abdominal legs and wings in the adult moth (Itikawa 1955). Lewis claimed that such genes arose by random duplication fol-lowing mutation, and hypothesized the physiological mechanisms as in the lactose operon of *E. coli*.

It is not yet fully understood how the complex loci, which usually involve posi-tion effects and pseudoalleleism, operate in chemical terms. Until the proteins or enzymes which the genes produce are recognized, we must assume that complex loci act as the histidine gene cluster in the bacterial system (chap. 2). Nevertheless, we know that, merely by an allelic change, two-winged insects produce four-winged offspring. This genetic change has occurred spontaneously in a laboratory stock; thus, there is no reason for it not to occur in natural populations. Indeed, such events may have occurred throughout the entire evolutionary journey of the many species.

Homoeotic variations occur in different species of vertebrates. In mice, the fact that the third molar is present in some strains and absent in others (Grüneberg 1963) was referred to as quasi-continuous variation due to single-gene differences. Variations are also described in the number of thoracic and lumbar vertebrae among inbred mouse strains (Green 1962). Inheritance does not quite follow the Mendelian pattern of segregation, and it was therefore concluded that as threshold character the variation is affected by a small number of genes.

Exactly the same type of variations have been observed during the development of inbred domestic rabbit lines. In rabbits, with the exception of a few minor types, the 13-6 and 12-7 thoracic and lumbar vertebrae combinations were most com-mon (Chai 1970). The thoracic and lumbar border shift can be perceived from the distribution of the last ribs (fig. 5.18). The differences between the inbred lines clearly indicated the influence of genes, but inheritance could not fully satisfy the Mendelian segregation ratio.

Various duplications have been observed and studied in different vertebrates, such as radius duplication in chickens (Landauer 1956), neural tube duplication

---

4. Pseudoalleles are two or more genes that are related in function, mapping at sites so close that they show a low frequency of crossing-over. They usually exhibit position effects.

in mice (Dunn and Caspari 1945), and body duplication in rabbits (Chai and Crary 1971). Of course most of these mutants cannot survive, but such great biological variation can occur genetically.

There are other types of well-known single-gene mutations causing discontinuous variations. Some of these may be of evolutionary importance, and some appear to be absolutely deleterious and have no evolutionary consequences.

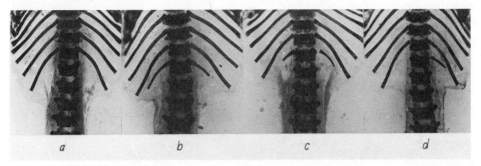

Fig. 5.18. Homeotic variations in the last pair of ribs in rabbits, ventral view: (*a*), normal; (*b* and *c*), asymmetrically reduced; and (*d*), left rib reduced and right absent. (Reprinted, by permission, from Chai, 1970.)

In maize, a mutant gene, corn grass (*Cg*), produces profound effects on the architecture of the plant. The development of young normal and corn grass plants diverges several days after seed germination, both in respect to the plastochron interval and the size of the apical meristem (fig. 5.19). The plastochron interval is a measure of the rate of leaf production and becomes shorter as the plant grows. It was found that the effect of the *Cg* gene on the rate of synthesis or replication of DNA in certain regions was greater in corn grass than in a normal plant. The gene exerts its effects by bringing about a higher frequency of mitosis in the flanks of the apical meristem and a lower frequency in its more distal regions, resulting in a plant different in shape and size from normal maize (Stebbins 1967).

Many genes influencing both size and shape have been discovered in mice. A classical case is Snell's dwarf (*dw*), affecting the development of the pituitary gland. The dwarf mice (fig. 5.16) have juvenile body proportions, myxedema, some accumulation of fat, and attain a size of about one-fourth to one-third that of normal littermates. It was discovered that the dwarf mice lacked the specific cell types responsible for growth hormone secretion and possibly also the cells for thyrotrophin. A similar type of dwarf has been reported that, although due to a different gene, (*df*), was comparable in effect to *dw*. These two genes are nonallelic and located on different linkage groups, but their histological and gross morphological effects are produced at almost the same time and stage of development. Other genes that produce retardation of body growth are miniature (Bennett 1961), stubby, achondroplasia, and so on (Lane and Dickie 1968), and although the causes are not quite as well known as those in the dwarf mutants, each is due to a specific gene and produces a characteristic body form. For instance, achon-droplastic mice are short in body, and the limbs, tail, and even the head give the general appearance of being small and disproportionate. Comparable single-gene effects are known for pituitary dwarfism and achondroplastic dwarfisism in man. But all these genes cause harmful effects in any environment. They are different from polygenes, producing effects of discontinuous types, and are constantly influenced against by natural selection.

Fig. 5.19. Plants of corn grass (*left*) and normal maize (*right*), at twenty-five days, when differentiation of embryonic tassels is beginning. These plants are sibs, segregating from the backcross (W23/oh51A/Cg/+). (Reprinted, by permission, from G. L. Stebbins, *Develop. Biol.* 1(1967 suppl.):113–35.)

However, there are genes that cause major biological changes, and the carriers of the gene appear to survive, sometimes even more vigorously than those that carry the wild-type gene allele, one example being the *t* locus in mice. It is a complex locus and more than thirteen alleles are present in different wild populations (Dunn 1956). Embryos heterozygous for the dominant gene (*T*/+) lose the distal part of the tail but survive; homozygotes (*T*/*T*) lose the entire notochord and die about midway through gestation. All recessive alleles produce morphologically normal mice when they are heterozygotes with the wild allele (*t*/+) and with the *T* allele (*t*/*T*,), tailless and viable. The presence of these alleles in relatively high frequencies in the natural population is believed to be due to the possibility of greater fitness of the heterozygotes. It is difficult to say whether this is an indication of the possibility of evolution toward tailless or short-tailed mice. If it is, this would be an example of macroevolution.

In summary, gene and chromosome mutations can cause large biological changes on a scale corresponding to the level of variations between species or genera or even higher taxonomic order. We recognize the fact that most of these mutations are probably deleterious for one reason or another and would thus be lost; however, there is the possibility that a small number of them may prove advantageous and therefore favored by natural selection. In the latter case, they would become able to establish themselves in the populations and contribute to the macroscopic level of evolutionary changes. Therefore, it may be fair to say that microevolution is a smooth transition and comparatively easy to achieve, while macroevolution, although rare, may also occur.

COMPARATIVE SEROLOGY AND IMMUNE RESPONSE

One of the most impressive physiological bits of evidence in Darwin's theory that similar species have been formed by descent with modification from a common ancestor is that of comparative serology and immune response. When a small amount of the blood serum of an animal is injected into an animal of a different species, the donor serum act as an antigen and cause the production of antibodies in the receiving animal that react with the serum of the donor animal. Such antigen-antibody reactions are highly specific. Yet the specificity is not complete, for serum immunized against one species may react with species closely related, although in ever-decreasing degrees as the relationships become more distant. This phenomenon has been known about for at least forty years, as demonstrated by immunization of guinea pigs with different genera of salamander. Much of such evidence was also shown in different species of primates and in insects and plants by the immunization of experimental animals with the proteins of these different animal and plant species (see Dodson 1952).

The technology in immunology has been much refined in recent years. A worthwhile experiment is the immunization of rabbits against the purified serum albumin of different primates, including that of *Homo sapiens*. The fundamental principle is the same as the immunization of guinea pigs against salamander serum. It was claimed, however, that albumin is a protein whose chemical structure is coded by specific genes. The antigenic differences and similarities are highly proportional to the differences in individual amino acid residues (Sarich and Wilson 1967). Therefore, such immunologic differences between the albumin of the different primates corresponds to that of the nucleotide base sequence of the gene. It can be seen that the results (table 5.9) are very close in their phyletic relationships.

STRUCTURAL REDUCTION

Against the background of the genetics of development and the action of various genes listed above, we can properly introduce the subject of structural reduction. As Romer (1949) stated, it generally refers to simplification of structure by degenerative changes of tissues, a phenomenon of evolutionary importance. The problem of reduction has been discussed recently against the theory of selection by Brace (1963) and Holloway (1967).

TABLE 5.9 The Immunological Distances between Different Primates

Between Primates	Estimated Divergence (million years)	Total Number of Generations Since Divergence (millions)	Albumin Immunological Distance‡
Chimpanzee-man	14	2.5	7
OWM*-hominoid	37	11.0	30
NWM†-hominoid	50	19.3	54
Prosimian-hominoid	65	45.3	123
OWM-NWM	50	21.4	59
Lemur-loris	65	54.4	90
Lemur-tarsier	65	54.4	126
Loris-tarsier	65	54.4	105

SOURCE: Reprinted from Lovejoy and Meindl 1972, by permission of the American Association of Physical Anthropologists.
*Old World monkeys.
†New World monkeys.
‡Obtained by multiplying mutation rate by total number of generations since divergence.

The concept of structural reduction is somewhat vague. It is considered here as meaning a reduction in either size or number of organs or parts of the body. An animal, such as the mouse, with a shortened tail or no tail at all, is a case of reduction. This can be affected by a single gene, such as the T alleles. The poll condition (hornlessness) in cattle is another case and is also due to a single gene. Actually, by close examination, it can be found that the organ is not completely absent in some cases. In the poll condition, some of the cattle still show elevated cornified development; cornification is always present at the corresponding location.

The absence of the heavy eyebrow ridges and the massive jaws of the Neanderthal man in modern man is generally regarded as a size reduction. In this case, it was probably not caused by a single gene. Similarly, a comparison of different breeds of dogs or horses reveals striking differences in body size. The small breeds may be regarded as having an overall reduction in structure. Most of these differences are caused by multiple factors or polygenes. Wright (1932a) assumed that there are general size factors that affect the overall size of the body and local size factors that affect certain parts.

The evolution of the horse from a three-toed to a single-hoofed animal may be regarded as a reduction in number. The reduction apparently has accompanied a more full development of the single hoof. However, the genetic mechanism must be similar to that of the poll condition in cattle, since a cornified spot (the vestigial toe) on the inside of each front leg is still present in the modern horse. These indicate that the structure remains, although not in a fully developed state. They suggest that genes that code for the different cellular substances may still be present. It is probably the regulating system in the genome that has suppressed the continuing expression of these structural genes.

Saunders and Fallon (1966) studied the limb development in fowl. They observed that in the chick embryo, the leg paddles appear after approximately sixty hours of incubation, then grow in length but retain their paddle-shaped outlines until the end of the fourth day. At this time, massive cell necrosis sweeps through the length of the limb in the soft tissues between the digits until their outlines are formed. The death and resorption of cells in these regions appears to contribute to the morphogenetic movements that model the contours of the individual digits. During the development of limbs in the duck, the homologous zones show no significant necrosis, but slight degeneration is observed between digits, except between the first and second, resulting in a web structure (fig. 5.20). Similar observations have been reported for mouse, mole, and human embryos (Milaire 1963; Menkes, Deleanu, and Ilies 1965). In my laboratory, necrosis occurred in the middle of limb buds of mice instead of being interdigital at later stages of development, resulting in the disappearance of the middle digits. This condition (fig. 5.21) is influenced by a single gene (Chai and Stevens 1972). In summary, structural reduction can be affected by a single gene or a number of genes likely to be of the regulatory type. On the assumption that ontogeny is a recapitulation of phylogeny, the selective growth and necrosis in the tissues of organisms during development may be analogous to the process of structural reduction that occurred in the phyletic lines.

FUNCTIONAL REDUCTION

Functional reduction here refers to partial or complete loss of the ability to produce a biochemical substance necessary for the maintenance of life at one time in phyletic evolution. An example in this case is the variation in the capacity of biosynthesis of ascorbic acid (vitamin C) among species. The reversible oxidation-reduction of ascorbic acid is related to many metabolic systems (White et al. 1959). Deficiency causes basically functional failure in cells of mesenchymal origin (an embryonic tissue in the mesoderm forming the connective tissues and the blood and lymphatic vessels in the body), producing a syndrome known as scurvy. The disease is characterized by sore and spongy gums, loosening of the teeth, and impaired capillary integrity and it results in subcutaneous hemorrhages, edema, joint pain, loss of appetite, and anemia.

Insects, other invertebrates, and fish are unable to synthesize ascorbic acid, but amphibians, reptiles, birds, and most of the mammals are able to do so. In amphibians and reptiles, the biosynthesis of ascorbic acid occurs in the kidney; in birds and mammals, it occurs in the liver. Apparently about 300 to 340 million years ago this function evolved in the kidney of amphibians and shifted later to the liver in the higher species.

The biosynthesis of ascorbic acid from glucose involves a few major steps (fig. 5.22), and each step requires at least one specific enzyme. Variation in the enzyme production can affect the final output. This explains the large variations in the rate of ascorbic synthesis among species as determined in vitro (table 5.10).

It has been generally known that humans, monkeys, guinea pigs, and flying mammals cannot synthesize ascorbic acid. Such functional failure, known to be

a

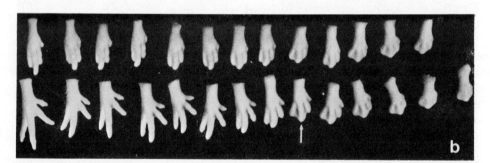

b

Fig. 5.20. (*a*), The pattern of necrosis in leg primordia of chick (*left*) and duck (*right*). Dots show the locations and the extent of necrosis in the limb buds, resulting in the differences in toe separation between the chick and duck. (*b*), Normal development of the right foot of the chick embryo serves as control (*top*); embryos killed at 6½ days (*far left*) and at six-hour intervals thereafter. Right feet from embryos that received 7.5μg of Janus green in the amniotic fluid at 6½ days of incubation and were killed simultaneously with their controls (*bottom*); all digits are joined by soft tissues. (Reprinted, by permission, from Saunders and Fallon, 1966.)

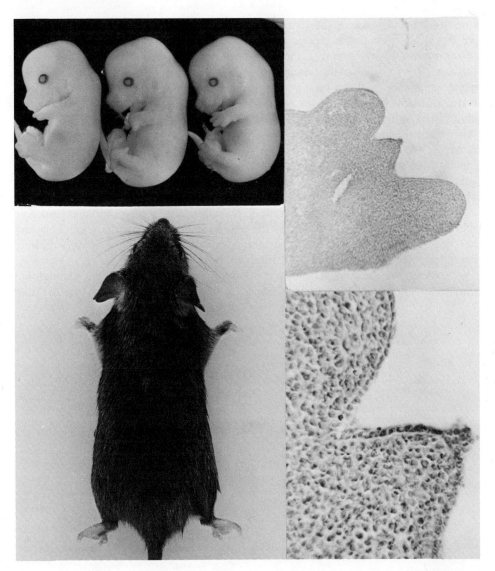

Fig. 5.21. Digital aplasia in mice due to a dominant autosomal gene. *Top left*, 15½-day-old embryos; the one on the right is normal and the other two have digital aplasia; except for the hind limb of the one on the left, both the front and hind feet are deformed. *Bottom left*, an adult digital aplasia mouse. Notice the lobster claw appearance of the hind feet, and the single digit in the front feet. *Right*, photomicrographs taken from a tangential section of a limb of a 12½-day-old digital aplasia embryo, showing the cell necrosis and retarded development of the middle digit. Top right × 65; bottom right × 260. Data from Chai and Stevens 1972.

caused by the absence of the terminal enzyme, L-gulonolactone oxidase (Chatterjee 1973), suggests that a genetic change affecting the enzyme production occurred
in the course of evolution of these species. Rajalakshmi, Deodhar, and Rama-
krishan (1965) studied the vitamin C balance in lactating women and found a
negative balance between intake and output in the milk and urine. If this proves
to be a general phenomenon in women, one has to assume that man still has the
ability to synthesize vitamin C but not under ordinary circumstances. Moreover,
this functional impairment is not caused by a change at the structural level, but
rather at the regulational level. Gene activation is closely related to endocrine
function. In the present case, it is possible that elevation in the level of certain
gonadotrophic hormones during lactation may activate genes not expressed under
ordinary circumstances. We need no further consideration on the genetic nature
at this point other than the fact itself. It reflects that a secure source of ascorbic
acid had been present in the diet of these species at a time in the past. This fact
has been taken as evidence for the assumption that man's early ancestors were
living in tropical environments (Livingstone 1971).

A comparable case on gene activation is the switch of beta chain hemoglobin
synthesis from the A type to the C type in sheep and goats, as previously dis-

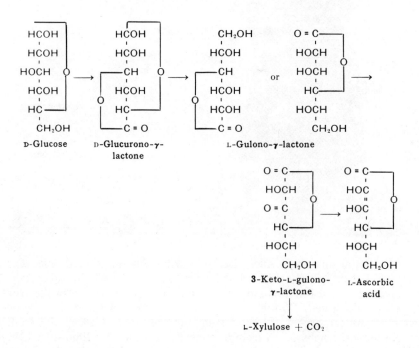

Fig. 5.22. The synthesis of L-ascorbic acid from 3-glucose. (Reprinted, by permission, from White et al., 1959.)

cussed. The particular point of emphasis here is that the switching mechanism, demonstrated both in vivo and in vitro (Adamson and Stamatoyannoupoulos 1974) by altering the level or concentration of erythropoietin, is hormone dependent. As indirect evidence, this supports the above interpretation that ascorbic acid synthesis, occurring in lactating women, may also be accomplished through alteration of the level of hormones concerning lactation. Both these cases seem to imply that functional reduction does not necessarily mean that the organisms have lost the genetic ability forever, but rather that the genes are still present,

TABLE 5.10 In Vitro Biosynthesis of Ascorbic Acid

Animals	Ascorbic Acid Synthesized (μg/mg protein/hr)	
	Kidney	Liver
Amphibians		
Toad (*Bufo melanostictus*)	144 ± 10	0
Frog (*Rana tigrina*)	115 ± 10	0
Reptiles		
Turtle (*Lissemys punctata*)	98 ± 8	0
Bloodsucker (*Caloter versicolor*)	50 ± 5	0
House lizard (*Hemidactylus flaviviridis*)*	46 ± 6	0
Common Indian monitor (*Varanus monitor*)	32 ± 4	0
Angani (*Mabuya carinata*)	25 ± 4	0
Snake (*Natrix piscator*)	18 ± 2	0
Tortoise (*Testudo elegans*)	14 ± 2	0
Mammals		
Goat	0	68 ± 6
Cow	0	50 ± 6
Sheep	0	43 ± 4
Rat	0	38 ± 4
Mouse	0	35 ± 4
Squirrel	0	30 ± 4
Gerbil	0	26 ± 4
Rabbit	0	23 ± 2
Cat	0	5 ± 1
Dog	0	5 ± 1
Guinea pig	0	0

SOURCE: Reprinted, by permission, from I. B. Chatterjee, *Science* 182(1973):1271–72. © 1973 by the American Association for the Advancement of Science.

NOTE: Biosynthesis by incubation of L-gulono-1, 4-lactone with the microsomal fractions from kidney or liver of animal of different species. The values are the means ± standard deviations obtained from a minimum of eight animals for each species.

*Kidneys from twelve house lizards were pooled for one determination, and four such determinations were made.

though not expressed under ordinary circumstances. In times of stress or under unusual conditions, these genes can be reactivated by hormonal regulation.

As far as the mechanisms of hormonal regulation on gene transcription and translation are concerned, we know very little. We should, nevertheless, appreciate the fact that, as gene regulators, hormones play an important role in evolution. When a large amount of genetic information is developed, such as in higher organisms, evolution of selective usage of this information (gene regulation) for adaptation as well as progressive modifications becomes essential.

SUMMARY

Organisms evolving from primitive to advanced forms call for not only more and newer genetic information, but also for its selective use. Structural genes encode genetic information; its realization depends on inducers and repressors, either endogenous (the products of regulatory genes) or exogenous (physical or chemical in nature). Differentiation and morphogenesis are genetically programmed; single-gene substitutions or certain environmental changes can alter or interrupt developmental processes, causing structural and functional changes. Microalterations and macroalterations can be accomplished by gene substitutions.

The success of a change depends on the genetic and external environment of the organism as well as population size and structure. These major areas will be discussed in the next two chapters.

the fate of a gene

Suppose a sexual organism is the carrier of a new and favorable gene. For one reason or another this individual may not leave any offspring. It may not have found a mate; it, or its mate may have been sterile; or the fertilized eggs may not have developed fully. If any one of these chance events occurred, the gene, although useful, would become completely lost, and no evolutionary progress would have been made.

If an individual produces fertile offspring, how much a useful gene would do for the survival of the organism would depend partly on its individual action and partly on its cooperation with its fellow genes. For instance, at a given locus, a gene, say A_1, may do better with A_2 than with the same allele A_1. For genes at a different locus, A_1 may do better with B_1 than with B_2. These are called gene-by-gene interactions. The former is allelic and the latter nonallelic. The selection value of any gene is altered by other genes.

When a genotype has a better chance of survival and reproduction than others, one may call this the survival of the fittest. But this individual will pass its genes on in haploid portions that will combine with haploid portions from other individuals. We can have no prior knowledge as to the fitness of the new zygotes thus formed.

All these uncertainties exist because, first of all, a gene cannot survive and spread autonomously in the population; its propagation comes about through the organism that is its carrier, itself conditioned by environment, population size, structure, and its own breeding practices. Second, even within the same environment, the selection value of the gene itself and how well it cooperates with its fellow genes are both factors that come into play. Evolution is based essentially on changes in gene frequencies, and gene frequency changes are events affected by many factors. Thus the likelihood of loss or survival of a gene becomes a population problem, and the relevant theories fall within the scope of population genetics. It is through the independent work of Wright, Fisher, and Haldane that Darwin's theory of natural selection has been laid on a Mendelian foundation—from Darwinism to neo-Darwinism.

POPULATION

A population of sexual organisms is a group of individuals having common characteristics and occupying a given space at a given time, an aggregate of individuals of different ages and sexes with particular rates concerning birth, death, and migration, and associated with each other by interbreeding for reproduction. In each generation, parental genes, temporarily dispersed and reassembled in their offspring, create a shuffling and recombining of alleles according to the law of Mendel. Wright referred to such a breeding unit as a Mendelian population; it was defined by Dobzhansky (1950b) as "a reproductive community of individuals who share in a common gene pool." The population is referred to as a panmictic unit if the gametes combine to form zygotes entirely by chance.

The combination of Mendelian inheritance with sexual reproduction greatly promotes evolution. Sexual reproduction, a biological process, is essential for shuffling the genes. Under the Mendelian system an enormous number of genotypes can be produced by the recombination of a relatively small number of genes. The potential number of genotypic combinations is limited by the size of the population itself. The property of dominance provides the potential of preserving hidden genetic variations, which can be released by the demands of a changing environment. Overdominance ($Aa > AA$) confers hybrid vigor and adaptive flexibility. With the interaction between loci or epistasis, organisms are equipped with another type of genetic variation. Moreover, recombination (associating in individuals in new ways two or more genes for which the parents differ) makes it possible for favorable mutant genes that arise in different members of the population to appear in a single individual. The genetic flexibility maintained by sexual reproduction is a conservative device for assisting species survival and evolutionary progress.

RANDOM MATING

In a population of bisexual organisms, the mating of an individual of one sex with any member of the opposite sex with an equal probability of fertility is called random mating, and the population so maintained is said to be panmictic. The consequences of deviations from random mating are essential elements contributing to evolution. As a starting point we need to understand the theory and a few basic concepts in connection with the random breeding system.

In a population of N diploid individuals with two alleles, A_1 and A_2, at a particular locus, N_{11} = the frequency of A_1A_1 individuals, N_{12} = the frequency of A_1A_2 individuals, and N_{22} = the frequency of A_2A_2 individuals. The proportion of the A_1 allele, or the frequency of A_1 in the population, is

$$p = \frac{2N_{11} + N_{12}}{2N} ;$$

similarly that of A_2 is   **6.1**

$$q = \frac{2N_{22} + N_{12}}{2N}$$

since each of the N_{11} individuals has two A_1 alleles, N_{12} has one A_1 and one A_2 allele, and N_{22} has two A_2 alleles.

In a panmictic population with p for the gene frequency of A_1 and q for A_2, where $p + q = 1$, the proportions of the three genotypes with respect to this locus in the population are

$$(p + q)^2 = p^2 + 2pq + q^2, \qquad\qquad \textbf{6.2}$$

where, p^2, $2pq$, and q^2 are genotypic frequencies for A_1A_1, A_1A_2, and A_2A_2, respectively.

Under random mating, the frequency of each type of mating depends on the frequencies of the different genotypes. For the frequencies of the three genotypes, let $d = p^2$, $h = 2pq$, and $r = q^2$, where $d + h + r = 1$. The proportion of each mating type equals the product of the genotypic frequencies. Table 6.1 shows the frequencies of genotypes for each mating type and the sum for all. The overall genotypic frequencies are the same for parental and offspring generations, demonstrating that under random mating the proportions of different genotypes remain in equilibrium, that is, constant from generation to generation. This state is referred to as the Hardy-Weinberg equilibrium. It is also demonstrable that equilibrium can be reached in one generation of random mating, regardless of the initial frequencies of the different genotypes (Li 1955).

TABLE 6.1 Frequencies of Mating Types in the Parental Generation and of Genotypes in the Offspring Generation under Random Mating

	Frequency of Mating	Offspring		
Type of Mating		A_1A_1	A_1A_2	A_2A_2
$A_1A_1 \times A_1A_1$	d^2	d^2	...	...
$A_1A_1 \times A_1A_2$	$2dh$	dh	dh	...
$A_1A_2 \times A_1A_2$	h^2	$\frac{1}{4}h^2$	$\frac{1}{2}h^2$	$\frac{1}{4}h^2$
$A_1A_1 \times A_2A_2$	$2dr$	...	$2dr$	...
$A_1A_2 \times A_2A_2$	$2hr$	...	hr	hr
$A_2A_2 \times A_2A_2$	r^2	...	...	r^2
Total	$(d+h+r)^2$	$(d+\frac{1}{2}h)^2$	$2(d+\frac{1}{2}h)(\frac{1}{2}h+r)$	$(\frac{1}{2}h+r)^2$
		p^2	$2pq$	q^2

NOTE: Homozygous dominant, d; heterozygote, h; homozygous recessive, r.

When the heterozygote is identifiable, equation 6.1 gives the exact gene frequency for the sample and the best estimate for the population, regardless of whether or not it is at equilibrium. Equation 6.2 requires the assumption of equilibrium for computing the gene frequency, and it is used when the heterozygote cannot be identified. Thus when the heterozygote is identifiable, the validity of this assumption can be tested by comparing the two estimates, using the χ^2 test with one degree of freedom. Whether or not mating in the population is at random with respect to the gene under investigation can thus be determined. The M-N blood types in man may be useful for checking, since the heterozygotes can be distinguished from both of the homozygotes. The observed and expected frequencies, based on the Hardy-Weinberg equilibrium, have been tabulated for five populations (table 6.2). The observed and expected frequencies showed no significant differences in the samples from the U.S. white, U.S. Indian, and Australian aborigine populations; but the differences are significant for S-Leut and L-Leut populations (two divisions of an endogenous religious sect in the United States) due to an excess of heterozygotes (MN). The original authors offered no logical explanations for the departure from Hardy-Weinberg equilibrium for the last two populations, but the general situation seems to suggest that gene frequency may be unable to maintain equilibrium in small populations due to selection which favors survival of heterozygotes. For the other three populations, the results suggest no significant departure from random with respect to the M-N blood type locus.

TABLE 6.2 The Distributions of MN Blood Types for Different Populations

Population	Total		M	MN	N	χ^2	Gene Frequency for M
White (United States)	6,129	obs.	1,789	3,039	1,303		0.54*
Navajo Indians (United States)	361	obs.	305	52	4		0.92*
Australian aborigines	730	obs.	22	216	492		0.18*
S-Leut (North America)	3,171	obs.	1,287	1,544	340	15.166‡	0.65†
		cal.	(1,336)	(1,445)	(391)		
L-Leut (North America)	2,563	obs.	1,083	1,220	260	9.671‡	0.66†
		cal.	(1,116)	(1,150)	(296)		

NOTE: Calculated = cal.; observed = obs.
*Mettler and Gregg 1969; calculated values are not listed.
†Steinberg et al. 1967
‡$P < .01$.

INBREEDING

Inbreeding means mating between individuals more closely related to each other than the average individuals in the population. The degree of relationship between individuals in a population depends on the size of the population. In a population of bisexual organisms each individual has four grandparents, eight great-grandparents, and so on, and 2^n ancestors n generations back. In a finite population, there is a value of n at which 2^n exceeds the population size. Therefore, any pair of individuals in a population must be related to each other through one or more ancestors in the past; the smaller the population size in the past, the less remote the common ancestors and the greater the number of common ancestors. Thus mating at random within small populations leads automatically to a certain amount of inbreeding. In natural populations, especially human, inbreeding may be compounded with assortative matings (matings between individuals phenotypically alike with respect to a certain character or characters) (Lewontin, Kirk, and Crow 1968). Inbreeding without selection will not change the gene frequencies but the genotypic frequencies, leading to gene fixation in individual demes.

The level of inbreeding may be determined on an individual as well as on a populational basis. For measuring the amount of inbreeding for an individual from its pedigree, Wright's (1921) path coefficient method is generally used:

$$F = \Sigma(1/2)^n (1 + F_A), \qquad\qquad 6.3$$

where the summation is over all possible paths, n is the number of individuals excluding x in each loop (fig. 6.1), and F_A is the inbreeding coefficient of the common ancestor for each path and is calculated similarly. The value F is generally referred to as Wright's inbreeding coefficient.

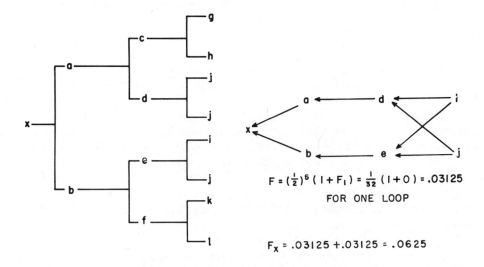

$$F = (\tfrac{1}{2})^5 (1 + F_1) = \tfrac{1}{32}(1+0) = .03125$$

FOR ONE LOOP

$$F_x = .03125 + .03125 = .0625$$

Fig. 6.1. Path diagrams for first cousins' marriage, and the calculation of Wright's inbreeding coefficient, F. This pedigree represents the first-cousin mating.

The inbreeding coefficient of an individual is the probability that the two alleles of one locus of an individual are identical by descent. It can also be worked out on probability theory. The method is due in essence to Malecot (1948). In the previous example, the probability that d and e receive the same allele from i is 1/2. The probability of receiving different alleles for the same locus is also 1/2, if there is no previous inbreeding of i. However, if there is previous inbreeding of i, the probability of two alleles passed randomly to d and e being identical is $1/2 + \tfrac{1}{2}F_i = \tfrac{1}{2}(1 + F_i)$. The probability of d passing a certain gene to a is 1/2, and the same from a to x; the joint probability of passing the same gene from i to x is, therefore, $(\tfrac{1}{2})^2$. Similarly, the probability of a certain gene passing from i through e and b to x is $(\tfrac{1}{2})^2$. The joint probability of x receiving two identical alleles from i is, therefore, the probability that d and e receive the identical allele from i, times the probability that this allele is passed on from d to x, times the probability that this allele is passed on from e to x, which equals

$$\tfrac{1}{2}(1 + F_i)\,(\tfrac{1}{2})^2\,(\tfrac{1}{2})^2 = (\tfrac{1}{2})^{2+2}\,\tfrac{1}{2}\,(1 + F_i).$$

The x individual can also receive two identical alleles from j on a similar probability basis. Therefore, the probability that x receives two identical alleles from either i or j is $\Sigma[(\tfrac{1}{2})^{2+2}\,\tfrac{1}{2}(1 + F_i)]$. This can be expressed as

$$F_x = \Sigma[(\tfrac{1}{2})^{n_1+n_2+1}\,(1 + F_i)], \qquad\qquad 6.4$$

where n_1 and n_2 are the number of ancestors excluding the common ancestor for each path. This is the form commonly used in most texts. The above equation can also be reduced to

$$F_x = \Sigma[(\tfrac{1}{2})^n\,(1 + F_i)],$$

where $n = n_1 + n_2 + 1$, which is the same as equation 6.3. The methods of computing an inbreeding coefficient based on pedigrees is referred to as a genealogical model (Yasuda and Morton 1967).

In consideration of inbreeding in populations, as Crow and Kimura (1970) pointed out, the effect of finite population size can be visualized as $2N$ allelic genes for the N parents, the two genes drawn at one time, with replacement, keeping N constant. The results will be that some genes are left out, while others are drawn more than once.

In a finite population, two gametes of a zygote have a chance of $1/2N$ of carrying two identical genes, and $1 - 1/2N$ of carrying different parental genes. When additional inbreeding occurs, caused by mating between more closely related than randomly chosen individuals, the probability of $1 - 1/2N$ genes being identical is F_{t-1}, the inbreeding coefficient for an average individual in the previous generation, where t is the generation number. Therefore, the general formula for computing the inbreeding coefficient is

$$F_t = \frac{1}{2N} + \left(1 - \frac{1}{2N}\right) F_{t-1}. \qquad\qquad 6.5$$

The above equation shows the effect of restriction of population size on inbreeding even though mating is random within the population. The consequence of this is the increase of homozygosity and of homogeneity of individuals in the population. The decrease of heterozygosity is by a fraction of $1/2N$ each generation, indicating the effect of population size on gene fixation.

Having discussed the two different models of inbreeding separately, we may now discuss their joint occurrence, as often observed in hierarchical populations, that is, a large population separated into a number of isolates, frequently seen in natural populations. Let S represent the isolate or subpopulation; T, the large or total population; and I, an individual of the isolate. F_{IS} is defined as the inbreeding coefficient due to the probability of two alleles of a locus being derived from the same gene of a common ancestor within the isolate, and F_{ST} that due to the probability of two alleles of a locus chosen at random from the isolate, both being descended from a single gene in the isolate. F_{IT} will be the inbreeding coefficient due to the overall probability of identity of the alleles of a locus in an individual I. Thus F_{IS} is the inbreeding coefficient corresponding to F of the pedigree model (eq. 6.3). The average of all the F_{ST} for individuals of the isolates corresponds to F of equation 6.5. The probability of nonidentity is the product of $(1 - F_{IS})$ and $(1 - F_{ST})$. The term $(1 - F_{IS})$ is the probability that two allelic genes are not from an identical gene in a common ancestor, and the term $(1 - F_{ST})$ is the probability that two allelic genes are not identical due to being randomly chosen from within the isolate. Therefore, according to Crow and Kimura (1970), $1 - F_{IT} = (1 - F_{ST}) (1 - F_{IS})$, and the relationship between the three different coefficients of inbreeding can be written as

$$F_{IT} = F_{ST} + (1 - F_{ST})F_{IS}. \qquad \textbf{6.6}$$

Steinberg et al. (1967) analyzed the rate of inbreeding in the S-Leut and L-Leut colonies, using only people with complete ancestry records. The members in the present colonies are descendants of sixty-three remote ancestors whose parentage traces back to those who first settled in North America. They computed the inbreeding coefficient for each mating F_{IT}, and the average inbreeding coefficient for each sect. They also estimated by computer the coefficient of inbreeding (F_{ST}) due to overall random pairing. The results are given in table 6.3. The surprising discovery was made that the coefficient of inbreeding F_{IT} was smaller than F_{ST} due to the fact that the marriage of close relatives was prohibited. This gives negative values for F_{IS}. In other words, in ordinary random mating some marriages between relatives would be included by chance, but in these sects marriages between close relatives were strictly avoided. We notice that the coefficient of inbreeding of 0.0311 for the L-leut is about equal to that (0.0313) resulting from marriages of first cousins once removed. Therefore, for natural populations of this size without restriction of close relative matings, this is the lowest level of inbreeding that can be expected.

TABLE 6.3 The Average Observed Inbreeding Coefficient and the Inbreeding Coefficient by Random Pairing in the S-Leut and L-Leut Colonies

Colonies	Number of Matings	Observed (F_{IT})	Random Pairing (F_{ST})	F_{IS}
S-leut	664	0.0211	0.0248	−0.0038
L-leut	618	0.0255	0.0311	−0.0058

SOURCE: Data from Steinberg et al. 1967.

In treating the hierarchical populations, gene frequencies (p) are different among the isolates. Wright (1951) used merely the gene frequencies for the isolates to estimate the coefficient of inbreeding:

$$F = \frac{\sigma^2}{\bar{p}(1 - \bar{p})}, \qquad\qquad 6.7$$

where σ^2 is the variance of the gene frequency among the isolates and $\bar{p}$ is the average gene frequency for the isolates. It can be seen that the larger the difference in gene frequency between the isolates, the greater the inbreeding coefficient. As an example, following are the gene frequencies for phenylthiocarbamide taste deficiency in eight Taiwan aboriginal populations (Chai 1967):

Atayal	Saisiat	Bunun	Tsou	Paiwan	Rukai	Puyuma	Ami	σ^2	$\bar{p}$	F
0.37	0.37	0.27	0.14	0.14	0.15	0.20	0.22	0.0092	0.232	0.052

The F so computed is the average for the individuals in the subpopulations, and disregards the variations in population sizes and the genealogical effects that can be different among them. Although this method requires the assumption of no natural selection on the gene under investigation, it has the virtue of simplicity in application.

In summary, inbreeding can come about either through the mating of related individuals at a higher frequency than would occur by chance, or the mating of related individuals as necessitated by finite population size. In the former case, if random mating is resumed, the inbreeding coefficient will drop to zero in one generation (Crow and Kimura 1970). But in the latter case, the inbreeding coefficient can be reduced only by outcrossing between isolated populations. (Matings of closely related individuals are used to advantage in laboratory and domestic populations, and are of medical interest in human populations.) It is the restriction of population size due to either social or geographical barriers that causes continual increase of inbreeding as indicated by equation 6.5.

Effect of Inbreeding

The principal effect of inbreeding on individual members of a population is in the increase of the frequency of homozygotes. Thus, inbreeding tends to expose recessive alleles that would otherwise go unnoticed when carried in the heterozygous state. In our human population, to cite one example, the parents of both the painter Toulouse-Lautrec, and of the writer, John Ruskin, were cousins. Both men were dwarfs. It can be assumed that there are genes causing the development of dwarfism as well as influencing mental ability. Being relatives, the parents tend to have a greater chance of carrying the same recessive genes than do two average members of the population. Thus their offspring have a greater chance of being homozygotes for these genes. Without knowing further family histories, such an interpretation must be considered speculative.

Natural populations generally carry large numbers of unfavorable recessive genes. In humans the frequency of development of some diseases from marriage between cousins is shown in table 6.4. However, in a large study of first-cousin marriage in Japanese populations, much less conclusive evidence of the effects of inbreeding was found (Schull and Neel 1965). In other species, the development of various abnormalities in laboratory animals was reported by Wright (1960) in guinea pigs and by Chai (1969) in rabbits. These laboratory populations are under continuous and intense inbreeding (usually sib mating), which is not encountered in natural populations. The effect of inbreeding on physiological fitness in humans is shown in table 6.5, and on reproductive fitness in domestic rabbits in figure 6.2. Other examples are numerous but will not be cited here.

What are the long-range effects of inbreeding in evolution? These depend on population size and intensity of breeding. With intensive inbreeding, a small population with a relatively large number of detrimental recessive genes can undergo losses so severe that the population size will diminish generation after generation

TABLE 6.4 Mortalities among Offspring of Marriages between Unrelated Persons and between First Cousins

		Unrelated		First Cousins	
		Total	Mortality	Total	Mortality
Deaths	Period	No.	(%)	No.	(%)
United States					
Young children	Before 1858	837	16.0	2,778	22.9
Before age 20	1700-1900	3,184	11.6	672	16.8
Children 0-10 years	1920-1956	167	2.4	209	8.1
France*					
Stillbirth, neonatal	1919-1950	2,745	3.9	743	9.3
Infantile, juvenile,					
and later	1919-1950	515	9.6	674	14.3
Japan*					
Stillbirth	1948-1954	63,145	1.5	2,798	1.6
Infants, to 1 month	1948-1954	63,145	1.8	2,798	2.8
Infants, during 9	1948-1954	17,331	4.7	822	6.6
months					
Children 1-8 years	1948-1954	544	1.5	326	4.6

SOURCE: Reprinted, by permission, from Stern, 1960.
*Deaths of children without visible major malformations.

TABLE 6.5 Frequencies of Disease and of Physical and Mental Defects among Offspring of Marriages between Unrelated Persons and between First Cousins

	Unrelated		First Cousins	
Population	No.	Affected (%)	No.	Affected (%)
France*	833	3.5	144	12.8
Japan†	63,796	1.02	2,846	1.69
Sweden‡	165	4	218	16
United States§	163	9.82	192	16.15

SOURCE: Reprinted, by permission, from Stern, 1960.
*Morbihan and Loir-et-Cher. Children in completed families from marriages 1919–25.
†Hiroshima, Kure, and Nagasaki. Children born 1948–54.
‡Three parishes, North Sweden. All cousin marriages registered 1947. The percentages of affected are estimates after various corrections applied to the data.
§Chicago. Children born 1920–56.

until, while the intensity of inbreeding increases because the population becomes smaller, the population eventually becomes extinct.

On the other hand, mild inbreeding in a fairly large population may be beneficial, serving to eliminate deleterious recessives and thus increase the fitness of the population. In the meantime, it will cause an increase in frequency of the favorable genes, some of which may eventually approach complete fixation. This, in fact, is a genetic process of adaptation.

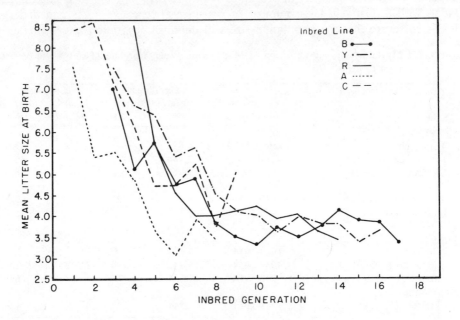

Fig. 6.2. The effect of inbreeding on litter size in five lines of rabbits. Data from Chai 1969.

MUTATION

As discussed in chapter 2, any changes along the DNA nucleotide base sequence, such as deletion, addition, or replacement of any one or more of the nucleotide base pairs, are mutations. A gene consists of, say, one thousand nucleotide base pairs; theoretically, numerous mutations can be observed at one locus. We do observe different numbers of mutations for different loci, ranging from a single to very large series of multiple alleles. In some loci the number of alleles is in the order of tens or even hundreds, such as in the red blood cell antigens of cattle. But in most cases the mutations observed consisted of only one or a few. We still lack understanding of why there is such a wide range.

A gene mutating from the wild to a different type is said to have undergone a forward mutation, but one reverting to its original form is designated a reverse or backward mutation. The frequency or rate of backward mutation is apt to be less than that of forward mutation. Sometimes, backward mutations have not been discernible; in many cases it is difficult to say which gene is of the wild type. Mutation is generally regarded as an unpredictable event, but in a large population each mutational event seems to recur regularly and with characteristic frequency.

Suppose gene A_1, with the original frequency of p_0, mutates to A_2 with a rate of u. After one generation the gene frequency of A_1 will decrease to p_1 so that

$$p_1 = p_0 - up_0 = p_0 (1 - u).$$

With one more generation, the gene frequency of A_1 is changed from p_1 to p_2 at the same rate as the change from p_0 to p_1. Thus

$$p_2 = p_1 (1 - u) = p_0 (1 - u)^2.$$

After n generations the gene frequency of A_1 will be

$$p_n = p_0 (1 - u)^n, \qquad\qquad\qquad \textbf{6.8}$$

so we expect that the gene A_1 will eventually be completely replaced by gene A_2. Since u is, in general, quite small in comparison to 1, the rate $(1 - u)^n$ can be written as approximately e^{-un} (e being the base of natural logarithm 2.718. . .). The above equation can thus be written as

$$p_n = p_0 e^{-un}.$$

In order to see, under mutation pressure alone, the number of generations required to decrease the initial gene frequency to one-half, the above equation can be reduced to $1/2 = e^{-un}$, or $ln (1/2)/u = n$, $n = 0.69/u$. For a gene with a mutation rate of 10^{-5}, approximately 69,000 generations are required for reducing the initial gene frequency to one-half. So it becomes obvious that without other pressures, changing gene frequency in any noticeable amount is an extremely slow process.

In the case of two alleles in reverse mutation, A_1 mutates to A_2 at a rate of u, and A_2 to A_1 at the rate of v. Suppose that the initial gene frequency for A_1 is p and that for A_2 is $1 - p$. After one generation there is a gain of $v(1 - p)$ and a loss of (up) in the frequency of A_1. The change in gene frequency of A_1 is

$$\Delta p = v(1 - p) - up. \qquad\qquad\qquad \textbf{6.9}$$

If these mutations are the only disturbing force, an equilibrium will be reached at a certain gene frequency. The gene frequency for equilibrium can be determined by putting $\Delta p = 0$. Thus

$$v(1 - p) - up = 0$$
$$p = v/(u + v). \qquad\qquad\qquad \textbf{6.10}$$

For example, if the mutation rate from A_1 to A_2 is 10^{-5} and that from A_2 to A_1 is 10^{-6}, at equilibrium the gene frequency for A_1 would be $5/55 = 0.09091$, or approximately 0.1.

The fixation of advantageous mutant genes is a key factor in evolution. The success or failure of survival of a new gene in a population depends on both chance and natural selection. Chance is involved in segregation and meiosis and in the number of offspring that each individual in the population contributes to the next generation. A mutant gene, even if it is favored by natural selection, cannot be assured of its survival or fixation in the population.

The probability of fixation of a mutant gene was investigated earlier by Haldane (1927) and Fisher (1930) and later by Wright (1942) and Kimura and Ohta (1969). Wright (1969) calculated the probabilities of fixation for each condition of dominance (table 6.6), the key factors affecting fixation being the degree of dominance and population size, in addition to the selective value of the mutant gene itself. When the selective value of the gene is high, and it possesses complete dominance, the probability of fixation is greatest. But in general, when the selective value is small, the chance of fixation is very low, no matter what the degree of dominance. Zero selection coefficient implies that the mutation is neutral in effect, and negative coefficient implies that the mutation is harmful. Obviously, the probability of fixation for a neutral mutation relates only to population size. Notice that a harmful mutation is less likely to be fixed, but it is not impossible. Depending on population size, once a harmful gene is fixed it reduces the fitness of the population and contributes to the possibility of extinction.

TABLE 6.6 Probability of Fixation of a Mutation A_2 at Various Values of the Parameter s

	Genotype	(1)	(2)	(3)	(4)
	A_2A_2	$1 + s$	$1 + s$	$1 + s$	1
	A_1A_2	1	$1 + s/2$	$1 + s$	$1 + s$
s	A_1A_1	1	1	1	1
Large		$1.1\sqrt{(s/2N)}$	s	$2s$	0
$4/(2N)$		$2.3/(2N)$	$4.1/(2N)$	$6.6/(2N)$	$3.2/(2N)$
$1/(2N)$		$1.3/(2N)$	$1.6/(2N)$	$1.9/(2N)$	$1.4/(2N)$
0		$1/(2N)$	$1/(2N)$	$1/(2N)$	$1/(2N)$
$-1/(2N)$		$0.70/(2N)$	$0.58/(2N)$	$0.49/(2N)$	$0.71/(2N)$
$-4/(2N)$		$0.12/(2N)$	$0.075/(2N)$	$0.042/(2N)$	$0.23/(2N)$

SOURCE: Reprinted, by permission, from S. Wright. © 1969 by The University of Chicago.
NOTE: (1), Recessive mutation; (2), mutation of partial dominance; (3), dominant mutation; and, (4), mutation conferring hybrid vigor.

Once the probability of fixation of a mutant gene is known, the time it will take to reach fixation can be estimated. This essential question of the rate of fixation is concerned with the biological properties of the gene and the population size, and it will form the basis of discussion in chapter 7. We merely offer here the simple solution for a neutral gene that occurred once at a given moment in a population size of N (Kimura and Ohta 1969):

$$t_{fixation} \approx 4N_e,\qquad\qquad\textbf{6.11}$$

where t is the number of generations required for fixation and N_e the effective number of individuals in the population. For a mutant gene favored by natural selection, the generation time will be relatively shortened. (For details see Wright [1969] and Kimura and Ohta [1969].) The existence of neutral genes in populations has been recently proposed as an issue by King and Jukes (1969) and Kimura (1970). Their hypothesis is based on statistical analysis of amino acid frequencies among 5,492 residues in fifty-three vertebrate polypeptides. The main supporting evidence is the fact that frequencies for the different amino acids align fairly closely with those in random permutations of nucleic acid bases. The numbers of the amino acid substitutions in the variable portions of polypeptides fit the expectations based on the Poisson distribution (a statistical description for random events). It is possible that some mutations which produced biological changes from the original genes are so small that the part played by natural selection cannot be perceived. On the other hand, one can see from equation 6.11 there is a chance of fixation for a neutral gene.

One of the arguments against the hypothesis is that each amino acid has a unique chemical structure and specific properties. Thus each may play a specific role in the function of the protein (Smith 1968). Another lies in the procedure of the statistical analysis (Clarke 1970), and a possible source of bias is the exclusion from the analysis of codons that do not fit the assumption of randomness (Dobzhansky 1970). Neutral mutation is one of the most recent interesting hypotheses resulting from the molecular genetic studies, and its confirmation or rejection requires further investigation on protein chemistry as well as more statistical data on amino acids of different proteins. However, one point has been established: the rate of fixation for mutation varies greatly among proteins (figs. 2.17 and table 2.10).

Corbin and Uzzell (1970) estimated that among the total number of mutations, 79% were deleterious and were eliminated by chance and selection. This calculation was based on the average of times of fixation per nucleotide substitution. They obtained 111×10^6 years as the average rate of substitution for thirteen positions of the fibrinopeptides A and B, and assumed that mutations at these positions were neutral. They then calculated the average fixation rate of 540×10^6 for the alpha and beta hemoglobins, cytochrome c, and insulins A and B, for a total of 442 homologous amino acid positions. The differences in fixation rate between the fibrinopeptides and others result from the differences in the nature of these mutations. They divided the fixation rate for fibrinopeptides by that of other proteins, ($111 \times 10^6/540 \times 10^6 = 0.21$), and then multiplied by 100 to convert their figure to a percentage. They arrived at 21% for an estimate of the proportion of mutations for alpha and beta hemoglobins and so on, as favored by natural selection. This means that the time required for the substitution of any one amino acid at any of the thirteen positions of the fibrinopeptides is only 21% of that required for the substitution, on the average, of any one at any position of the other proteins. They reasoned that the longer time required for the latter proteins is because mutation causing amino acid substitution had occurred for some of these positions but was lost due to deleterious effects. The percentage of these mutations is therefore $100 - 21 = 79$. Producing an estimate for the proportion of deleterious mutations is very difficult, influenced as it is by many factors, with the question of selective neutrality a still unresolved issue. Nevertheless, this estimate remains perhaps the most accurate possible under present conditions from data of natural populations.

Among the protein molecules, cytochrome c has been most extensively analyzed, since it is an essential enzyme for all species and thereby has necessitated the maintenance of gene homology throughout phylogeny. The number of differences in amino acids between species is illustrated in figure 6.3. The theories regarding the construction of phylogenic trees based on amino acid variations differ among investigators. My purpose here is to show the magnitude of the correlation of the fixation of mutations in each phyletic line. Certainly there are difficulties in the construction of phylogenetic trees. For example, why do the primates separate from mammalian lines before the marsupial kangaroo? Why does the shark show a closer relationship with the lamprey than with the tuna? We need to realize that such trees are merely based on a protein molecule determined by a single gene in a number of contemporary species. The merit of any protein sequence analysis lies in the complete registration of the events of gene fixation in the past.

SELECTION

In any population, some genotypes leave more offspring than others, due either to greater fertility or to lesser mortality. This process is referred to as selection, and its occurrence in natural populations is in accordance with the differences in fitness among genotypes.

Complete Selection

When the dominant effect of an allele is complete and both the heterozygote and homozygote dominant either die or fail to reproduce, resulting in the next generation being composed entirely of the homozygous recessive individuals, complete selection is said to have taken place. Dominant lethals are examples of this. Therefore, when complete selection is against a dominant allele, the gene will be completely dropped from the population in one generation.

But when selection is against a recessive gene, elimination will require a longer time. As in each generation only homozygous recessive individuals are eliminated, recessive alleles carried by the heterozygotes remain hidden and are not affected. The heterozygotes, when mated among themselves, will produce homozygous recessives that will drop out of the population in the following generation. Therefore the frequency of the recessive gene decreases in each generation. After many generations, the frequency will be extremely low.

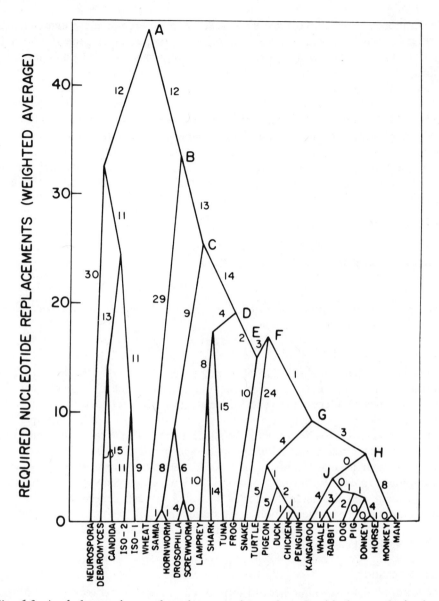

Fig. 6.3. A phylogenetic tree based on numbers of nucleotide base substitution in cytochrome *c* gene for twenty-nine eukaryotic species. Numbers on segments are the actual number of nucleotide base replacements required to account for the descent of the present sequences. The method of construction is based on the cluster analysis technique. (Reprinted, by permission, from W. M. Fitch and E. Margoliash, *Evolutionary Biol.* 4(1970):67–109.)

In an infinite population, a recessive gene cannot be completely eliminated by selection for the following reasons: Let p be the frequency of gene A and q be that of a, and $q = 1 - p$. For a diallelic locus in a random-bred population, the frequencies for the three different genotypes are as follows:

Genotypes	AA	Aa	aa
Frequency	p^2	$2pq$	q^2

When the aa are completely eliminated by selection in each generation, only heterozygous Aa individuals will produce the a gametes. The total gametic contribution of a by Aa individuals is pq, half of its genotypic frequency, since each Aa will produce by chance 50% A gametes and 50% a gametes. The total gametic contribution will be changed to $1 - q^2$. Therefore the gene frequency after each generation of selection will be

$$q_1 = pq/(1 - q^2) = (1 - q)q/[(1 + q)(1 - q)] = q/(1 + q).$$

It can be easily shown that following n generations of complete selection against the recessive gene, the gene frequency will be

$$q_n = q_0/(1 + nq_0), \qquad \textbf{6.12}$$

where q_0 is the initial gene frequency. The change of gene frequency per generation is

$$\Delta q = q/(1 + q) - q = -q^2/(1 + q). \qquad \textbf{6.13}$$

The above equations show that after many generations of selection, the gene frequency will approach zero asymptotically. Secondly, the change of gene frequency becomes smaller as the gene frequency (q) becomes lower.

The effect of selection against a recessive gene is shown by data collected from laboratory populations of *Drosophila* (Wallace 1963). The data are plotted in figure 6.4, showing a case of selection against a recessive lethal gene. The experiment was started with a gene frequency of 0.5. At generation zero, all the flies were heterozygotes. In each following generation there were two genotypes, the homozygote dominant and the heterozygote, as classified in each generation by progeny tests. The data for each generation agree quite well with theoretical expectations, but when all the generations are considered there is a tendency toward a slight selection against heterozygotes.

Partial Selection

In natural populations, differences in fertility and mortality between genotypes are often relative; that is, some genotypes leave less offspring than others but are not completely sterile or nonviable. This may be referred to as partial selection, and its intensity is expressed as the coefficient of selection, s, which is the fraction of reduction in the gametic contribution of one homozygote compared with the other. The contribution of the favored genotype may be taken as 1, and that of the unfavored as $1 - s$. These values represent the fitness of the genotypes. It is under-

stood that the fitness of a genotype with respect to a particular locus is not necessarily the same for each individual, depending somewhat on the other genes of the
individual and its environment. For the sake of mathematical treatment, however,
we assume the genetic background and environment to be the same.

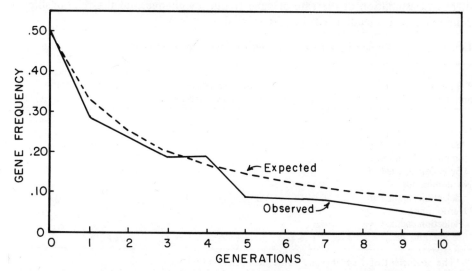

Fig. 6.4. Selection against an autosomal recessive lethal gene in *Drosophila melano-*
gaster. The difference between the observed and expected frequencies is explained
by a slight selection pressure against the heterzygotes. Data from Wallace 1963.

The effect of selection is complicated by the dominance relationship between
the alleles of a gene. Most mutant genes are completely recessive to the wild type
in their visible effects; that is, the phenotype of Aa is identical to AA. But the fitness of the heterozygote may not be the same as the wild homozygote. If the fitness
of Aa is less than AA and greater than aa, the relationship is called no dominance
or partial dominance with respect to fitness. (In some genes the phenotype of the
heterozygote is intermediate. In the case of isozymes, the heterozygotes show the
gene products for each gene; this relationship is called codominance with respect
to the isozyme.) When the fitness of Aa is greater than AA, this condition is called
overdominance. Therefore, we have to treat the effect of selection according to
the different dominant relationships.

We can work out the change of gene frequency of each case similarly to that for
the complete selection against the recessive gene. For example, in the case of partial selection against the completely recessive gene, the total gametic contribution
to the recessive gene will be $(1 - s)q^2 + q(1 + q)$ in the next generation, and
the new total gametic contribution will be $1 - sq^2$. Therefore the gene frequency
of a in the next generation will be $[(1 - s)q^2 + q(1 + q)]/(1 - sq^2)$. Sub-
tracting q, the gene frequency in the previous generation from the above quantity,

and by simplification, we obtain $-sq^2(1 - q)/(1 - sq^2)$ as the change in gene frequency. The formulas that can be worked out for the change of gene frequency under each condition of dominance and selection are given in table 6.7. The denominator for each is omitted for simplification. This omission only causes a very slight overestimation of the rate of gene frequency change, but the relationship of dominance and selection will be appreciated more readily.

TABLE 6.7 Gene Frequency Change after One Generation of Selection

Type of Dominance and Selection	Initial Gene Frequencies and Fitness of each Genotype*			Gene Frequency Change for a (approximation)†
	AA	Aa	aa	
Complete dominance Selection against a	1	1	$1 - s$	$-sq^2(1 - q)$
Complete dominance Selection against A	$1 - s$	$1 - s$	1	$+sq^2(1 - q)$
No dominance Selection against a	1	$1 - \dfrac{1}{2}s$	$1 - s$	$-\dfrac{1}{2}sq(1 - q)$
Overdominance Selection against AA and aa	$1 - s_1$	1	$1 - s_2$	$+pq(s_1p - s_2q)$

*The coefficient of s can be any value between 0 and 1.
†The denominator $1 - sq^2$ for the expression in each line is omitted (see text for explanation).

Let us examine the effect of selection on the gene frequency under each dominant condition. This can best be illustrated by the graphs in figure 6.5, which show the relative change of gene frequencies at different values of q. In the case of complete dominance, selection for a dominant gene is most effective at intermediate gene frequencies (about 0.3) and less effective at higher or lower frequencies. To completely eliminate an unfavorable recessive gene or to completely fix a favorable dominant gene in natural populations takes a long time. Selection for a recessive gene is very ineffective when the recessive gene is rare, and almost entirely hidden in the heterozygous state. The most effective selection for a recessive gene is at the gene frequency of about 0.7 and, for a partially dominant gene, at a gene frequency of 0.5.

How many generations would be needed to change a gene frequency from one specific value to another would depend on both the initial gene frequency and the selection coefficient. The theory was worked out earlier by Haldane (1932) and, more recently, by Ewens (1967) and Crow and Kimura (1970) on the basis of large populations and constant selection coefficients for the entire period. These assumptions are hardly realistic and result, therefore, in some unrealistic values. For example, with a selection coefficient of .001, a change of gene frequency from .99 to .99999 for selection of a dominant gene, or from .00001 to .01 for a recessive gene, requires about 100 million generations. Both Crow and Kimura (1970) and Ewens (1967) were aware of this; Crow and Kimura (1970) stated that when a gene frequency approaches 0 or 1, random fluctuation in gene frequency can carry the gene to loss or fixation in a shorter period of time than calculated. This may well be the case, but the assumption of a large population and constant selection coefficient may not hold. Although we do not have records covering this aspect of the evolutionary period, populations such as demes or tribes could have been rather small. Based on these calculated values, the truth is that evolutionary progress would have been too slow to arrive at its present stage. It is because of this fact that theoreticians (Eden 1967) have challenged the validity of the Darwinian

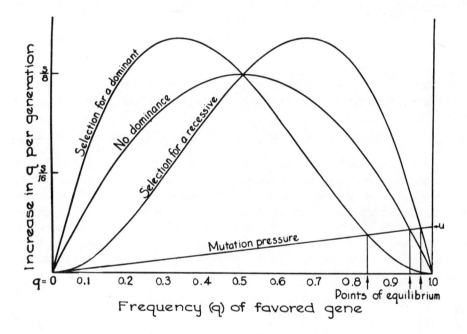

Fig. 6.5. Rate of change in gene frequency under constant selection, *s*, which is opposed by a constant mutation rate, *u*; drawn with *u* equal to 0.03*s*, which is a rather weak selection. The height of the curved lines indicates the rates at which selection would change the frequency, *q*, of the desired gene under the three conditions specified for dominance. The height of the straight line, "mutation pressure," indicates the rate at which mutation would change gene frequency in the absence of selection. The difference between the heights of the curved lines and the straight line indicates the net rate at which gene frequency is changed by selection and opposing mutation. Arrows indicate gene frequencies at which selection pressure and opposing mutation pressure are equal. (Reprinted, by permission, from *Animal Breeding Plans* by Jay L. Lush, 3d ed. © 1945 by Iowa State University Press, Ames, Iowa.)

theory of evolution. Perhaps a more practical way of looking at the problem would be to realize that population size, structure, and selection coefficient vary with time. One needs to keep in mind the fact that the rate of gene frequency changes due to selection depends on the gene frequencies themselves (fig. 6.5).

JOINT EFFECT OF MUTATION AND SELECTION

Having discussed the consequences of mutation and selection separately, we must now consider them in conjunction. In actual populations these two genetic processes operate simultaneously, and more or less in opposite directions. Their magnitude of effect is different with respect to the gene frequency. The mutation of a particular gene is most effective in increasing its frequency when its frequency is low, because there are more genes of the other type of mutate. But selection would be least effective at low gene frequency. Therefore, eventually a gene frequency will be arrived at where the effect of selection is exactly counterbalanced by that of mutation. The point of equilibrium depends on the intensity of selection and mutation rate. To find the gene frequency at the point of equilibrium, we can merely combine equation 6.9, the mutation rate of change, with that resulting from selection (table 6.7, line 1)

$$u(1 - q) - vq = sq^2 (1 - q)/(1 - sq^2). \qquad \textbf{6.14}$$

The equation thus becomes too complex to reveal the actual relationships between selection, mutation, and gene frequency but, as we are interested primarily in low frequencies, we can, according to Falconer (1960), safely make approximations for different conditions of dominance and selection. Since low gene frequencies greatly diminish the chance for backward mutation (remember the mutation rate from $A \to a$ is u and from $a \to A$ is v), we can leave out both vq and sq^2 in the denominator of the right side of the equation and arrive at an adequate simplification consisting of

$$u(1 - q) \cong sq^2 (1 - q)$$
$$q \cong (u/s)^{1/2}. \qquad \textbf{6.15}$$

For selection against a gene with no dominance, we have

$$u(1 - q) - vq = \frac{1}{2} sq(1 - q)/(1 - sq^2) \text{ (see table 6.7, line 3)}.$$

We arrive at

$$q \cong u/s. \qquad \textbf{6.16}$$

In selection against a completely dominant gene, the frequency of the dominant gene is $1 - q$, and the mutation rate from it is u. The term $u(1 - q)$ in equation 6.14 becomes negligible. At equilibrium,

$$vq \cong sq^2 (1 - q)$$
$$q(1 - q) \cong v/s. \qquad \textbf{6.17}$$

What the above three equations reveal is that in each case of selection and mutation, the change in gene frequency will eventually reach zero, where selection pressure and mutation pressure were exactly counterbalanced, and the gene will survive in the population at such a frequency in spite of having been selected against. The frequency may fluctuate slightly in each generation, but it will continue to oscillate around the point of equilibrium.

To illustrate the relationship between mutation and selection, the mutation pressure is plotted against gene frequency in figure 6.5. The mutation pressure increases at a constant rate with the increase in gene frequency. When the gene frequency is low, the mutation pressure is low, simply because there are so few genes in the population to mutate. The effect of selection pressure on the change of gene frequency varies between different dominant conditions, but it is a common characteristic that after gene frequency decreases to a certain level following selection, selection pressure decreases. The interception of the line for mutation pressure on the curve for selection pressure gives the point of equilibrium gene frequency, according to the specified coefficient of selection and mutation rate.

If we accept that the percentage of mutants in a natural population results from the joint effect of mutation and selection, we can, for the present, consider gene frequency as representing a state of equilibrium as a result of counteraction between selection pressure and mutation rate. The frequency of dominant dwarfism (chondrodystrophy) in man has been estimated at 10.7×10^{-5} (Haldane 1949). The fitness of such dwarfs $(1 - s)$ was estimated at 0.196 from the number of children produced by the dwarfs compared with their normal sibs (Falconer 1960). The mutation rate can be estimated by using equation 6.17, $q(1 - q) \cong v/s$, the quantity on the left side of the equation being the percentage of heterozygotes, so that the mutation rate emerges as 4.3×10^{-5}. Using this example, we can estimate the mutation rate for recessive and dominant genes with known detrimental effects in any natural population, if the fitness can be determined.

Although fitness values may be difficult to estimate in some cases, we may accept the gene frequency for many hereditary single gene diseases in the human populations as representing the state of equilibrium. The difference in incidence for the same hereditary disease between populations is therefore not due to a difference in mutation rate but to a difference in the selection coefficient, reflecting environmental variations.

SELECTION FOR HETEROZYGOTES

There remains to be explored a most interesting phenomenon of selection, that is, the fitness of a heterozygote proving to be greater than that of both homozygotes. This circumstance is referred to as overdominance and is considered to be a genetic basis of heterosis (the superiority of the heterozygotes over the homozygotes with respect to one or more traits). When selection favors the heterozygote, gene frequency tends to reach equilibrium at an intermediate value. When two or more alleles are present in a population at the same time in such a proportion that the rarest of them cannot be maintained by merely recurrent mutation, this type of variation is referred to as polymorphism[1] (Ford 1964). Hopkinson and Harris (1971) consider a genetic variation polymorphic when the frequency of the rarest gene at equilibrium reaches a level as high as 0.02.

While selection favors the heterozygote over both homozygotes, there may be a difference in fitness between the two homozygotes themselves, so that different selection coefficients are assigned to them (table 6.7). The change in gene frequency per generation depending on selection is given in the same table. The condition for equilibrium is that there be no further gene frequency change following selection. Therefore $\Delta q \cong pq(s_1 p - s_2 p) = 0$, which can be written as $\Delta q \cong s_1 p - s_2 q$. The gene frequency at the point of equilibrium is

$$s_1 p = s_2 q. \qquad\qquad \textbf{6.18}$$

By substitution, we obtain equilibrium gene frequencies as follows:

$$p = \frac{s_2}{s_1 + s_2} \qquad\qquad q = \frac{s_1}{s_1 + s_2} \qquad\qquad \textbf{6.19}$$

When q is greater than its equilibrium value and p is therefore less, $s_1 p$ will be less than $s_2 q$ and Δq will be negative (s_1 and s_2 are constant). Then q will decrease. On the other hand, if q is less than its equilibrium value, it will increase. Thus, at any value between 0 and 1, selection changes the gene frequencies toward the values of equilibrium. In the case of three or more alleles, equilibrium will be maintained similarly, providing the heterozygote of any pair of alleles is superior in fitness to either of the homozygotes of that pair (Kimura 1956b). The gene frequencies at the point of equilibrium depend on their relative selective values.

The presence of genes at intermediate frequencies is common in many species. Would all of them be due to selection in favor of heterozygotes? This question cannot presently be clearly answered.

Polymorphism is present in practically all species, for example the shell color variations in the land snails *Cepaea nemoralis*, tracing back to the Neolithic period

1. Cavalli-Sforza and Bodmer (1971) proposed a simplified definition for polymorphism: "the occurrence in the same population of two or more alleles at one locus, each with appreciable frequency" and assigned an arbitrary value of 0.01 for appreciable frequency. They pointed out that it is difficult to accept Ford's definition unequivocally on the ground that the cause of polymorphism is often unknown. It is true that contributing factors to polymorphism are difficult to determine. But a gene frequency of 0.01 or above implies that it is unlikely to be maintained by recurrent mutation alone, because mutation rates are generally in the order of 10^{-3} to 10^{-6} (see table 2.9). It is also true that as a definition the description of Cavalli-Sforza and Bodmer is brief and clear. It is questionable, however, whether an arbitrary value of 0.01 should be given without theoretical ground. For instance, if one found genetic variations with gene frequencies on the border line, say between 0.001 to 0.01, one would hesitate to conclude that they are not polymorphism. Indeed, a gene frequency in this range may represent a transient polymorphism for either a new gene on the way to being established or an old gene on the way to dropping out of the population.

(Cain and Sheppard 1954). The blood group differences in many species are well known, and recently the t-locus (brachyury) polymorphism (Dunn, Beasley, and Tinker 1960) and the H-2 histocompatibility locus in the mouse (Snell and Stimpfling 1966) have been disclosed. There are numerous cases of isozyme variations from nonvertebrate and vertebrate species.

In humans a survey of electrophoretic variants was carried out (Hopkinson and Harris 1971) on some twenty arbitrarily chosen enzymes in both European and Negro populations. The variant enzymes and the incidence of heterozygotes for each are shown in table 6.8. Among them, three structural loci specify phospho-

TABLE 6.8 Estimates of Average Heterozygosity per Locus from a Survey of Twenty Arbitarily Chosen Enzymes

	Incidence of Heterozygotes	
	Europeans	Negroes
Red blood cell acid phosphatase	0.51	0.28
Phosphoglucomutase		
Locus PGM_1	0.35	0.33
Locus PGM_3	0.38	0.47
Adenylate kinase	0.09	...
Peptidase A	...	0.16
Peptidase D (prolidase)	0.02	0.10
Adenosine deaminase	0.11	0.06
Average heterozygosity (detected electrophoretically) per locus, assuming twenty-six loci were screened	0.056	0.054
Average heterozygosity per locus for alleles determining all structural enzyme and protein variants (that is, electrophoretic variants $\times 3$)	0.168	0.162

SOURCE: Reprinted, by permission, from D. A. Hopkinson and H. Harris, *Ann Rev. Genet.* 5(1971):5–32.

glucomutase, and in four of the other enzymes two structural loci are implicated. Therefore, in this survey twenty-six loci were covered and an average heterozygosity per locus was computed. Hopkinson and Harris pointed out that it seems unlikely that more than one-third of all structural variants of enzymes can be detected electrophoretically, and suggested, therefore, that the average heterozygosity may be about 0.16 and that the percentage of polymorphism in these two populations, given as about 23 (or $[6 \times 100]/26$), may be an underestimate. Notice that although the averages are about the same for Europeans and Negroes, there are discrepancies in the incidence of heterozygotes for some individual isozyme types between the two populations. The populations are rather heterogeneous, but the discrepancies do suggest differences in the selection coefficient for some enzymes between populations.

There has been a continuing search for the mechanism that determines heterozygote superiority. A well-known example in human populations is sickle-cell anemia, a disease caused by a single gene. Homozygotes for the gene have abnormal hemoglobin and are fatal in most cases. Heterozygotes are found with 35% or more abnormal hemoglobin, but do not suffer from anemia. It was also found that the heterozygotes were more resistant to malaria. This is believed to be the reason for the high frequency of the sickle-cell gene in certain populations. In American Negroes, the frequency of individuals with the disease is approximately 9%. In African populations, in which malaria is prevalent, the frequency is as high as 40% (Allison 1955). Among the 40%, it is estimated that 2.9% were homozygotes. Thus for the whole population, it is estimated that 38.84% are heterozygotes and 1.16% are sickle-cell homozygotes. Thus the estimated gene frequency for sickle-cell anemia is 0.206 (eq. 6.1). The fitness of the homozygote has been estimated at 0.25 relative to the heterozygote based on their respective viability and fertility (Falconer 1960). Thus the selection coefficient against sickle-cell homozygotes is 0.75, and that against the normal homozygotes, 0.197 (eq. 6.18). The relative fitness of the heterozygotes is 1.24 (or $1/[1 - 0.197]$) times that of normal homozygotes, and is five times that of sickle-cell homozygotes. The difference in the frequency of sickle disease between the American and African Negro populations indicates the effects of environments on the selective value of the different genotypes.

The physiological basis of heterosis has remained an intriguing problem through this century, the basic biochemical and physiological mechanisms being diverse. It involves essentially allelic (overdominant) or nonallelic (epistatic) gene interaction, or both, and in some cases it is intimately related to the problem concerning the maintenance of polymorphism and to the problems of genetic load to be discussed later. Evidence tending to foster physiological explanations is still lacking, but with modern biochemical and immunological techniques some interesting facts are emerging, and a few cases are discussed below.

The first evidence of hybrid products of allelic interaction came from the studies of red blood cell antigens in the dove family. Irwin and Cumley (1945) crossbred the pearlneck (*Streptopelia chinensis*) and the ringdove (*S. risoria*) of the family Columbidae. The antiserum prepared against cells of the F_1 hybrids could not be exhausted of antibodies by absorption with the red blood cells of the two parents. That is, after absorption there still remained some antibodies that agglutinated F_1 cells. This shows that there are antigens peculiar to the hybrid but present in neither parent. Further analysis showed that the hybrid substances contained two or three fractions due to specific gene interactions.

A clearer example of hybrid products was obtained from the study of polymorphism for alcohol dehydrogenase in maize by Schwartz and Laughner (1969) in which four alleles for alcohol dehydrogenase were described. Studies on the physicochemical properties of the isozymes specified by the various alleles revealed that the isozyme produced by one homozygote was stable but less active at a high pH, and that those of the other three were unstable but relatively active. Hybrids consisting of the allele of the homozygote stable at a high pH and any one of the other three alleles produced an enzyme that was demonstrated on starch gel electrophoresis. It contained one subunit of each parent and was both stable and active at a high pH and, as the authors pointed out, more advantageous than the homozygotes to the effective growth and development of the organism.

In hybrid enzymes due to allelic gene interactions reported in the fruit fly, F_1 hybrids produced an electrophoretically intermediate phosphatase in addition to both the parental varieties (Beckman and Johnson 1964); similar results were obtained in alcohol dehydrogenase (Johnson and Denniston 1964).

It is often difficult to classify a gene variant as polymorphic or mutant. We are correct in saying that genes affecting the ABO blood groups are of the polymorphic type, and those causing ketopolyuria and dwarfism indicate mutant types. But what about color blindness in man? At present, it is considered that a gene with a frequency above 0.02 should be classified as polymorphic. We cannot too complacently say that color blindness is maintained in the population as a polymorphism, because we know that those with such defects are disadvantaged; but the incidence in some populations (Kuo 1967) is so high that it cannot be explained by mutation alone. Color-blindness genes may have herotic or pleiotropic effects. The selection coefficient of a gene varies in different environments.

A list of rare alleles for the isozymes was reported (table 6.9), but the incidences are too low to be classified as indicating polymorphisms. However, we do not know whether some of them may be on their way out, or just beginning to spread in the population.

TABLE 6.9 Incidence in Europeans of Heterozygotes for Rare Alleles (Gene Frequency < 0.01) Determining Electrophoretic Variants of Enzymes

Enzyme	Approximate No. of Unrelated Individuals Screened	No. of Different Rare Variants	Total Incidence of Rare Heterozygotes per 1,000
Phosphoglucomutase PGM*	6,500	5	1.4
Phosphoglucomutase PGM_2	6,500	3	0.6
Peptidase A	6,000	4	1.0
Peptidase B	4,500	3	2.5
Peptidase D*	3,000	2	1.3
Adenylate kinase*	5,000	1	0.2
Adenosine deaminase*	4,000	2	0.5
Indophenol oxidase	6,500	1	0.8
NADH diaphorase	2,700	4	8.1
Triose phosphate isomerase	2,000	2	1.0
Nucleoside phosphorylase	1,600	3	1.9
Phosphohexose isomerase	1,500	1	0.6

SOURCE: Reprinted, by permission, from D. A. Hopkinson and H. Harris, *Ann Rev. Genet.* 5(1971):5–32.

*Indicates polymorphism (that is, two or more alleles with gene frequency > 0.01), which also occurs in Europeans.

LINKAGE AND EPISTASIS

One essential role of chromosomes is that genes on a single chromosome tend to be transmitted as a group to the zygotes of the next generation. Such genes are referred to as being linked, and the association itself is called linkage.

Picture two linked loci with alleles A and a at one locus and B and b in the other. With respect to the fitness of the organism, A is advantageous in the presence of B, but disadvantageous in the presence of b, and vice versa. Such interaction between genes is called epistasis in a broad sense. Natural selection will favor the gametes AB and ab, but recombination in heterozygotes AB/ab will produce the less favored gametes Ab and aB. Thus the distribution of different gametic types in the population depends on the closeness of the linkage and the extent of epistatic effect.

The importance of linkage in evolution has long been recognized (Fisher 1930), and recently the problem of linkage and gene interaction has been reexamined by a number of investigators, notably Kimura (1956a); Lewontin and Kojima (1960); Bodmer and Parsons (1962); and Wright (1969). The general conclusions are that the problem is complex and the evolutionary consequences, depending on the conditions of linkage and gene interactions, unpredictable. We shall discuss this subject briefly in order to see its importance in evolution.

Assuming two linked loci with two alleles, AB and ab, for each, four types of gametes will be produced. The gametic frequencies and fitness can be written as follows:

Gamete	ab	Ab	aB	AB
Frequency	p_1	p_2	p_3	p_4
Fitness	m_1	m_2	m_3	m_4

The alleles in combinations of AB and ab are said to be in coupling phase, and those of Ab and aB in repulsion phase.

The measurement of departure from linkage equilibrium may be made by estimating the ratio of the zygotic frequency of AB/ab to that of Ab/aB. Say that Z represents the ratio

$$Z = \frac{p_1 \, p_4}{p_2 \, p_3} \qquad\qquad 6.20$$

when the product of zygotic frequency of the coupling phase equals that of the repulsion phase, $Z = 1$, and this state is called linkage equilibrium. In the absence of any selection pressure, the value of Z will continue to approach unity with succeeding generations. When there is selection pressure, Z will approach a value either greater or smaller than unity. The attainment of such a state is called linkage disequilibrium. When the ratio approaches a positive value greater than unity as a limit, the condition is referred to as positive disequilibrium. This would lead to an excess of ab and AB chromosomes, that is, the chromosomes in the coupling phase would be favored by natural selection. When the ratio approaches a value less than unity as a limit, the condition is referred to as negative disequilibrium. This would lead to an excess of aB and Ab chromosomes of the repulsion phase. Both these states have been referred to as quasi equilibrium. When the frequency of crossing-over, which is measured by c $(0 < c < 0.5)$, is smaller than the epistatic effect (measured by $E = m_1 - m_2 - m_3 + m_4$), there being no upper or lower limit for Z leads to the eventual fixation of one type or the other. The mathematical method for finding departure from linkage equilibrium as selection proceeds is to take the derivative of Z. The expression of Z is in terms of gametic frequencies, epistasis, frequency of crossing-over, and the fitness of gametes. A rather elaborate derivation is given in Crow and Kimura (1970, p. 198).

Examples illustrating the effects of the departure of different variables from linkage equilibrium follow:

Genotype	A-B-	aaB-	A-bb	$aabb$	Crossing-over Frequency
Fitness (1)	1.00	0.99	0.98	1.02	0.20
Fitness (2)	1.00	0.95	0.95	1.10	0.01

The first example shows a case of linkage disequilibrium with arbitrary values for the fitness of the genotypes involving two linked loci. Assuming the frequency of crossing-over to be 0.20, in this case the epistatic effect would be relatively small. As time proceeds the Z value approaches 1.29 as an upper limit of quasi equilibrium (fig. 6.6) and the coupling phase of chromosomes will prevail. In natural populations, this would correspond to selection favoring the extremes of the distribution.

In the second case, the linkage is closer (crossing-over frequency, 0.01), fitness for each genotype is slightly different from the first case, and the Z values continue to change. The rate of change is greater when the Z values are larger. This leads to fixation of the ab chromosome. The change of frequencies for the different chromosomes after a number of generations is shown in table 6.10.

TABLE 6.10 Changes in Chromosome Frequencies and Gametic Phase Unbalance (Z) in a Diploid Model with Close Linkage ($c = 0.01$)

Generation	Chromosome Frequencies			Z
	AB	$Ab = aB$	ab	
0	0.250	0.250	0.250	1.00
10	0.278	0.218	0.268	1.67
20	0.294	0.185	0.336	2.87
40	0.263	0.114	0.508	10.02
80	0.025	0.007	0.961	483.00

SOURCE: Reprinted, by permission, from Crow and Kimura, 1970.
NOTE: Fitnesses are $W_{aabb} = 1.10$, $W_{A-bb} = W_{aaB} = 0.95$; $W_{A-B-} = 1.00$.

The evolution of mimicry in the butterfly can be considered a case of selection of closely linked genes with strong epistasis. Mimicry gains protection for an individual from its enemies by resemblance to a type of impalatability. Mimetic resemblance must be so striking that it misleads the enemy. It has been suggested that mimicry was affected by a series of genes and it must have been evolved gradually (Ford 1953). The perfection of the mimicry in *Papilio dardanus* was dependent on a whole gene complex (Clarke and Sheppard 1960). When this coadapted genotype was disrupted by hybridization to a race, the pattern became less effective. It was claimed that in the evolution of mimicry a new mutant resemblance to some model may appear, and then be enhanced by a selection of modifiers. Clarke and Sheppard also suggested that the combination of a limited number of characters accounts for the various patterns. The genes responsible for these characteristics are very closely linked, the whole unit being a supergene. This case is very similar to the second arbitrary example given above, with the eventual consequence being that a specific chromosome giving the best mimic effect became most frequent. But there were large numbers of loci involved.

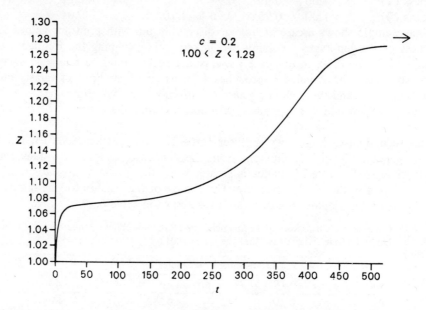

Fig. 6.6. The fast approach to quasi-linkage equilibrium and the slow change in Z thereafter when epistasis is small and linkage is loose. The fitness values are given in the text. (Reprinted, by permission, from Crow and Kimura, 1970.)

The above examples are offered to show that unless linkage is close, or the epistatic effect large, it may be expected that epistasis would have little effect on selection. This is a general conclusion reached by both Crow and Kimura (1970) and Wright (1969). The patterns of evolutionary changes brought about by gene interaction are complex and inconsistent (Kojima 1965). When more than two loci are involved, the situation would be more complex (Lewontin and Kojima 1960).

MATHER'S CONCEPT OF HETEROZYGOSITY AND LINKAGE IN EVOLUTION

The importance of linkage and epistasis in evolution has long been recognized, beginning perhaps with Fisher (1930) and continuing with investigations by Ford (1953), Sheppard (1959), and others on polymorphism and linkage. The concept of Mather (1943) is that in natural populations selection tends to build up an alteration of plus and minus genes[2] on chromosomes. Thus evolution tends to minimize variability but maximizes potential variability. Mather referred particularly to quantitative characters.

In most quantitative traits, variations, if measured on appropriate scales, are generally distributed symmetrically and are bell-shaped, suggesting natural selection in favor of intermediates. Wright (1935) called them the optimum genotypes. Such a process of selection has been referred to as centripetal selection (Simpson 1953), stabilizing selection, or normalizing selection (Waddington 1960) (fig. 6.7). Natural selection favors genotypes close to the mean of the distribution and tends to reject those away from the mean. The genotypes close to the mean consist more of heterozygotes than of homozygotes. It was based on this idea that Lerner (1954) proposed the concept of genetic homeostasis. He pointed out that heterozygotes confer the homeostatic property and possess buffering ability for normal growth and development under varying environments, whereas homozygotes in general lack this capacity.

Generally speaking, quantitative characters are affected by a large number of genes. These genes act independently and/or interact among themselves to affect the character under investigation. Thus when many genes are involved, the distribution of the variation usually gives a continuous type. This is distinct from the discrete type of distribution, which is bimodal for a single-gene (two alleles) trait with dominance, and trimodal with no dominance for a random-bred population. The genes responsible for continually varying characters are referred to as multiple factors. Mather (1943) termed them polygenes and believed that each of their effects was small, additive, or interactive, as discussed in chapter 5. He once claimed an association between polygenes and heterochromatin; this concept has lacked good supporting evidence, and for that reason has not been generally accepted. But the term *polygene* has been widely adopted and is used loosely to refer to conditions that cannot be explained by single-gene inheritance.

As I see it, the variation in a quantitative trait is the result of variations in many biological systems, each of these systems being influenced by one or more genes. Each contributes a specific physiological effect through an individual biological system to the trait under investigation. Spickett (1963) has studied the individual effects of a few polygenes influencing the bristle number in *Drosophila*. They can

---

2. This expression is generally used for genes that affect quantitative traits; with reference to the mean of the population, those that cause quantitative increase of the trait under study are referred to as plus genes and those that cause quantitative decrease are referred to as minus genes.

be either structural genes or regulator genes. It is probable that, as an example, isozyme variations are a type of polygenic variation. Hybrid enzymes or substances that differ from the parental types and have specific biological properties are the basis of genetic homeostasis.

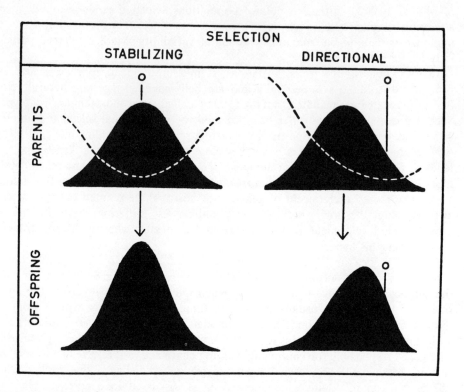

Fig. 6.7. Stabilizing and directional selection. The solid figures represent the distribution of the character in the population and the broken curves the relative intensity of selection against the various expressions of the characters, *o* being the optimum expression in the sense of suffering the least action of selection. In stabilizing selection, *o* is at the average expression: the action of selection is to narrow the average of expression of the character. Birth weight (fig. 6.9) is an example of such selection. In directional selection, *o* is not at the average expression, and the selection tends to shift this average toward the *o*. This illustration shows that selection against different genotypes is relative, which presumably occurs in natural populations. (Reprinted, by permission, from Mather, 1964.)

Since polygenes are normally present in populations, and their effects are small, they are essentially the genetic units responsible for microevolution. Mather has extensively discussed the polygenic concept in conjunction with linkage. A simple example of two closely linked loci, each with two alleles, will illustrate his major thesis. We will disregard all other loci, or assume they have identical genes. We want to show why intermediate types are favored by natural selection, how linkage affects this condition, and how the population carries the hidden genetic variation.

The intermediate genotypes of the parents, and most of the genotypes in the off-spring generation for two closely linked genes, are

Genotype	Parents	Offspring
1	$\dfrac{Ab}{aB}$ $\rightarrow$	$\dfrac{Ab}{Ab}, \dfrac{aB}{Ab}, \dfrac{Ab}{aB}, \dfrac{aB}{aB}$
2	$\dfrac{AB}{ab}$ $\rightarrow$	$\dfrac{AB}{AB}, \dfrac{AB}{ab}, \dfrac{ab}{AB}, \dfrac{ab}{ab}$
3	$\dfrac{Ab}{Ab}$ $\rightarrow$	$\dfrac{Ab}{Ab}$
4	$\dfrac{aB}{aB}$ $\rightarrow$	$\dfrac{aB}{aB}$

Genotype 1 produces all intermediate types and 50% of them are homozygotes. Genotype 2 produces 50% intermediate types and 50% are double homozygotes of the extreme types. Genotypes 3 and 4 produce only intermediate types but these are all homozygotes. They do not possess the homeostatic property of the heterozygotes. We can see that the repulsion phase of genotype 1 produces relatively more offspring favored by natural selection, especially when the linkage is tight, than any other type.

In case there are environmental changes favoring either one or the other extreme type in the distributions, these genes can be released by crossing-over to produce double homozygotes AB/AB and ab/ab, or the other types AB/Ab, aB/ab, and so on. Thus the aB/Ab genotype has minimum variability in a stable environment, but all the variability in reserve can be released if there is a demand.

Directional selection experiments showed that after the plateau that followed the beginning generations of selection there were suddenly sharp responses (fig. 6.8). It has been speculated that this condition was due to the recombination of closely linked genes, as in the above case. But this was not proved until the experiment of Thoday and Boam (1961), in which striking individual gene effects were observed and later the genes were identified. The experiment involved a series of lines of *D. melanogaster* selected for increased sternopleural bristle number. The results are shown in figure 6.8. Due to some variations in selection procedure among the lines at some generations, the number of generations of selection is not shown in the figure. They are adjusted so that the lines are in correspondence. Three lines (*dp* 1, *dp* 2, and *dp* 6) showed markedly similar responses. Each of them had a fairly good response at the beginning generations of selection, then the lines plateaued for a number of generations. Thereafter, there were sharp and accelerated responses in all the three lines, as shown in the middle portion of the curves. The fourth line (*vg* 4) began with a relatively high bristle number and showed slight responses for a large number of generations. Then there were accelerated responses at generations corresponding with the other three lines, reaching a level greater than those of the three other lines. These accelerated responses, shown in the middle of the curves, could not be explained by random mutation, because mutation would have been unlikely to occur simultaneously in all the lines. It has recently been possible to show that some of the accelerated responses were due to two linked loci at about 28 to 32 crossing-over units at the chromosome III affecting bristle number (Thoday 1961).

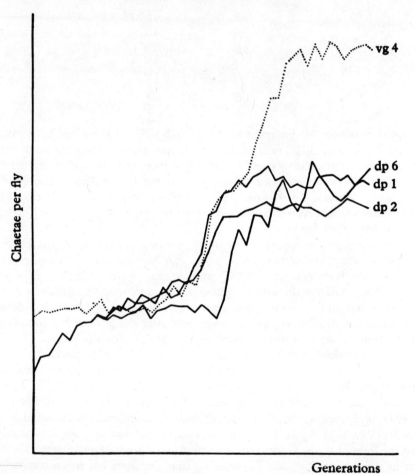

Fig. 6.8. Accelerated responses to selection in *Drosophila.* The solid curves show the similarity of response of three lines, *dp* 1, *dp* 2, and *dp* 6. The dotted curve shows the response of a fourth line, *vg* 4, plotted so that its accelerated response coincides with those of the dp lines. Note the accelerated responses in the middle portion of the curves for all the lines and the plateau at later generations. See text for detailed explanation. (Reprinted, by permission, from J. M. Thoday and T. B. Boam, *Genet. Res.* 2(1961):161–76.)

The most plausible explanation for the sharp responses after the plateau is that the genes were arranged in a complex of a type similar to the Ab/aB. The concealed genetic variability was released by recombination under continued selection pressure and was called heterozygosity potential by Mather (1943). Many instances of genes possessing some similar effects on similar biological systems are closely linked. There is evidence for this in both lower and higher organisms, as stated in chapter 2. We can see that linkage is a biological property playing an important role in evolution.

In natural populations, examples of selection in favor of intermediate types are many. Human newborns showing extremes of body weight are greater mortality risks than are those of average weight, as is well known (fig. 6.9). But to prove that the intermediate weight is due to linked polygenes in the repulsion phase is extremely difficult.

GENETIC LOAD

The term *genetic load* was introduced by Muller (1950) when he discussed the problems of past and future damage to the genetic material of man. His major concern was with radiation-induced mutations whose damage he feared might reach a dangerously high level before it was recognized. The term has been used since then to refer to genetic consequences involving reduction of fitness due to genetic variations carried in the population. Thus its meaning has been widened and some mathematical theory has been developed. The genetic load in a population (Crow 1958, 1963) is expressed as

$$L = \frac{W_{max} - \overline{W}}{W_{max}},$$

where W_{max} represents the fitness of the most favorable genotype and $\overline{W}$ the average fitness of the population affected by a specific genetic locus. Accordingly, there are two major kinds of genetic load: mutational and segregational. Mutational load refers to the reduction of fitness due to genes that are detrimental in any genetic and natural environment. Segregational load refers to the reduction of fitness due to heterotic (overdominant) genes rendered homozygous. Some other types of genetic loads are discussed by Mayr (1963).

Mutational Load

Let s be the selection coefficient, and h a constant so that $0 < h < 1$, while the relative fitness of the three different genotypes and frequencies in the presence of inbreeding (Crow and Kimura) are

Genotype	AA	Aa	aa
Fitness	W_{max}	$W_{max}(1 - hs)$	$W_{max}(1 - s)$
Frequency	$p^2(1 - F) + pF$	$2pq(1 - F)$	$q^2(1 - F) + qF$

where F is the inbreeding coefficient. For simplifying the derivation, let W_{max} equal 1 in this case. The average fitness of the population is, therefore, the sum of the products of the genotypic frequencies times their respective fitness. The general expression for the mutational load is, therefore,

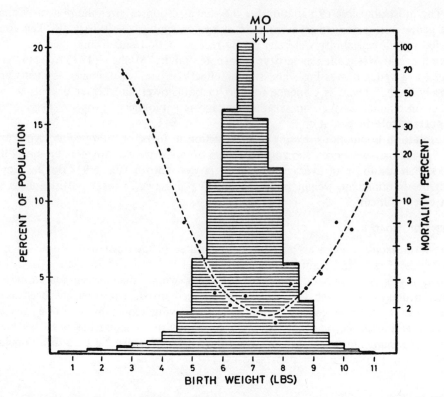

Fig. 6.9. The distribution of birth weight, a polygenic character, among 13,730 children and the early mortality of the various birth weight classes. The hatched histogram shows the proportions of the population falling into the various classes with respect to birth weight. The broken line is the curve of mortality in relation to birth weight, the values actually observed for the classes of birth weight being represented by the points to which the curve is an approximation. The percentage mortality is set out as a logarithmic scale for ease of representation. Mean birth weight, M; birth weight associated with the lowest mortality, hence the optimum weight, O. (Reprinted, by permission, from Mather, 1964.)

$$L = \frac{W_{max} - \overline{W}}{W_{max}} = hs[2\,pq(1 - F)] + s[q^2(1 - F) + qF]. \qquad \textbf{6.21}$$

In a completely inbred population ($F = 1$), the expression for mutational load is simplified as

$$L_i = sq, \qquad \textbf{6.22}$$

where *i* refers to an inbred population.

In a random mating population ($F = 0$), the mutational load is

$$L_0 = 2hsupq + sq^2, \qquad \textbf{6.23}$$

where u is the mutation rate from A to a. The value of the first term of the right-hand side of the equation is very small and can be dropped out. When the mutant gene is completely recessive, $h = 0$. According to equations 6.15 and 6.23,

$$L_0 = sq^2 = s\frac{u}{s} = u \qquad\qquad \textbf{6.24}$$

since $q = (u/s)^{1/2}$. Hence, the mutational load equals the mutation rate in the case of complete recessiveness as was first noted by Haldane (1937). In the case of partial dominance $(h = 0.5)$,

$$L = \frac{2u}{1 + u}. \qquad\qquad \textbf{6.25}$$

In other degrees of partial dominance, unless h is very close to 0, the mutational load is about $2u$, that is, two times the mutation rate. The mutational load in the case of partial dominance is greater than with complete dominance. This is because, in addition to the homozygous recessive class, the fitness of the heterozygote is also reduced. The mutational load is additive. That is, the total mutational load for a number of loci in a population is the sum of the individual loads.

It has been estimated that in the fruit fly one-fourth or one-third of each chromosome contains at least one lethal or semilethal gene (Dobzhansky 1957). In man, an average of 1.5 to 2.5 lethal equivalents per gamete or 3 to 5 per zygote has been reported (Morton, Crow, and Muller 1956) based on data collected in two French populations (Morbihan and Loir-et-Cher) with different degrees of consanguinity in each. One lethal equivalent was defined as a group of recessive mutant genes present in the individuals of a finite population in such numbers that they caused an average of one zygotic death. That is, one lethal equivalent corresponds to one lethal mutant or to two mutants each with a 50% probability of causing death, and so forth.

Segregational Load

Segregational load refers to the reduction of fitness of the population when a heterozygote is favored by natural selection. The decreased fitness arises from the fact that in a Mendelian population there will be inferior homozygotes rising by segregation. Assuming s and t to be the selection coefficients against the homozygotes AA and aa, respectively, the fitness and genotype frequencies of the three genotypes will be as follows:

Genotype	AA	Aa	aa
Frequency	p^2	$2pq$	q^2
Fitness	$1 - s$	1	$1 - t$

The heterozygote has the maximum fitness and is assigned the score of one and is the reference genotype for evaluation of population fitness (Crow 1963). According to the definition of genetic load, the segregational load is similarly calculated as for the mutational load:

$$L = 1 - [p^2(1 - s) + 2pq + q^2(1 - t)].$$

Since in a random-bred population, $p^2 + 2pq + q^2 = 1$,

$$L = sp^2 + tq^2. \qquad \qquad \textbf{6.26}$$

For a number of independent loci, the total segregational load is roughly the sum of the individual loads.

For example, in the African populations cited earlier we have calculated the gene frequency and selection coefficient of sickle-cell anemia as 0.206 and 0.75, respectively, and for the normal allele, 0.794 and 0.197. The segregation load $L = 0.75(0.206)^2 + 0.197(0.794)^2 = 0.156$. The fitness of the population is therefore reduced to $1 - L = 0.844$.

In the house mouse (*M. musculus*), a high percent of the wild population carries more than thirteen lethal and male-sterile alleles at *t* locus on the seventeenth chromosome. The homozygote recessives are lethal or sterile in males. There is a high transmission ratio (0.96) (Dunn, Beasley, and Tinker 1960) of the recessive alleles in male heterozygotes. A lethal allele favored by such a transmission advantage should attain a very high equilibrium value. The relative fitness of the normal homozygote and the heterozygotes tends to be in favor of the latter and thus further increases the frequency of the lethal alleles. The distribution of the *t* alleles in many of the populations of *M. musculus* found in different parts of North America is shown in table 6.11. If we disregard the abnormal male transmission ratio and assume that the relative fitness between the normal homozygotes and heterozygotes is 0.9:1, we arrive at an estimated segregational load of 0.102 ($Sp^2 + tq^2 = 0.1[0.81]^2 + [0.19]^2 = 0.102$). The average fitness of the populations is about 0.9 ($1 - 0.102 = 0.898$), but mouse populations are nevertheless surviving throughout the world with this large genetic load.

TABLE 6.11 Frequencies of Different Genotypes with Respect to the *T* Locus in Natural Populations of the House Mouse (*M. musculus*)

Populations	Total Number of Mice	Normal Homozygotes	Heterozygotes	Lethel Gene Frequency
16 states, United States	140	62	78	0.28*
Alberta	176	134	42	0.12†
Total	316	196	120	0.19

*Dunn, Beasley, and Tinker 1960.
†Anderson 1964.

In criticism of the theory and application of segregational load (Sanghvi 1963; Li 1963; and Bruce 1970), Li (1963) gave the example of a population containing 1,000 *AA* individuals, in which a favorable mutation arose. That is, one of the individuals was *Aa*. If the heterozygote conferred a 2% advantage in fitness over the homozygote, then the fitness of the population increased by 0.002; but the population had a segregation load of 0.0195 in comparison with 0 for the population without the advantageous mutation. Li points out that the more beneficial the mutation, the greater the genetic load. The population with the new mutation, rather than with the "optimum" genotype, is at a disadvantage. It was argued (Sanghvi 1963; Li 1963) that in calculating the genetic load the reference genotype should be the most favorable homozygote, not the heterozygote genotype. Brues (1969) questioned whether the segregation load is a load at all. In the same sense, she used sickle-cell anemia as an example. If two populations live in the same malarial environment, the population with a low percentage of sickling gene of 0.1 has its size increased by 1.25% as compared with the population without the sickle-cell gene. Thus the population with the segregation load is more successful than the normal population.

For the calculation of a segregational load for a large number of loci, there is a further problem. For instance, if one heterotic locus contributes to 0.01 of the segregational load, for 100 such loci there will be $0.01 \times 100 = 1$, which means that the fitness of the population will be zero. In natural populations there are very high percentages of heterozygous loci, among which there must be a high percentage of heterotic genes. How can populations carry these enormous genetic loads?

Explanations have been offered by various investigators (Sved, Reed, and Bodmer 1967; Milkman 1967; Crow and Kimura 1970), some of which are hypothetical and require supporting evidence.

The problem regarding the theory of segregational load is associated with assignment of fitness values to the genotypes. It seems logical to assume that the best homozygote, not the heterozygote, should be used as the reference and be assigned a fitness of 1 (Sanghvi 1963). The heterozygote with heterotic effect would then have a value greater than 1. In a population with heterotic loci, the fitness gained by the heterozygote would tend to compensate that lost by the homozygote. Heterosis is a genetic phenomenon advantageous to the population. The average fitness of the population should not be reduced by its presence.

Summary

The Mendelian system is capable of producing an enormous number of genotypes by recombination of a relatively small number of genes, the number of potential combinations being surprisingly great. The resulting variations offer freedom for natural selection to operate. Fisher's (1930) fundamental theory of natural selection states: "The rate of increase in fitness of any organism at any time is equal to its genetic variance at that time."

Gene mutation supplies fresh material to the gene pool. Many of the mutant genes are lost by chance or eliminated by natural selection. Those left, if favored by natural selection, are retained in the population, and maintained according to the theory of selection. Those that are closely linked may form a gene complex, causing improvement of certain characters.

In general, the frequencies of alleles at different loci and the direction of natural selection operating in populations living in a more or less constant environment are closely connected. Genes at very low frequencies are generally those selected against, but they are maintained in the population by recurrent mutation and contribute to the mutational load.

Most of the genes in intermediate frequencies are maintained by some form of balanced selection. They are referred to as polymorphisms when their individual effects are recognizable and separable; when not, they are polygenes (Mather 1943). These genes are the basic units for the evolutionary shift and adaptation of the population concerned with the segregational load.

The correlation between allelic frequencies and selection forces may be disturbed in small populations due to genetic drift or inbreeding. However, in large populations, the gene frequencies will, in general, be such as to tend to maximize the population fitness at that time and space. Heterosis is one mechanism for the maintenance of polymorphism. We are beginning to understand its biochemical and physiological basis. There are other explanations for polymorphism, such as the neutral gene theory and evolutionary relics, which will be further discussed in the next chapter.

An organism needs to keep some of its genes in reserve for environmental changes and adverse direction of selection. The dominant property of genes provides such a mechanism. Recessive genes making for reduced fitness at a specific time can be carried in the population hidden by effects of dominant genes. When the environment changes, the relative fitness of the genes will be altered, and types previously unfavored may become favored by natural selection in the new environment. New combinations would occur and new variations would appear. Such a conservative process of evolution represents essentially the theory of heterozygosity potential of Mather (1943), whereby release of genetic variability occurs when there is need.

Evolution of Mendelian populations involves a continuous interaction of their gene pools and their environments. In this interaction system, population structure, such as size and breeding systems, and the properties of genes, such as dominance, linkage, and mutation, all play important roles. Populations are changing ceaselessly in time. The variations we observed represent the concerted action of all such factors at a given moment. Variations persist and, at a slow pace, genes continuously mutate and renew—become lost, fixed, or coexist with their allelic genes.

Seven

adaptation, speciation, and extinction

A few biologists in the eighteenth and early nineteenth centuries expressed the idea that all living species were derived by a process of branching off from ancestral lines. Darwin noted a "natural subordination of organic beings in groups under groups." On this hierarchical structure the race represents the outermost twig on branches delineating species, genus, family, order, class, phylum, and kingdom according to a taxonomic system of classification. For adaptation with further differentiation, this branching process represents speciation or phyletic evolution.

Evolution is a continuing process in time. What has been observed in fossils and in present living forms illustrates sections of it. The fate of a present living population or species cannot be predicted. However, we can make generalizations regarding cause and effect, based both on theory and on fact.

Fossil findings have disclosed that many species once in abundance in the history of evolution have disappeared. The phenomenon occurred rather generally and sometimes took the form of not only pruning some of the twigs of the phylogenetic tree, but chopping off large branches of it. Fossil records indicate that extinction was the final destination of almost all ancient species. The fate of a population or a species existing at present will be either extinction or further development, as a twig or branch of a tree either falls off or grows and branches.

Charles Darwin spent two decades gathering all the evidence he could find bearing on his theory of evolution, and concluded that all existing organisms evolved by successive progressive changes from previously existing forms as a consequence of natural selection. After the discovery of Mendelian inheritance, his theory of natural selection was put on a scientific basis in terms of genes and population genetics. We have discussed in previous chapters the physical and chemical properties of the gene, its changes, its effects, its transmission processes, and its fate. We can now attempt to relate genes and populations, and to elucidate the fundamental processes of speciation and extinction.

Population genetics is a separate discipline, and one we cannot treat in depth here. We can only review briefly a few fundamental theories as a guide for discussion relevant to speciation and extinction. Since the discovery of nucleic acid as the carrier of genetic information there has been a great advance in genetic knowledge, which, together with population genetic theories, has provided further illumination of the evolution of populations.

GENETIC THEOREM OF EVOLUTION

During the processes of transmission of genes from one generation to the next, many factors come into play and these may be grouped into two categories: random and nonrandom. The random categories of genetic drift and mutation and the nonrandom categories of immigration and selection operate jointly in natural populations. However, in order to see the individual effects of each, we shall have to assume that in some cases the effects of one or the other will be zero or negligible.

Random Genetic Drift

In natural populations, unfixed genes are always in the act of segregating. At each locus a certain number of homozygotes and a certain number of heterozygotes are present at a given time. Without assuming the factors of natural selection, mutation and migration, one would expect that the relative proportions of homozygotes and heterozygotes would vary between generations due to random breeding between individuals. However, after a large number of generations the population tends to consist exclusively of one or the other type of homozygous individuals, indicating the loss or fixation of some gene that originally segregated. The random fluctuation in gene frequency in time, together with the tendency toward genetic fixation of one gene, is called random genetic drift or the "Sewall Wright effect." The process was originally described by R. A. Fisher, but its value has since been disavowed by him. It is Wright who recognized its importance in evolution, and further developed the concept and its mathematical basis.

The significance of the theory of genetic drift in evolution has not been fully appreciated by many laboratory and field workers. In this respect, let us make use of an arbitrary example. Suppose we start with 100 cultures of fruit flies, each containing ten males and ten females randomly chosen from a population containing the genotypes AA, Aa, and aa in the frequencies of 0.25, 0.50, and 0.25, respectively. In each generation, the parents are removed before the eggs hatch. Then ten newly hatched males and ten females are randomly chosen as parents for the next generation, and so on for a large number of generations. Assuming the three types to be equally fertile and viable, we will find that in each generation the different genotypes vary in frequency between cultures. We will also find that they vary between generations for each series in each culture. The greater the number of intervening generations after the initial cultures, the greater the differences between the cultures will be. Eventually we may find that some cultures contain only flies of the AA genotypes, other only those of the aa genotype, and the rest will contain all three genotypes AA, Aa, and aa, but in proportions different from those in the original sample.

In addition to fluctuation in gene frequencies, the general tendency is to shift from intermediate values to either 0 or 1. Thus there is a tendency for heterozygosity to decrease. This is known as the "decay" of variation.

It can be seen that the loss or fixation of a gene as a fundamental of evolutionary change is due entirely to chance. If a gene that happens to be useful is fixed, the population benefits and the evolutionary process toward adaptation is accelerated. If the gene is not useful, or is harmful to the population, the fixation would become a handicap, and the only cure for the difficulty is by mutation or migration of genes from other populations. If many harmful genes become fixed, the population may face extinction, as is known by field biologists who have observed that a species very abundant one year may be difficult to find in some later year. Athough the reason for it may be complicated, random genetic drift can play an important role; in general, the smaller the population the greater is its effect. The basis for this process and its mathematical analysis follows.

Let us consider a population of N sexual organisms, with a gene frequency of p for A and q for a, where $p + q = 1$. Two gametes are drawn repeatedly from among the $2N$ gametes to produce the next generation. The gene frequency will vary, by chance, according to the binomial expansion

$$(p + q)^{2N}.$$

The $2N + 1$ possible values of q in the next generation of N individuals are 0, $1/2N$, $2/2N$, . . . , $(2N - 3)/2N$, $(2N - 2)/2N$, $(2N - 1)2N$, 1. According to the binomial theorem, the probability for any particular value of q, q_i, is

$$\frac{(2N)!}{(2Nq_i)!(2Np_i)!}\, q^{2Nq_i} p^{2Np_i},$$

where $p = 1 - q$.

The gene frequency varies as a result of random sampling in a population of finite size, and may increase or decrease at random. The variance of q in the first generation is

$$\sigma^2_{q_1} = \frac{q_0 p_0}{2N},$$

where p_0 and q_0 are the frequencies of A and a in the 0th generation, respectively. In the following generation the variance of q (Wright 1931) is

$$\sigma^2_{q_2} = \frac{q_0 p_0}{2N} [1 + (1 - \frac{1}{2N})].$$

The variance at the nth generation is

$$\sigma^2_{q_n} = \frac{q_0 p_0}{2N} [1 + (1 - \frac{1}{2N}) + (1 - \frac{1}{2N})^2 + \dots, (1 - \frac{1}{2N})^{n-1}]$$

$$\sigma^2_{q_n} = q_0 p_0 [1 - (1 - \frac{1}{2N})^n] \approx q_0 p_0 (1 - e^{\frac{-n}{2N}}) \qquad \textbf{7.1}$$

where $e = 2.718$, the natural logarithm base, and n represents the number of generations. The above equations show that in the early generations variances increase. As n becomes larger, variance approaches $q_o p_o$ as a limit. The probability distribution for q becomes a rectangular type and the decay of heterozygosis would be at the rate of $1/2N$ (fig. 7.1).

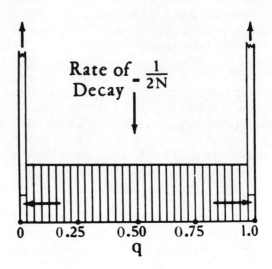

Fig. 7.1. Theoretical gene frequency distribution for small populations. The fixation and loss of genes are each proceeding at the rate $1/4N$ per generation, so that the total rate of decay is $1/2N$ in the absence of mutation, selection, or migration. (Reprinted, by permission, from Li, 1955. © 1955 by The University of Chicago.)

Say one started with a number of populations having intermediate values of q. If q moved toward either 0 or 1, immediate fixation of either A or a would occur. But the values of q for the rest of the populations would spread between 0 and 1 and, finally, the distribution curve would flatten (fig. 7.1). Then the rate of fixation for either A or a would be $1/4N$, and the probability of either loss or fixation of a gene $1/2N$. Eventually the heteroallelic populations would be entirely depleted, all becoming either AA or aa, with the ratio $q_o:p_o$. Note the importance of population size on the rate of gene fixation. (The rate of gene fixation here arrived at checks with equation 6.5 for the effect of population size on inbreeding. The inbreeding coefficient F is a measure of gene fixation. Here the term *genetic drift* is used in reference to the population with direction implication to evolution.)

The snail *Cepaea nemoralis* has three shell colors: yellow, pink, and brown, with yellow recessive. To study the effect of population size on distribution of the yellow gene, LaMotte (1959) observed the gene frequencies in populations of different sizes. Figure 7.2 shows the distribution of the gene frequency for yellow. The populations were divided into two groups: one of them between 500 and 1,000 individuals and the other between 3,000 and 10,000 individuals. The effect of population size on the rate of fixation can be seen. The smaller the population size, the greater the rate of decay of heterozygosity or polymorphism, or the effect of random genetic drift. The slightly skewed distribution patterns suggest some selection in favor of yellow. Without selection they would be similar to the theoretical type (symmetrical) (fig. 7.1).

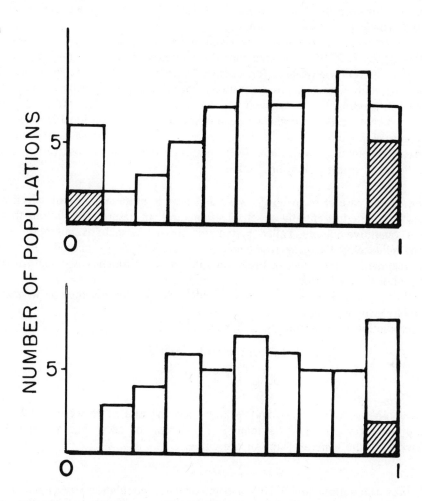

Fig. 7.2. Observed frequency distributions for the yellow gene in small (upper, $500 < N < 1,000$) and large populations (lower, $3,000 < N < 10,000$) of snails (*Cepaea nemoralis*). The hatched areas represent gene fixation for additional traits. (Reprinted, by permission, from M. LaMotte, *Cold Spring Harbor Symp. Quant. Biol.* 24 (1959): 65-86. © 1959.)

Systematic Pressures

It must be fully realized that, in nature, population size and the environment are more or less constantly varying; selection pressures also vary and immigration rates may differ at different times. Wright (1969) has pointed out that any mathematical formulas expressing this process in full would be so complicated as to be of little value. In treating the effects of individual systematic pressures, therefore, the functions for each individual pressure are illustrated chiefly to aid understanding of the genetic processes of evolution.

The gene frequency distributions are given under different specific conditions in the following. As Li (1955) pointed out, these relative frequencies may be interpreted from different viewpoints; for example, a change of gene frequency for a locus in a population over many generations; the distribution of gene frequencies for all loci in a population at a given generation, assuming all loci are subject to the same systematic pressures; and the gene frequency distribution for a given locus in a number of populations of the same size and subject to the same systematic pressures. These interpretations attempt to associate theory with observation, and will be discussed in the following section.

Mutation. In the case of recurrent forward and reverse mutation without selection, the genes are treated essentially as selectively neutral. In natural populations there are other operative factors, called systematic pressures, such as migration and selection, influencing the gene frequencies in one direction or another. For the present treatment of the effect of mutation rate on gene frquency, we shall disregard migration and selection.

The effect of mutation rate on gene frequency and fixation depends on its magnitude relative to population size. With respect to these two variables, the probability distribution of q, according to Wright (1931), is

$$\phi(q) = Cq^{U-1}(1-q)^{V-1} \qquad \textbf{7.2}$$

$$C = \frac{(U+V)!}{U!V!}$$

where $U = 4N\mu$ and $V = 4N\nu$, and μ and ν are the rate of forward and backward mutation, respectively. Assuming $\mu = \nu$, it can be seen that the quantity $4N\mu$ determines the general shape of the curve.

$$\phi(q) = C[q(1-q)]^{4N\mu-1} \qquad \textbf{7.3}$$

If we take a mutation rate of 10^{-4}, and the effective population size (number of breeding individuals in the population), N, of 2,500, $4N\mu = 4 \times 2,500 \times 10^{-4} = 1$. Therefore, when $N = 2,500$, $4N\mu = 1$, equation 7.2 is reduced to $\phi(q) = C$.

From the above example and equations, we can see that the magnitude of population size relative to that of the mutation rate is the determining factor for the general shape of the gene frequency distribution curve. In order to appreciate the gene frequency distribution in the problems of evolution, we now construct a few distribution curves under specified conditions (fig. 7.3), and examine the meaning of each. Under $\mu = v$, when U or V is equal to unity, the gene frequency distribution is rectangular, that is, each frequency is likely to be found in a population. When U or V is smaller than unity, the gene frequency distribution changes from rectangular to concave. Therefore, in populations we expect to find more gene frequencies close to either end of the distribution curve than to the midpoint of 0.5 and, thus, we expect to find populations with the gene either fixed or lost. (Note the general shape of the curve is very similar to the one described for the random genetic drift.) When the U or V value is greater than unity, the shape of the curve changes again from rectangular to convex and, with the further increase of the U or V values, to a bell shape. In this condition, we expect to find more populations with gene frequencies close to the midpoint of the distribution than to either end, and fewer populations with the gene either fixed or lost. Now we can see the results of counteraction between genetic drift and mutation pressure. Genetic drift continually pushes the population toward both ends of the gene distribution curve, whereas the mutation pressure acts in a reverse direction, to prevent fixation from occurring. The balance between them depends on population size and mutation rate.

In natural populations, such as those of insects, we may find in some species that the population size may be larger than 2,500. However, in higher species, such as mammals (except humans), the effective number of a breeding unit in most species hardly ever exceeds 2,500. It is often much smaller, Thus for a mutation rate of 10^{-5} and an effective population size of 2,500 or less, there is continuous genetic fixation. This conclusion must be kept in mind for our later discussion of phyletic evolution and extinction. Experimental results on the yellow gene in *C. nemoralis* confirm this.

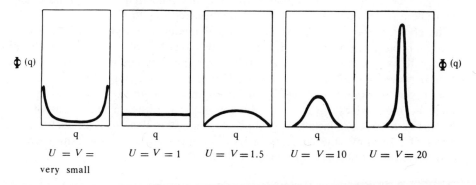

Fig. 7.3. Theoretical distribution of the gene frequency, q, under mutation pressure, assuming the same mutation rate for forward and backward mutations. See text for the specific condition for each graph. (Reprinted, by permission, from C. C. Li, 1955. © 1955 by The University of Chicago.)

Immigration. In a species or in a very large population subdivided into small groups, the groups will exchange individuals at a constant rate, m, the percentage of immigrants. Let q represent the gene frequency for gene A in one group, and $\bar{q}$, the average gene frequency of A for all groups. The rate of change of q in each subgroup per generation as affected by immigration is $\Delta q = -m (q - \bar{q})$. Immigration tends to pull the subgroup gene frequencies toward the general mean of the entire population, with an effect similar to that of mutation, that is, $m\bar{q}$ corresponds to μ and $m(1 - \bar{q})$ corresponds to ν. Thus, by substituting mq for μ and $m(1 - \bar{q})$ for ν in equation 7.2 the distribution of gene frequencies under migration pressure (Li 1955) is

$$\phi(q) = Cq^{4Nm\bar{q}-1} (1 - q)^{4Nm(1-\bar{q})-1}, \qquad \textbf{7.4}$$

where C is similar as that in equation 7.2, except here $U = 4Nm\bar{q}$, and $V = 4Nm$ $(1 - \bar{q})$; N, the number of individuals in the population; and m, the percentage of immigrants. When $q = 0.5$, the distribution is reduced to the following form:

$$\phi(q) = C[q(1 - q)]^{2Nm-1}. \qquad \textbf{7.5}$$

The distribution curves for the gene frequency at various values of $2Nm$ are the same as those in figure 7.3, assuming $q = 0.5$. The distribution changes from a bell-shaped to a U-shaped curve following the change of $2Nm$ from greater to less than 1. When $2Nm$ is less than 1, the curve is U-shaped and there is a high tendency for a gene to be fixed or lost within a subgroup, as discussed above. When $2Nm$ is larger than 1, the gene frequency for the subgroups tends to be close to the mean gene frequency of the whole population. This means that the lower the rate of migration, the higher the chance for gene fixation. When $2Nm = 1$, an exchange of one individual in each two generations and isolation between subgroups is considered to be rather strong. The effect of migration is exactly checked by that of random drift, thus the gene frequency has no change. When there is no migration, the above equation reduces to $C[q(1 - q)]^{-1}$.

The above theory of isolation is the island model of Wright (1946). When a population on a continent is subdivided into a number of subgroups, each subgroup will have a directly adjacent subgroup. Immigration would be more frequent between the adjacent subgroups than those farther apart—the "stepping-stone model" (Kimura and Weiss 1964). In this case, m in the above equation should be replaced by $m(1 - r)$, where r is the correlation coefficient of gene frequencies between adjacent subgroups. As r may be close to 1 in such circumstances (for example, the gene frequencies for them are about the same), even a large migration would lead to considerable fixation.

Selection. Gene frequency distributions under selection may be considered under different conditions of dominance (table 6.7). The general distribution equation for selection alone (Li 1955) is

$$\phi(q) = \frac{C\overline{W}^{2N}}{q(1-q)}, \qquad\qquad 7.6$$

where $\overline{W}$ is the average fitness, and C (see Li 1955 for its expression), q, and N have their usual meanings. The average fitness $\overline{W}$ for each type of dominance is given in table 6.7. In the case of partial dominance, $W_{AA} = 1$, $W_{Aa} = 1 - s$, and $W_{aa} = 1 - 2s$, where s is the selection coefficient; the average fitness of the population would be $\overline{W} = 1 - 2sq$. The gene frequency distribution for A can be expressed (Li 1955) as

$$\phi(q) = Ce^{-4Nsq}q^{-1}(1-q)^{-1}.^1 \qquad\qquad 7.7$$

As a numerical example, assuming a population with 1,000 effective breeding individuals, and a selection coefficient $2s = 0.0025$, the distribution curve for the gene frequency of A would be J-shaped (fig. 7.4). The greater the value of s, the higher the rate of fixation of the gene A. For smaller values of N or s, or both,

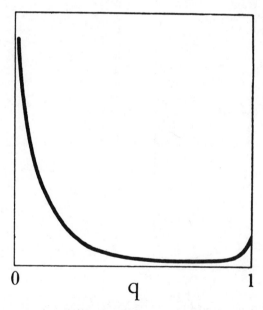

0   q   1

Fig. 7.4. Gene frequency distribution under selection and partial dominance. See text for population size and selection pressure. (Reprinted, by permission, from C. C. Li, 1955. © 1955 by The University of Chicago.)

1. The conversion of $\overline{W}^{2N}$ in equation 7.6 to e^{-4Nsq} in equation 7.7 is based on $\log_e(1 - \epsilon) = -\epsilon$, and $(1 - \epsilon)^{2N} = e^{-2N}$, when ϵ is much smaller than unity. Therefore, $\overline{W}^{2N} = (1 - 2sq) = e^{-4Nsq}$.

random genetic drift would have more effect, and the gene distribution curve would tend to approach the general forms of the curves given in figure 7.3.

For the selection in favor of heterozygotes (overdominance), the mathematical expression can be found in Li (1955), Wright (1969), or Crow and Kimura (1970). We merely illustrate here (fig. 7.5) the gene frequency changes with the genotypic fitness values of $W_{AA} = 1 - s$, $W_{Aa} = 1$, $W_{aa} = 1 - s$. Notice that there is practically no fixation when the selection coefficient is 0.01 or greater. When the fitness of the homozygotes is different, the distribution is asymmetrical and there is a greater chance for fixation of the comparatively favorable gene.

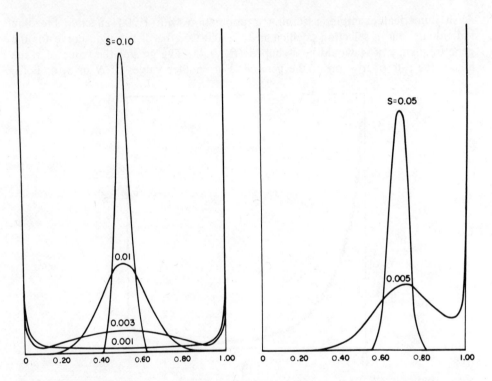

Fig. 7.5. Gene frequency distributions under selection for heterotic genes in populations of 1,000. *Left,* equal selection against both homozygotes ($s = 0.001$, 0.003, 0.01, or 0.1). *Right,* selection against one homozygote twice as much as against the other ($s = 0.005$ or 0.05). (Reprinted, by permission, from S. Wright, 1969. © 1969 by The University of Chicago.)

The theory of natural selection points to the fact that when selection pressure overwhelms that of random genetic drift, the changes in gene frequency are at its disposal: to increase the gene frequency to complete fixation when selection is in favor of a homozygote, or to hold it at a balanced state (polymorphism) when selection is in favor of the heterozygote. But when the selection pressure is not strong enough to counterbalance the random genetic drift, the latter takes over. This condition seems to bear on the issue of neutral mutation, as previously mentioned.

In natural populations, the three pressures—mutation, migration, and selection—are presumably operating jointly. We should treat these three pressures together. All the preceding discussions deal with only one locus with constant systematic pressures, assuming that the effects of genes at other loci are independent. This is rarely the case, since genes have pleiotropic and interactive effects, and fitness involves numerous genes. Even in the case of two loci, the systematic pressures may differ widely between them. Thus a population may be considered of large size for one gene, but small for the other. The mathematical treatment under these circumstances becomes very complex and is out of place in this book. Those who are interested in the theory may read the excellent texts of Li (1955), Wright (1969), and Crow and Kimura (1970).

We should now examine two examples. One is the result of an experiment and the other is from the fossil record. In natural populations, the finding that tooth size increases in the horse can be considered an example (fig. 7.6). Of course, during the long period of horse evolution, there were also basic increases in body size; but the proportion of tooth size increase was greater. The change from a browsing to a grazing life because of environmental change apparently was one of the major factors in this.

Suppose a population has two alleles, A_1 and A_2, at a single locus. If selection favors A_1, the frequency of A_1 increases and that of A_2 decreases. Immigration (inward migration) of individuals carrying A_1 would obviously accelerate the replacement of A_2 by A_1. However, if the immigrants carried A_2, the replacement of A_2 would be retarded, so that in this case immigration would operate against natural selection. The point of equilibrium for the gene frequency depends on the rate of immigration and selection pressures.

The joint effect of selection and migration is shown in an experiment involving sternopleural bristle numbers in *D. melanogaster* (Streamer and Pimental 1961) (fig. 7.7). Artificial selection for increasing the bristle number was made for eight generations, with animals having the highest number of bristles being used as parents in each generation. In two of the selection lines, immigrants from a random line were introduced. The effects of immigration on selection are shown in figure 7.7.

In summary, genetic drift may be looked upon as a type of centrifugal force; it tends to drive the population away from the center (heterozygosity) toward the periphery (homozygosity). Gene mutation and immigration represent "centripetal" forces that restrain the population from moving away from the center, thus restoring heterozygosity. Selection acts in the same way as genetic drift, for example, centrifugally when a homozygote is more favored than the other types by selection; when the heterozygote is more favored, however, the action is centripetal, resembling that of mutation and immigration.

Adaptive "Landscape"

The concept of an adaptive "landscape" describes the average fitness of populations for specific traits, according to the altitude of the terrain at a given point. Developed by Wright (1932b, 1970) as a model to explain the variations among populations within a species, its application to evolution has not been fully appreciated because of the difficulty of making illustrations for large numbers of genes. Let us begin the discussion with some simplified models.

Two-dimensional Landscape. This, the simplest landscape of all, we may take to represent changes in gene frequency at a diallelic locus at which the two alleles affect the relative fitness of a population. Let us take two instances, one involving a recessive lethal gene, and the other a case of polymorphism with better fitness of the heterozygotes than of any of the homozygotes.

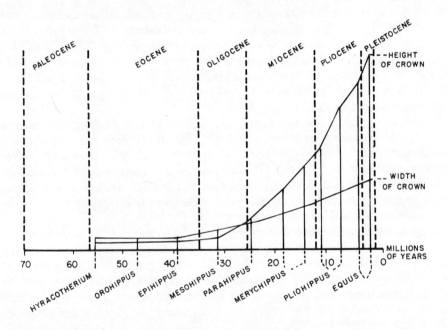

Fig. 7.6. The relative increase of the horse tooth size, measured by the crown width and height from Hyracotherieum through all the phylogenetic categories to Equus. It is assumed that the genetic basis for the horse tooth size is polygenic. The continuous increase represents the continuous fixation of polygenes through natural selection. Of course, other random and systematic pressures must also operate simultaneously. Apparently, selection in favor of large tooth size has been effective throughout the whole evolutionary history of the horse. (Reprinted, by permission, from Olson, 1965.)

In the case of a deleterious recessive, let p represent the frequency of A, and q the frequency of a, and $p + q = 1$. The average fitness of the population equals $1 - sq^2$ (see table 6.7). Assuming the other pressures to be negligible and the selection coefficient $s = 0.5$, the change of gene frequencies with the corresponding change in fitness is illustrated in figure 7.8. As the frequency of the recessive gene decreases, the fitness of the population increases. The curve in figure 7.8 represents a landscape without depth. Natural selection tends to push the popula-

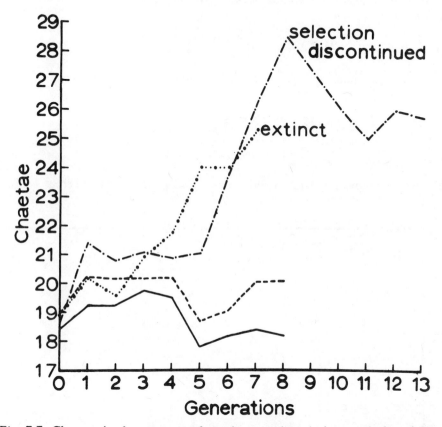

Fig. 7.7. Changes in the mean number of sternopleural chaetae during eight generations of selection in experiment 1. Selection was practiced each generation by retaining the 10 percent, four males and four females, with the highest number of chaetae as parents of the next generation. *Line A1*, (———), the unselected control line; *line B1*, (. . . . . .), the isolated line; *line C1*, (. - . - . -), two immigrants added each generation; *line D1*, (- - - - -), eight immigrants added each generation. Line C1 was continued five additional generations without selection. (Reprinted, by permission, from F. A. Streamer and D. Pimental, *Am. Naturalist* 95 (1961): 201-10. © 1961 by The University of Chicago.)

tion toward the peak of the hill and eventually reaches the maximum point, with a fitness of 1.0, and $q = 0$.

For a locus with heterozygote superiority, we may assume two alleles, A and a. The Aa individuals have higher fitness than the AA or aa homozygotes. We have shown a population with such a type of polymorphism (chap. 6). The average fitness of the population is $\overline{W} = 1 - sp^2 - tq^2$, where s and p are the selection coefficient and gene frequency of A, respectively, and t and q of a. For finding the maximum fitness of the population, we first write the derivative of the above equation, since $q = 1 - p$

$$\overline{W} = 1 - sp^2 - t + 2tp - tp^2$$
$$d\overline{W}/dp = -2sp + 2t - 2tp$$

and for the maximum value of $\overline{W}$, we set $d\overline{W}/dp = 0$, thus obtaining

$$p = \frac{t}{s + t}$$

$$q = \frac{s}{s + t}$$

These two equations are identical with equation 6.19, except that different symbols are used for the selection coefficients.

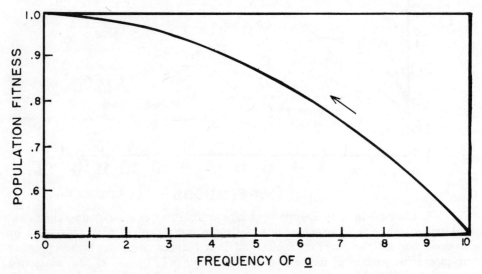

Fig. 7.8. Relation between the frequency of a recessive gene, a (with a selection coefficient, $s = 0.5$) and the fitness of the population, an illustration of a two-dimensional landscape. Arrow indicates gene frequency change toward maximum fitness of the population.

In this case it happens that the maximum fitness of the population is reached at the point of gene frequency equilibrium. If we let $s = 0.1$, and $t = 0.3$, the maximum fitness of the population occurs when $p = 0.825$ and $q = 0.175$. Natural selection will push the population toward the point of equilibrium that represents the maximum fitness of the population, in this case $\overline{W} = 1 - sp^2 - tp^2 = 1 - 0.1(0.825)^2 - 0.3(0.175)^2 = 0.923$. Figure 7.9 illustrates this polymorphic case.

Three-dimensional Landscape. We will now consider a three-dimensional landscape. (We will assume that the other genes have no influence and that there is no other genetic variation, an assumption that is not realistic because there are many loci in the segregation phase, but that will serve our purpose here.) Taking as an example chromosome polymorphism in the grasshopper (*Moraba scurra*), we find two chromosomes, CD and EF, each of which has two distinct cytological forms: a metacentric (V-shaped) and a telocentric (the centromere is at the end, rod-shaped) form. Although it may be assumed that the metacentric form is due to a pericentric inversion, let us consider the metacentric chromosome as the standard form (ST). The telocentric form is designated by Blundell (BL) for the CD chromosome and Tidbinbilla (TD) for the EF chromosome.

The observed frequency distribution of the combinations of each type of CD with each type of EF chromosome for 584 grasshoppers is given in table 7.1,

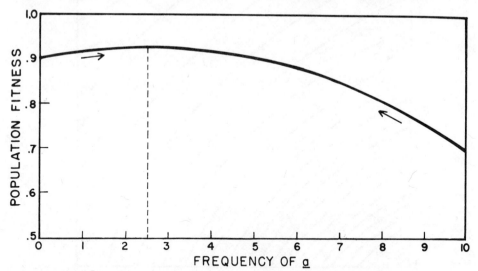

Fig. 7.9. Relation between the frequency of a heterotic allele ($AA < Aa > aa$) and the average fitness of a population. See text for assumed selection coefficients and point of equilibrium. Arrows indicate gene frequency change toward maximum fitness of the population.

there being a total of nine genotypes. The relative fitness of each genotype was estimated (table 7.2), and the difference between the observed and expected variations in joint frequency was used to estimate the relative fitness of the different combinations of chromosome types. Examining the fitness values, it should be noticed that there are interactions between the different types of chromosomes, so that given the fitness values for each of the nine chromosome types it is possible to work out an adaptive landscape, assuming a specific value for each chromosome frequency, then calculating the frequency for each of the nine different combinations. Each frequency is multiplied in fitness, and the sum of all nine products represents the fitness of a population having these specific gene frequencies. Lewontin and White (1960) calculated the fitness for each chromosome frequency from 0 to 100% by 5% increments. An idealized adaptive landscape based on their results is given in contour form (fig. 7.10). A three-dimensional form (fig. 7.11) was provided by Wallace (1968) to help in appreciating the contour map.

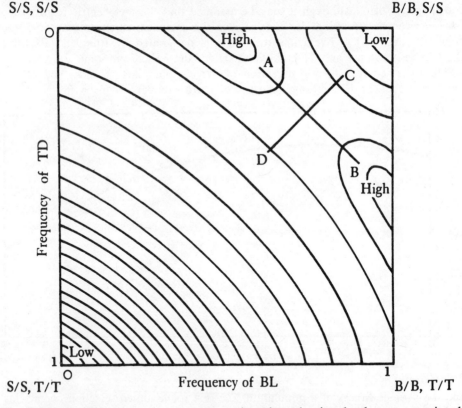

Fig. 7.10. Typical contour map representing the adaptive landscape associated with different frequencies of BL and TD chromosomes in population of *M. scurra*. (Reprinted, by permission, from Lewontin and White, 1960.)

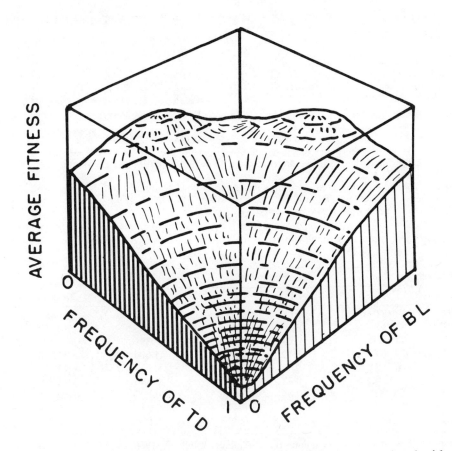

Fig. 7.11. Cutaway model of the three-dimensional landscape associated with different frequencies of BL and TD chromosomes in populations of *M. scurra.* (Reprinted, by permission, from Wallace, 1968. © 1968 by W. W. Norton and Company, Inc.)

The idealized adaptive landscape described by Lewontin and White is saddle-shaped, with the seat of the saddle higher than either side, but lower than the front or back horn. This particular topography covers all the fitness values for all possible frequencies of BL and TD. A population of grasshoppers sampled in Australia had a frequency of 87.5% for BL and 14.0% for TD, the fitness of this population falling directly on the seat of the saddle. Other populations also were found whose fitness centered in a very close neighborhood.

Why did population fitness not occupy the highest spots, that is, the front or back horn? Whether it is on the way to climbing toward the high points is not clear. The problem was reexamined by Allard and Wehrhahn (1964), who assumed some inbreeding instead of random mating. By assigning different inbreed-

TABLE 7.1 Estimated Relative Fitness of Nine Genotypes in the Grasshopper, *Moraba scurra*

| | Chromosome EF | Chromosome CD | | |
		ST/ST	ST/BL	BL/BL
Wombat	ST/ST	1.002	1.000	0.927
Royalla	ST/ST	0.842	1.000	0.808
Wombat	ST/TD	0.646	0.849	1.044
Royalla	ST/TD	0.636	0.997	0.974
Wombat	TD/TD	0.000	1.054	0.626
Royalla	TD/TD	0.393	0.682	0.916

SOURCE: Reprinted, by permission, from R. C. Lewontin and M. J. D. White, *Evolution* 14(1960):116–29.

NOTE: CD and EF represent two nonhomologous chromosome pairs; ST represents the standard gene arrangement; BL (Blundell) and TD (Tidbinbilla) are names given to telocentric forms of these chromosomes.

TABLE 7.2 Calculation of a Single Point on the Adaptive Landscape Formed by the Relative Fitness (see table 7.1)

| Genotype | | Frequency | | | |
CD	EF	CD	EF	Fitness	Product*
ST/ST	ST/ST	0.64	0.09	0.842	0.0485
ST/ST	ST/TD	0.64	0.42	0.636	0.1710
ST/ST	TD/TD	0.64	0.49	0.393	0.1232
ST/BL	ST/ST	0.32	0.09	1.000	0.0288
ST/BL	ST/TD	0.32	0.42	0.997	0.1340
ST/BL	TD/TD	0.32	0.49	0.682	0.1069
BL/BL	ST/ST	0.04	0.09	0.808	0.0029
BL/BL	ST/TD	0.04	0.42	0.974	0.0164
BL/BL	TD/TD	0.04	0.49	0.916	0.0180

SOURCE: Reprinted, by permission, from Wallace, 1968. © 1968 by W. W. Norton and Company, Inc.

NOTE: The point chosen is that for which the frequency of BL is 0.20 and that of TD is 0.70.

*Sum (average fitness) = 0.6497.

ing coefficients from 0.05 to 0.25, the landscape changes from a saddle shape to that of a single hill, with the highest elevation in the region of 80% to 90% for BL frequency and 10% to 20% for TD. Space is lacking for a detailed landscape reconstruction, but with these latter landscapes the populations were found to be actually on the hilltops.

Hyperdimensional Landscape. Neither the two-dimensional nor the three-dimensional landscape model is realistic, for in natural populations a large number of loci are segregating along with the added possibility of chromosomal polymorphisms. But for a landscape of more than three dimensions, the difficulties of constructing a map are multiplied, requiring a drawing of lines of various density such as that long used by Wright (1932b) (fig. 7.12). It may be imagined as a sketch of an aerial map. The closely spaced lines represent elevated hilltops and those further apart, low regions or valleys, merely a diagrammatic form of an extremely complex situation. For example, in human populations 16% of the loci are in the segregation phase (Hopkinson and Harris 1971), and in *D. pseudoobscura* 40% (Prakash, Lewontin, and Hubby 1969). If there are 10,000 loci for each species, the frequencies of genes for 1,600 loci in man and 4,000 loci in the fruit fly must be represented so that an adaptive landscape for each cannot be made in a realistic manner. But this does not mean that they do not exist.

Fig. 7.12. Adaptive landscape with several peaks (+) and valleys (−).

The above discussion assumes that all the populations of a species occupy the same environment. In reality this is not so, since in one geographical region there will be specific biotic and physical factors more or less different from those in others. Thus the same genotype may be more adaptive in one region than in another and the hills and valleys with respect to the gene frequencies could vary from one population to another. This adds another complication, but it does not necessarily destroy the landscape concept. Ford, however (1964, pp. 33–34) found little merit in the adaptive landscape concept, stating that peaks and valleys are not permanent, that peaks may lower while valleys may elevate. He viewed the adaptive topography as resembling a seascape more than a landscape since it varies in time and space. Apparently what he meant is that when the environment changes the adaptive topography will change, but this is exactly what we have been saying and it does not contradict the landscape model since all shapes change with time. The purpose of the adaptive landscape concept is to show that, among populations within a species, peaks and valleys represent the population fitness under gene equilibrium; that they change with respect to time and space but do hypothetically exist; and that populations move constantly toward peaks even though they may fail to reach them.

INTEGRATED GENE POOL

Our knowledge of the biological mechanisms by which genes affect fitness is very limited. We know, however, that genes produce additive and interactive effects. Consistent responses to directional selection in experiments indicate the presence of additive effects; heterozygote superiority is an expression of allelic interaction and epistasis is an expression of nonallelic interaction. The presence of such gene interactions has led to the theory of the integrated gene pool of Dobzhansky (1950a).

Suppose that in a population consisting entirely of A_1A_1 individuals, with a fitness value of W_1, A_2 arises by mutation. Although the A_2A_2 individuals have lower fitness than the A_1A_1 individuals, selection has caused an increase of the frequency of A_2, and the average fitness of the population due to the increased presence of A_1A_2 heterozygotes has now become greater than W_1. In other words, the fitness of a population reaches its maximum when a gene frequency equilibrium is reached. Thus selection maximizes the fitness of the above-described population even though handicapped A_2A_2 individuals became part of it. In the words of Dobzhansky (1950a), "the fitness of some individuals may be sacrificed for the fitness of the population as a whole," and he demonstrated (1955a) that such a condition applies to heterozygotes for different chromosomal rearrangements, as well as in epistasis (for example, interaction between genes at different loci) (Wallace and Vetukhiv 1955). Thus a Mendelian population represents a level of organic integration having specific laws of operation: the fitness of a population depends not so much on the fitness of its individual genes as on combinations that provide an integrated gene pool in the reproducing community. (The concept of integrated gene pool is associated with the theory of segregational load, as discussed in the above chapter, and of interactive load, as described by Mayr [1963].)

Wright (1970) recognized gene interactions in natural populations. Indeed, the multiple peaks in his adaptive landscape demonstrate the case. Note that the fitness for the different genotypes of the grasshopper (table 7.1) illustrates epistasis. It is the presence of epistasis that accounts for the multiple peaks of the landscape.

As pointed out at the beginning of this chapter, in actual populations many genes with high orders of interactions are involved. Mathematical expression for these would be exceedingly complicated and of little value (Wright 1969). Mayr (1959) criticized that this imposed a limitation on the mathematical models, and Haldane (1964) discussed the general application of mathematical theories. In essence, he said that different models and hypotheses serve for deducing evolutionary consequences from observed facts, and that the study of gene properties belongs to physiological genetics. Wright (1970), in a different way, pointed out that "evolution is, in short, a process for which it is unlikely that there will ever be long-term predictions." Mathematical theories of evolution thus attempt to explain past evolutionary processes. Evolution is guided by random and nonrandom forces. Even the nonrandom component, such as selection, is in a sense unpredictable. The direction of selection, in fact, will shift as the environment changes; a gene or genotype favored at present may be selected against in the future. The fate of a population is unpredictable.

EVOLVING LIVING POPULATIONS

With systematic pressures constantly operating, living populations are never static —they are ever evolving. When an interbreeding (Mendelian) population is subdivided geographically, the environment of different groups is different. Natural selection acts independently in each, pushing it toward its own adaptive peak. In this process the formation of group, subspecies, and then of species is found. The following large and local populations were examined with respect to some specific traits and characteristics of their distribution.

Large Populations

The world distribution of the ABO blood group system is better known than any other polymorphism in the human population. There are three major alleles at this locus: I^A, I^B, and I^O; six genotypes, and four phenotypes:

Phenotypes	Genotypes
A	$I^A I^A$, $I^A I^O$
B	$I^B I^B$, $I^B I^O$
AB	$I^A I^B$
O	$I^O I^O$

The presence of the antigens determined by these alleles on the red blood cells is by mixing cells and antisera. Group A contains subtypes A_1 and A_2. For the present discussion, however, we will disregard them.

The frequency distribution of I^B in the world is roughly displayed by the contour lines on the map (fig. 7.13). In dealing with human populations of such size and geographical distribution, there is a possibility that one population occupying the same geographic area consists of various interbreeding groups, such as castes, religion isolates, and so on. (This is one reason this map has been criticized; in view of difficulties in sampling human populations in such scale, however, it is useful for our discussion.) The gradual increase of gene I^B eastward and westward

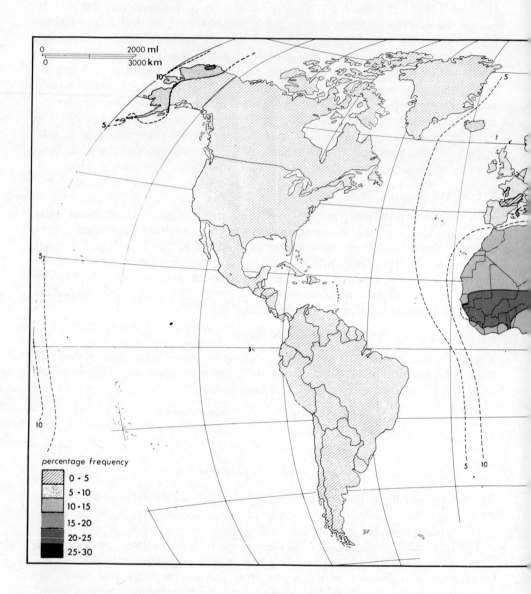

Fig. 7.13. Frequency of the gene *1*^B for blood group B in the world. (Reprinted, by permission, from Mourant, Kopec, and Domaniewska-Sobczak, 1976.)

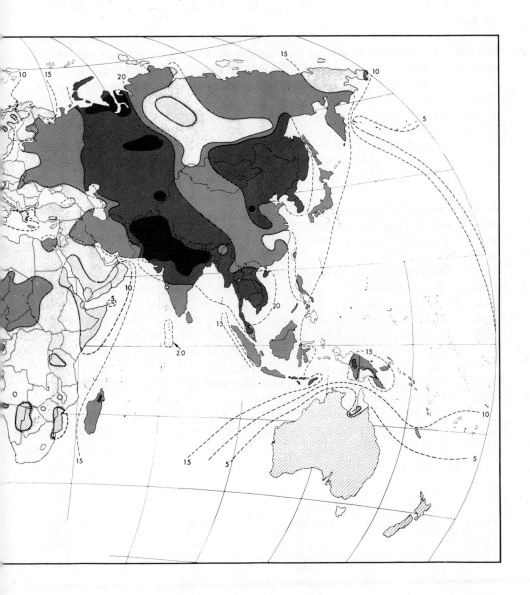

from the highest region of Central Asia and India is a geographical gradient. In North India, the frequency reaches as high as 30%, dropping somewhat in Southwest Asia and the East Indies. The gene is rare in Australians and in American Indians.

However, the distribution of I^O does not display either the same geographical relationships or the same genetic relationships as I^B. For instance, American Indians are considered to have been associated with the Mongoloids, since it is known that the ancestors of American Indians moved to the New World from Asia by way of the Bering Strait. Yet the low frequency of I^B and the high frequency of I^O in American Indians are in sharp contrast to those of Asiatic Mongolian populations.

The sharp difference in the frequency of I^B from those of the Mongoloids reflects possibly the strong effect of genetic drift. As previously pointed out, the smaller the population, the greater the effect of genetic drift. Of course, we have no knowledge of the size of the group of ancestors who presumably crossed the Bering Strait. In view of the geographical situation and the total world population at that early time (see chap. 8), one can reasonably assume that the number of migrants was not large. It is also possible that most of the migrants were carriers of gene alleles other than I^B.

We cannot completely disregard the effect of natural selection, however, as evidence is accumulating that blood group genes are associated with resistance to certain diseases. There is a correlation of the frequencies of certain ABO blood types with those of gastric and duodenal ulcer (table 7.3). The relationship of high frequency of type O and low frequency of type A in ulcer patients was derived from samples of Caucasian populations only, and these statistics are consistent between populations within and outside England. This relationship apparently holds in different geographical and biotic environments, although we cannot yet deduce whether this is due to the pleiotropic effects of the blood group gene, or of genes very closely linked with the blood group gene. Either way, the data do suggest that ABO blood group genes are correlated with fitness, at least in respect to some diseases.

TABLE 7.3 Frequencies of Blood Groups O and A in Duodenal Ulcer Patients and in Controls

Center	Size of Sample		Group O Frequency (%)		Group A Frequency (%)	
	Ulcer	Controls	Ulcer	Controls	Ulcer	Controls
London	946	10,000	56.6	45.8	32.9	42.2
Liverpool	1,059	15,377	59.6	48.9	29.0	39.1
Glasgow	1,642	5,898	57.7	53.9	31.5	32.3
Copenhagen	680	14,304	50.3	40.6	38.4	44.0
Vienna	1,160	10,000	41.0	36.3	41.3	44.2
Iowa	1,301	6,313	53.7	45.8	36.3	41.6

SOURCE: Reprinted from Roberts 1961 by permission of the Medical Department, The British Council.

There are wide speculations regarding relationships between blood groups and infectious diseases. In view of the fact that blood group substances are red blood cell surface antigens, it is possible that they may effect the process of adherence of bacteria to the wall of the red blood cells. Livingstone (1960) speculated that blood group A is antigenically similar to several *Diplococcus pneumoniae* organisms, hence persons with anti-A may have greater resistance to the bacteria than those with blood group A. The latter would not likely produce antibodies against their own system.

For comparison with the ABO blood groups as a single gene trait, let us examine some traits of polygenic basis in the same populations. A familiar example is the three basic dermal finger ridge patterns: arch, loop, and whorl. The percentages of arches and whorls for the world populations are given in figure 7.14. The population characteristics are quite clear: the Negro populations have the smallest percentage of whorls, the Caucasians are next, and Mongoloid populations have the greatest percentage. Such correlations are usually present for polygenic traits, because many genes are involved, and genetic drift affects one single gene more than a number of genes. We do not know the relative selective significance for finger ridge patterns or how genes affect the development of a specific pattern, although there are many congenital abnormalities involving different biological systems that can be diagnosed from specific variations in the dermal ridges (Penrose 1963). Nevertheless, the racial characteristic for the trait is clearly shown.

Tribal Populations

There are eight aboriginal tribes on the island of Taiwan. All of them inhabit mountain ranges with altitudes up to 3,000 meters, except two tribes on the east side of the island with populations living mostly at sea level (fig. 7.15). The origin of these aborigines is uncertain. They were present at least 4,000 years ago (Chai 1967), but the history of their separation into the present tribal groups is not known. The average altitudes at which they live are shown in table 7.4. It is believed that initially they populated the entire island, and then moved into the mountain ranges due to the influx of Chinese migrating from the mainland of China during different dynasties. The map shows that all except one of the tribal areas are bordered by the Chinese on the outer perimeters. Population sizes of the tribes range from 3,200 to 87,000 according to the Chinese government report for 1960.

TABLE 7.4 Average Altitudes at which the Taiwan Aboriginal Tribes Live

Tribe	Atayal	Saisiat	Bunun	Tsou	Paiwan	Rukai	Puyuma	Ami
Altitude (meters)	960	830	1,320	1,030	630	630	160	160

Nineteen anthropometric measurements, including the most important head and body measurements, were made for a total of 126 to 277 individuals in each tribe (Chai 1967). The average differences were calculated according to the method of Mahalanobis' squared distances (Rao 1955). This distance may be called bio-

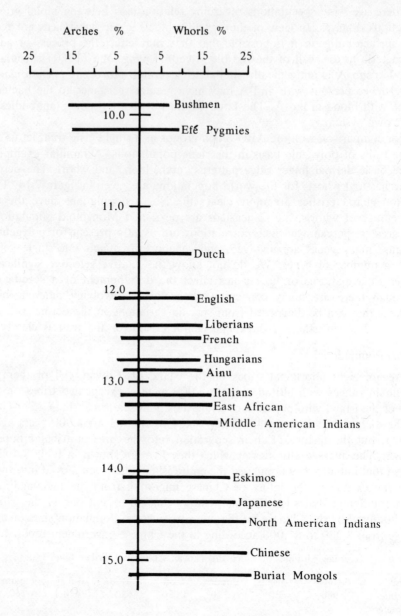

Fig. 7.14. The percentage of arches and whorls, horizontal scale, and the pattern intensity index for various human populations. (Redrawn from Harrison et al., 1964, modified from Cumins and Midlo, 1961.)

logical distance and is given in figure 7.16. The average differences are in fairly close agreement with the geographical distances between the tribes, taking into consideration some natural barriers, such as rivers, mountain peaks, and valleys.

Another set of observations for the same groups of people consists of dermatoglyphics of the fingers (the finger ridge patterns) and palms. The development of the dermatoglyphic ridges is completed in the fourth month of pregnancy and thereafter environment has no effect. They contrast, therefore, with the anthropometric measurements of the head and body, which are known to be influenced

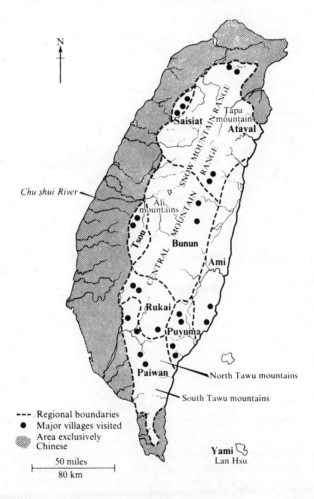

Fig. 7.15. Geographical distributions of eight Taiwan aboriginal tribes. (Reprinted, by permission, from Chai, 1972.)

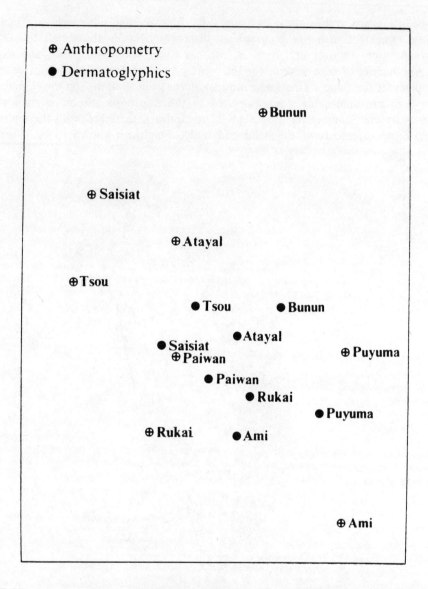

Fig. 7.16. A point diagram illustrating the biological distance between tribes as based on anthropometry and dermatoglyphics. The difficulty of projecting the distances in a hyperspace on a two-dimensional space is obvious, but the plots, being approximations, may be helpful as a visual aid. The distribution of tribes based on anthropometry seems in close approximation to their geographical localities. It may be imagined that the Bunun is on the mountaintop, and the Ami

by environment. Eleven dermatoglyphic observations on each hand were made for each individual. The detailed descriptions of the dermatoglyphic observations (Chai 1967) include the biological distances based on the dermatoglyphics in figure 7.16. The two sets of data reveal similarities and discrepancies. Each set of traits is influenced by a large number of genes. The most marked variation is the greater distance for the anthropometric measurements than that for the dermatoglyphics. It should be pointed out that the Mahalanobis analysis corrects for dependence between variables and that each variable is measured in standard deviation units. The effect of natural selection acts more on body and head measurements than on the dermal ridges of the hands. This conclusion is supported by the fact that relative chest girth was found to correlate with altitude of habitation (fig. 7.17). The larger biological distances for the anthropometric traits are therefore credited to natural selection (Chai 1972).

To summarize, natural selection may act more on some traits than on others; probably more on the ABO blood groups and on head and body forms than on dermatoglyphics in the human population. These examples demonstrate the effects of natural selection as a major force in creating genetic differences among groups, the basis of speciation.

ADAPTIVE RADIATION

In the following pages we will attempt to explain several concepts—speciation, phyletic evolution, and extinction—as the consequences of adaptive radiation.

Adaptive radiation refers to a process in which phyletic lines repeatedly split due to adaptation to different environments. When a population splits into separate groups or subpopulations, each population occupies some part of the new sphere and becomes a distinct type, narrowly and specifically adapted to its particular sector. Adaptive radiation for land vertebrates is illustrated by Romer (fig. 7.18).

Adaptive radiation is a consequence of random genetic drift and a response to three systematic pressures: mutation, migration, and selection. It may also be visualized as a way in which these factors impose themselves on adaptive landscapes, the peaks and valleys of which differ between populations with respect to gene frequencies. Evolution operates in individual gene pools in the domain of such landscapes. (Thus progressive evolution is one of the consequences of adaptive radiation, speciation is another.)

---

and Puyuma close to the sea level on the coast, and the remaining tribes at various intermediate topographical regions. The plots for the distances based on dermatoglyphics are even more difficult due to relative small differences. The relative differences in amount of spread in the space between the two sets of points can be seen. Note that in each set, the Atayal, and to some extent the Paiwan, seem to be close to the center of the distribution. (Reprinted, by permission, from Chai, 1971.)

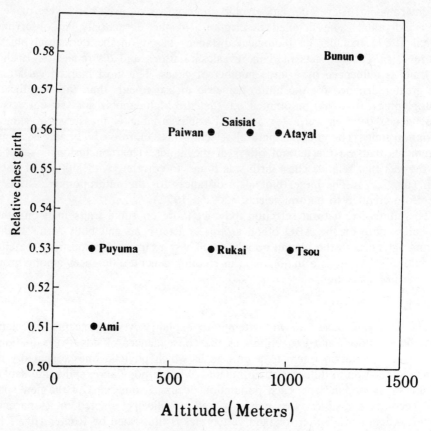

Fig. 7.17. The relationship of relative chest girth and altitude for eight Taiwan aboriginal tribes. (Reprinted, by permission, from Chai, 1972.)

Speciation

Lamarck, who spent almost his entire life classifying species of plants and animals, was perhaps the originator of the hypothesis that species evolve from races. He believed that new races and species result from inherited consequences of the use or disuse of organs. Darwin, on the other hand, believed that evolution was brought about by natural selection. There is no sharp division between their theories: both men agreed that races are incipient species. According to Mayr (1963), species are "groups of actually or potentially interbreeding populations, which are reproductively isolated from other groups."

The essential difference between local populations and species is, therefore, that the genetic systems of local populations are open, and those of species closed; an exchange of genes occurs between local populations but not between species. Consequently, when speciation occurs, independent evolution proceeds. From an evolutionary standpoint, Simpson (1961) stated that "an evolutionary species is a lineage (an ancestral-descendant sequence of populations) evolving separately from others and with its own unitary evolutionary role and tendencies."

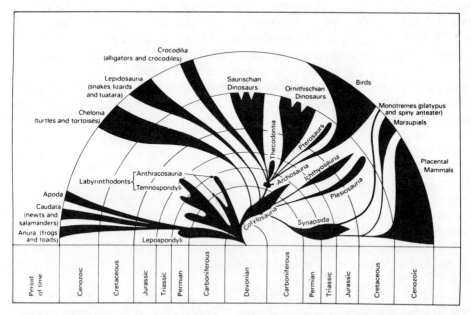

Fig. 7.18. The pattern of evolutionary relationship of the land vertebrates, showing adaptive radiation and extinction of various groups of vertebrates. (Reprinted, by permission, from Romer, 1968. © 1968 by Alfred S. Romer.)

A living species is generally divided into populations, with size as well as the extent of isolation varying between them. In an animal population, for example, some members tend to penetrate into surrounding areas as strays, travelers, or invaders, later to become permanent settlers, for better or worse. If the environment is favorable, they may establish a colony called a cline, or tribe, occupying an isolated territory that is often referred to as a niche; and they are then termed marginal populations with respect to the main population of the species. The early settlers of a marginal population of sexual organisms may be as small as one or a few pairs of males and females; they may become the founders of large populations, subspecies, or species.

The genetic structures of marginal and central populations differ: marginal founders do not carry all the genes that were present in the central populations, or at least not in the same frequency. In addition, their genes are not necessarily a random sample of coadapted gene pools. The smaller the number of founders, the greater the chance for a biased sample, thus breaking the cohesive system of genes, which must be resynthesized in order to cope with a new environment.

The process of resynthesis operates differently from synthesis in a central population. Natural selection tends to favor homozygotes rather than heterozygotes at the majority of loci, except those whose heterozygosity is essential for fitness in general (in a specific niche). This is because the environmental variation within the niche is of a rather narrow range in comparison with that of central populations inhabiting larger areas. Moreover the tendency toward selection in favor of homozygotes is reinforced by inbreeding, due to small population size (eq. 6.5).

Thus the fixation of a large number of genes and chromosomal structures results in a relatively high degree of homozygosity for the total genotype of the organisms in marginal populations as compared to that in the central. Biologically, organisms in a marginal population show, in general, a rigid and perhaps high level of adaptation to their specific environment. This genetic shift that leads to fixation of individual genes J. B. S. Haldane called the principal unit process in evolution.

The reduction of heterozygosity eliminates the continuous production of undesirable homozygotes, which would be a heavy drain on small populations. Consequently, a marginal population will be relatively free of deleterious genes, such as were concealed under dominance and/or balanced polymorphisms in the central populations; thus they are relatively free of both mutational and segregational loads. The tendency toward a decline of polymorphism in small (marginal) populations, and its extensive presence in large ones, supports this hypothesis (Carson 1959).

In addition to selection for adaptation to the particular environment, we must keep in mind the effect of random genetic drift, the effect of which is greater, the smaller the population. We should realize therefore that many genes fixed in isolated small populations may not be necessarily adaptive. However, the effect on gene fixation in general does accelerate speciation.

The same process of fixation or favoring of homozygosity applies to various forms of chromosomal structural variations, such as inversion, translocation, and fusion, which are often deleterious. Probably most of them are eliminated by either chance or natural selection, perhaps even more frequently than with mutant genes. However, we must also expect a small percentage of chromosome variants to confer additional fitness on the organisms. These chromosomes will recombine freely and may eventually reach fixation (Dobzhansky 1955b) as mutant genes favored by selection. In certain cases, such as that involving duplication of segments and translocation, there is a large increase in the amount of genetic material, signifying a big step in evolution. This may not occur as often as do single-gene mutations, but is of great importance. The model discussed for single genes can be similarly applied to this case (Wright 1970).

Evolution guided by selection of favorable genes has obviously led to an advancement from less differentiated to highly differentiated forms, having greater effect in large rather than in small populations; this results from the fact that in a large or a central population, a large percentage of heterozygote advantage is constantly present (Carson 1959) due to intragenic interaction (hybrid vigor) and intergenic interaction (epistasis). This, in turn, is partly due to the great number of possible mutations, and also to the environment's larger and more variable areas. When a new gene arrives, of course, in most cases, it may be lost. If not, it may fit into one part or another of the environment of a large population.

When populations of varying size within a species are isolated, each will assume its own independent course of evolution and, taking advantage of whatever genetic heterogeneity may be available and whatever new mutations may arrive, it will move toward its own adaptive peak. Because of basic and environmental (geographical and spatial) differences between gene pools, the genetic distance between them eventually widens until interbreeding becomes impossible and independent species are formed. But how do they achieve isolation in the beginning?

We have discussed the formation of marginal or small populations by splitting from a central or large population. In general, there are three major causes of isolation: spatial, environmental, and reproductive.

1. *Spatial isolation.* Two or more populations of a species may be capable of interbreeding, although they live so far apart that gene exchange between them is reduced by distance.

Since organisms inhabit both land and water, geological changes such as the sinking or rising of a land mass can split a population. The formation of the Bering Strait created a barrier for the land animals inhabiting this area; the rise of the isthmus of Central America separated the sea animals of that region. It has been inferred that South America was originally joined with Australia through the Antartic, in a single land mass. Thus marsupials may have reached Australia from South America via Antartica and have moved northward, becoming differentiated, specialized, and isolated from all placental mammals that evolved elsewhere (Jardine and McKenzie 1972). There is evidence that alligators and magnolia trees once inhabited a range reaching from South China to the southeastern United States, a range cut by Pleistocene glaciation (Mettler and Gregg 1969). The historical development of different fauna in Latin America resulted from geological evolution and climatic changes, as discussed by Simpson (1950). Many similar examples can be cited, examples important for their effect on species formation.

2. *Environmental differences.* Two or more populations may be adapted to different environments in the same or different territory, and can interbreed, although their hybrids may be limited by the availability of habitats.

In natural populations, the spatial and environmental components of isolation usually go together. Environment varies with latitude, longitude, and altitude; different geographical areas differ not only spatially but also ecologically. For example, Texas frogs are to a large extent infertile when mated with Vermont frogs, whereas closely adjacent frog populations are interfertile. Many examples of spatial and environmental isolation have been given by Grant (1963).

3. *Reproductive isolation.* Organisms of a species may be unable to interbreed for various reasons. With respect to their zygotes, there may be seasonal differ-

ences in mating, incompatible mating behavior, or physiological differences in the reproductive systems. There are cases where normal mating and fertilization occur, but the fertilized eggs fail to develop because of chromosomal differences. Cases of hybrid inviability or sterility due to chromosomal or genic incompatibility in either the F_1 or F_2 hybrids have also been reported. In reproductive isolation, we need to make a clear distinction between cause and effect from the point of view of speciation. Many of the above-stated mechanisms (Grant 1963; Mettler and Gregg 1969) are the *result*, not the cause, of speciation. For example, a mule produced by the mating of a horse and an ass is generally infertile, and the mating itself represents a species cross.

As a mechanism causing speciation, reproductive isolation between individuals within a population can result from gene or chromosomal mutations. Gene mutations may lead to physiological or behavioral differences so that assortative matings may then take place between individuals having the mutation as one group and between those without as another. Chromosomal mutation may cause meiotic difficulties in heterozygotes with the consequence of hybrid sterility. Either case can lead to a split of the population. It was reported, for instance, that one mutation in *Panaxia dominula* influenced its mating preference (Sheppard 1952) and that the ebony mutant in *D. melanogaster* mates more successfully in the dark than in light (Rendel 1951). In the case of yellow, mutant males of *D. melanogaster* meet with reduced success in fertilizing normal females (Bastock 1956), its effects being deleterious to mating success. But it could be incorporated into a population if the mutant concerned were useful in any other way. Then selection of other divergences associated with differences in sexual behavior might eventually bring about the sexual isolation of the population.

Lewis (1953) has described the geographical locations of the seven-, eight-, and nine-chromosome species of the genus *Clarkia* as found in Northern California above San Francisco, in San Francisco, and further south, respectively. Species with larger chromosome numbers tend to inhabit regions of increasing dryness. Lewis points out that an extra chromosome would hardly improve the fitness of an individual in a long-established population, but he suggests that an extra chromosome might help to promote marginal adaptive changes. The possibility exists that such chromosomal differences occur before the formation of these different species. Chromosomal mutations (rearrangements) can be a primary factor for speciation. The case of chromosomal fusion in feral mice discussed in chapter 4 is another example.

Sympatric[2] species are more likely to be produced by geographical isolation (Grant 1963; Mayr 1942). Large genetic differences gradually become established

2. Related species occupying a single territory without interfering with each other are referred to as sympatric, such as zebras, wildebeests, and giraffes that feed together on an African plain. The counterpart of sympatric species is allopatric species, which refers to species occupying different territories. If they were inhabiting the same territory, they would compete with each other for food and one might eliminate, drive out, or absorb the other.

between them so that when they again inhabit the same locality they do not inter-breed. It has also been suggested that if reproductive isolation arose through gene or chromosomal mutation in a population as discussed above, there is a possibility that a sympatric species could be formed without geographical isolation. This represents a minority view (White 1973) that many investigators dispute.

Evolutionary Divergence and Phyletic Evolution

In *On the Origin of Species by Means of Natural Selection*, Darwin pointed out that "organic beings have been found to resemble each other in descending degrees, so that they can be classed in groups under groups." The fact that all members of one order share some common feature, as do similar members of one family, one genus, or one species, is well established. According to Darwin, the result of evolutionary divergence is a fact of nature and not an artifact of classification. The reason is that more common or similar genes are present between groups of the same phylogenetic branch than exist between them and groups of another branch.

The above-discussed speciation can, in fact, be considered as the beginning of divergent evolution. Grant (1963) stated that "the evolutionary motion is linear when measured against time for a single population, but is relative when measured against the trends in other related populations." Darwin termed the two evolutionary modes as "descent with modification" and "the origin of species." Modern evolutionists refer to them as phyletic evolution and evolutionary divergence (Grant 1963). This disagreement is merely a matter of difference in reference points (fig. 7.19).

Organic evolution improves the adaptability of organisms through functional and structural modification. Such adaptive changes may be regarded as constituting evolutionary progress. In the course of evolution, however, because of chance or environmental differences, the structural and functional changes made by one population (or one phyletic line) are apt to differ from those made by another. Such differences can be considered as evolutionary divergence. It is believed, for instance, that mammals and birds evolved from reptiles. Some differences in structure and function between each group and the reptiles may be expressed in terms of evolutionary progress, while some between birds and mammals furnish examples of evolutionary divergence. Phyletic evolution concerns evolutionary progress and divergence resulting from the joint effects of random genetic drift, mutation, and selection.

Certain geneticists have theorized that evolutionary progress along phyletic lines results from changes in gene frequencies or in the combinations of genes. This is an oversimplification derived at a time when clear understanding of gene mutation was lacking. Allelic forms of a gene (such as those occurring in polymorphisms) can coexist in the same population simultaneously; they are different from the homologous genes present in different species, in most cases in the number of nucleotide base differences of a gene. The differences between allelic genes may be due to one mutational event, whereas what distinguishes homologous genes from each other may involve an accumulation of many such differences. Fixation of an allele in a population constitutes only a single step in phyletic evolution.

It should now be clear that the change of a gene in one species to a copy identical to the homologous gene in another species is virtually impossible. For instance, for mutation by nucleotide substitution it is not likely, in the sequence of the two

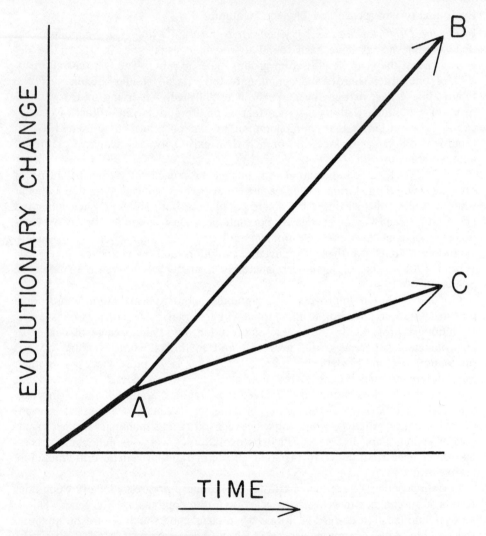

Fig. 7.19. Phyletic evolution and evolutionary divergence. A is the ancestor population of populations B and C. The differences between A and B, and A and C, are referred to as phylectic evolution; that between B and C, evolutionary divergence.

homologous genes, to have the substitution at the same position with the same nucleotide. Thus, once species are formed, nonidentical mutations continue to accumulate with evolution, and the species are carried further apart. This is the reason why phyletic evolution follows the pattern of radiation. There is, of course, a selection factor involved. But mutations of the nonidentical type, which supply the raw material for evolution, are the cause. This fact will not be fully appreci- ated until we understand the chemical structure of a gene.

When a new advantageous gene occurs by mutation in a population, it depends on selection for its establishment in the population. After establishment, the new gene may mutate again to a more advantageous form, and selection again helps to establish the newer type. For the same gene (locus), mutation and selection oper- ate alternatively and continually, as long as it is essential in the course of phyletic evolution. The fact of this continuous process of gene evolution has been revealed from the amino acid sequence analysis of proteins. If we accept that the rabbit, as a mammal, is more advanced (or more highly differentiated) than the bullfrog, as an amphibian, some of the eleven amino acid differences in cytochrome c (table 2.4) between rabbit and bullfrog can be credited to evolutionary progress. As another example, the growth hormone (GH) contains a large number of amino acids with species specificity and is probably coded by a single gene. It has been discovered that primates do not respond to cattle GH, and that rats do not respond to fish GH, but that fish do respond to mammalian GH. Again, if we accept the fact that primates are more advanced than cattle, and rats more advanced than fish, the differences in GH between members of two groups constitute evolutionary progress. Note that GH from higher species is effective in lower species, but not vice versa. This indicates that, while preserving certain basic properties, GH evolves by adding some species features evidently not present in its earlier carriers.

The evolution of genes includes all basic mutational processes: nucleotide base substitution, gene duplication, and deletion. The evolution of cytochrome c in- volves primarily nucleotide base substitution; the evolution of GH probably in- volves both nucleotide base substitution and duplication. The latter supposition is based on the fact that GH was found to be effective after enzymatic digestion had removed 25% of its total number of amino acid sequence.

Perhaps a prime example of the effect of gene duplication on phyletic evolution is the hemoglobin gene discussed in chapter 2. Invertebrates may have a gene similar to the primitive globin gene, but they probably do not have any of the hemoglobin genes. Among the vertebrates, the less advanced species lack some of the genes present in the advanced species. Here the presence or absence of a gene refers to its complete nucleotide base sequence.

We discussed the evolution of genes in chapter 2, giving different examples, and there is no need for further emphasis here. However, we do need to become aware of the presence of gene inactivation. Its importance in ontogeny appears to imply its involvement in phyletic evolution. We have cited some examples of sex chromosome inactivation and given several speculative cases involving the inactivation of autosomal genes. We can imagine that during the course of phyletic evolution there must be genes that, although completely obsolete, are nevertheless still carried by the organisms.

Evolutionary divergence resulting from natural selection is obvious and needs no elaboration. However, the random elements of evolution, such as genetic drift and mutation, must also play important roles, causing adaptive changes through modification of one or many biological systems. Assuming that two loci homozygous for A and B, respectively, are present in populations 1 and 2, we may conjecture the following evolutionary process:

1. Mutation a arises in population 1 and is established by selection; thus the individuals of population 1 now have the genotype $aa\ BB$. Mutation b occurs in population 2, contributing approximately the same adaptive effects but through different biological alterations. Thus the individuals of population 2 now have the genotype $AA\ bb$.

2. Populations 1 and 2 now have identical adaptive abilities, although they differ in certain biological aspects determined by their respective mutated genes.

3. If mutation b (instead of mutation a) had occurred in population 1, that population would then have consisted of individuals having an $AA\ bb$ genotype; in this case, population 1 would still be able to meet environmental demands, that is, the $AA\ bb$ genotype would become established through natural selection.

Thus the evolutionary divergence between two such populations can be seen to have arisen fortuitously.

Random genetic drift plays an even more important role in genetic divergence. Its effect is compounded with those of mutation and selection and cannot easily be distinguished from them. We may, however, visualize its importance from work done in laboratory populations. Beginning from a common population from which a few replicate lines are derived, one selects for a particular quantitative genetic trait. These lines are kept in environments as nearly identical as possible and subjected to the same selection pressure (that is, the same number of breeders are chosen from approximately the same number of progeny in each generation). After a number of generations, one may find that these lines differ in the character under selection as well as in some other correlated or uncorrelated traits. Such differences, arising mainly from sampling errors in choosing the breeders from each generation, correspond to the results of random genetic drift in natural populations. Divergence between inbred lines that originated from a single population and were kept in like environments will later offer an illustration of this fact in a rabbit population.

Extinction

Extinction is a common evolutionary phenomenon. As Romer stated, "the major question is not so much why animals become extinct as why a certain favored few survive; for extinction is the common lot, survival the exception." In the course of evolution, relatively few species have left descendants. The present continental reptiles, birds, and mammals have actually been derived from a limited number of the Mesozoic assemblage of species. As a crude estimate, 99% of the tetrapod genera surviving in the early middle Meoszoic period left no descendants.

For a concrete example of widespread extinction, the evolution of reptiles in the Mesozoic era provides perhaps one of the most dramatic. The Mesozoic periods are the Triassic, Jurassic, and Cretaceous, covering a total of some 125 million years. During this time, changes occurred in the configuration of continents and ocean waters, in temperatures and in climates. The beginning of the Triassic period marked the beginning of the successful radiation of vertebrates onto land. A welter of animals came into being, some leading to the more recent kinds of reptiles. The great radiation of land reptiles flourished at first, but was later decimated through widespread extinction (fig. 7.18); some lines declined well before the end of the Cretaceous period, while others disappeared completely.

Two general trends in the evolution of reptiles appear to have led to their extinction: an increase in size and the development of a high degree of specialization. Each kind became adapted to a restricted way of life, and each progressive step in morphological development led to greater specialization than before. As a consequence, diverging lines of stocks with distinct biological characteristics were generated. Besides their remarkable size, unusual anatomical characteristics developed, such as the large backplates of the stegosaurs, the head frills and horns of the ceratopsians, the 6-inch-thick skull roofs of armored dinosaurs, and the hollow, ornate crests of the duckbills. Not long after the appearance of these odd features, certain stocks became extinct (Romer 1949). Similar evolutionary phenomena, varying in extent, occurred in other categories of organisms, including plants, invertebrates, and certain other vertebrates.

The extinction of a population or species in different time periods may be viewed as a consequence of genetic and environmental (physical and biotic) interactions. During the Mesozoic era, the seas were withdrawing after a period of flooding the continents; climates were also changing, and it has been claimed that there was a rise in temperature as well as great bursts of radioactivity. Physical changes cause biotic changes. Organisms having high genetic specialization that developed in a particular environment faced changing conditions with which they could not cope. The upset of a delicate balance in ecological communities was one major factor. Failure in competition for common food supply and the disappearance of plants or animals on which a group customarily fed was another. Regarding the reptiles, it is claimed that mammals were increasing in numbers and kinds and that they devoured many reptilian eggs. We are not going to discuss various environmental factors here; instead we will seek certain genetic interpretations for the phenomenon.

The genetic system or the gene pool of a population is not static. Random genetic drift and selection tend to force each locus to fixation, while mutation and immigration operate in an opposite direction. The balance between these two pressures determines the state of the genetic system or the percentage of homozygosity (see figs. 7.3, 7.4, and 7.5). It is assumed that the pressure of drift and selection outweighs that of immigration and mutation based on the fact of speciation. Thus Mendelian populations, in general, are continually moving toward gene fixation. This is especially true when a species is formed; hence immigration is shut off and mutation is the only pressure countering fixation. The usefulness of heterozygosity for preserving the genetic potential maintaining a homeostatic genetic system has been discussed in chapter 6 and needs no further elaboration. When a genetic system reached a state with a high percentage of homozygosity, the biological systems became highly specialized and adaptable only to a narrow environment. Furthermore, the genetic system is so rigid that there is no sufficient reserve (heterozygosity) for making genetic changes. Consequently, the population or species declines and faces extinction when the environment changes.

The situation is comparable in certain ways to what occurs in laboratory populations under directional selection. It has frequently been observed in artificial selection that, after a number of generations, the fitness of a population decreases, although the character under selection has no direct relationship with fitness. The general interpretation for such effects is either that the genes under selection have pleiotropic effects or that undesirable linked genes have been selected. In most cases, definite evidence supporting such interpretations remains lacking. However, it is certain that a general increase in homozygosity occurs in the selected populations because of the limited number of individuals used as breeders in each generation (see eq. 6.5). In this respect, the genetic systems of the artificially selected populations bear some similarity to those of natural populations and develop an overspecialization similar to that of the vanished reptiles. The population has drifted into an evolutionary blind alley.

From years of experience in rabbit inbreeding and observation of family extinction in inbred colonies, I believe that such an inbreeding system is perhaps analogous to evolutionary radiation in natural populations. The family differentiation, expansion, and extinction of these inbred colonies are delineated in figure 7.20 from breeding and mortality records. It can be seen that the families descended from a randomly bred stock as radiation of natural populations from a common origin. Genetic differences between the rabbit families were found in the loci that could be tested. Many families became extinct before the tenth generation of inbreeding; at the time of this writing, families A_2 and C_2 are declining and will probably become extinct in the near future. The direct causes of extinction are sterility and early mortality, which lead eventually to complete termination. Geneticists called such degenerative changes inbred depression. They can hardly be traced to any single-gene effects; the increase of homozygosity is generally recognized as the basic cause. Thus the phenomenon and genetic basis of extinction in inbred rabbit colonies are very similar to those in natural populations.

It can be argued that in natural populations the amount of inbreeding may never reach the degree attained in intensely inbred laboratory populations. It is possible that the maximum extent of inbreeding found in natural populations may not be incompatible with survival in a rather static environment for a long period of time. But once the environment has altered, especially if drastically, genetic changes by mutation or recombination of genes that are not fixed may not be rapid enough to meet this challenge. This is probably a basic factor responsible for the wholesale extinction of species in evolutionary history, a reason for which Wright (1970) predicated that the fate of each species is extinction.

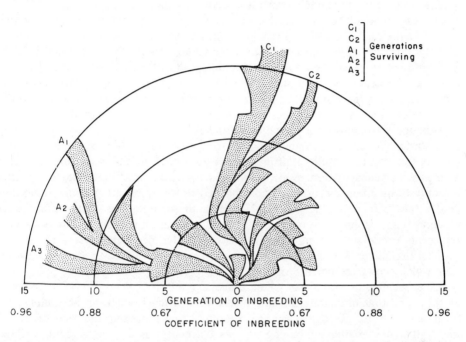

Fig. 7.20. The branching patterns and extinct families in an inbred rabbit stock of the Jackson Laboratory. It is similar to the pattern of evolution of vertebrate radiation shown in figure 7.18.

Summary

The history of evolution is essentially a history of speciation, adaptation, and extinction, the consequences of mutation and natural selection. Organic evolution has its characteristics and definite patterns, among them the phenomenon of irreversibility. Dollo's law of irreversibility, formulated in 1896 or earlier (Simpson 1953; Rensch 1960), states essentially that no species ever returns to any one of its previous states.

Dobzhansky (1970) discussed the validity of this proposition from the standpoint of the magnitude of genetic change. Mutation *is* reversible, for example, in rodents: an albino gene can, in fact, mutate back to full color in a single step.

But the difference between a species at a given moment and its condition at some previous ancestral level involves many genes, each of which may differ in many mutational respects. For instance, when two loci are involved, the chance of both mutating back to a previous form presumably implies two mutation rates. Assuming the mutation rate is the same, the probability of reversibility is v^n for a mutation rate of v and n genes. For a mutation rate of 10^{-5}, the probabality of reverse is 10^{-10} for two genes, and 10^{-15} for three genes. The chance for three genes to mutate back to their original form thus becomes almost impossible.

The above calculation is based on a condition of one mutational step for one gene. But the farther apart a species is from its ancestors, the greater the number of mutations that have occurred in the same gene. This has been revealed by protein sequence analyses. The number of amino acid differences between homologus proteins represents the minimum number of mutational steps occurring in the same gene. Thus the probability of evolutionary reversibility of any species to the genotype of even its most immediate ancestor is virtually impossible.

Although we regard organic evolution as being an orderly progression, life may have started by random combinations of the nucleotides. It was in the development of function that some orderliness was established through selection. Mutation as a process tends to destroy order so that selection must continue to operate, not only to maintain the originally established order by rejecting unfavorable mutations but to increase the order by keeping those mutations that confer better function. Thus evolution from the protozoa to the highest forms, such as mammals, represents essentially a continuous increase in orderliness. On this basis, we can say that organic evolution is a decrease of entropy, in contrast to inorganic evolution, which shows an increase of entropy.

It has been assumed that the whole universe might be running toward maximum entropy, and that ultimately the whole cosmos will level out to an absolutely uniform temperature. Recently some astronomers have evoked the "steady state" theory that new matter containing new usable energy is constantly being created in voids of space between stars and galaxies, so that the dreadful prospect usually believed to be in store for the universe may not be realized. Whatever the outcome will be, it remains distant. But the physical and biotic environments are constantly changing, and the environment in which the ancestors of any species once lived will no longer exist at some point in the future. Thus, even if genotypes identical with those of their early ancestors should appear, the recurring genotypes could not survive. In examining the two conditions necessary for repetition of a previous genetic stage, we find that the recurrence of the ancestral genotypes is most improbable, and that the ancestral environment would no longer exist. The probability of evolutionary reversal, then, becomes practically zero. Natural law seems to hold that organic evolution goes from higher to lower entropy, and inorganic evolution from lower to higher, with neither process reversible.

"Evolution is a creative process" in somewhat the same sense as "carving a statue or painting a picture," in the words of Dobzhansky (1970). An art work is unique. The evolution of a species is also unique; in fact, each phyletic line in the history of evolution is unique. Evolution from the very beginning of life functions through trial and error from the moment a gene is started. Life evolving from very primitive to more sophisticated forms has constantly involved the concerted action of mutation and selection. As in the writing of a text, mutation is reversing, and selection is editing. A gene corresponds to a sentence, and all the genes of an organism together represent the text. But there is a certain difference between the task of revising a text and that of bringing about a change in genes. The alteration of one nucleotide base by one mutation in one generation is analogous to a change of one, and only one, word in one sentence *at a time*. And just as the changed word has to fit both the sentence and the whole text, so a single mutation, in order to survive, must fit its entire genetic and exernal environment. Biological revision is comparable to having to follow an editorial rule that each sentence must be increased in descriptability and sophistication throughout the entire text. This indeed describes what actually happens when the change in an organ brings about evolutionary progress and speciation.

The evolution of life is largely the history of adaptive radiation. In the evolutionary course of a plant or animal population, or of a group of many species, a time arrives when certain groups or species of organisms rise to a peak in variety and abundance and then decline. Some of the remnants die out, others may again acquire new genes and new environments that lead to a second radiation. Then another tide of propagation and differentiation arises. But the organisms produced differ from their ancestors in some genes and in certain functions and structures. Old species are continually becoming extinct while new ones are generated, carrying a number of the ancestral genes. Genes evolve, so do organisms.

Eight

the character of human evolution

If we consider the human species in terms of its genetic apparatus, we find that it differs little from any other eukaryotic organisms. In effect, modern man and the other organisms still share identical and homologous genes. Human evolution has, however, taken a profoundly different course ever since the development of culture. If we define evolution as successful adaptation to the environment, we can see that man's environment now includes not only the physical and biotic spheres common to all animal species, but a cultural sphere as well. Since its inception, culture has served to guide the course of life; as it has progressed, it has assumed a proportionately greater influence, becoming in modern societies perhaps the leading force of natural selection.

Human evolution is a broad subject and we are unfortunately unable to give it a detailed treatment here. Because of the uniqueness of the phenomenon of cultural intervention in man's evolution—one not present in any other species—we will discuss this evolution in terms of a few essential biological characteristics, such as bipedalism, expansion of brain size, development of self-consciousness, and the interplay between biological and cultural evolution.

A Brief History of HOMO SAPIENS

Man belongs to the kingdom Animalia, class Mammalia, order Primates, and genus *homo* and species *Homo sapiens*. According to the Linnaean system of nomenclature, with slight modifications, we have the following full classification (Simpson 1972):

Kingdom Animalia
Phylum Chordata
Class Mammalia
Subclass Theria
Infraclass Eutheria
Order Primates
Suborder Anthropoidea
Superfamily Hominoidea
Family Hominidae
Genus *Homo*
Species *Homo sapiens*

The similarities between man and other mammals are obvious when man is compared with other advanced members of the primate order, for example, gorillas and chimpanzees. This was clearly pointed out by Darwin and Huxley, who based their comparison on gross anatomical features. Now this can be reinforced by evidence from studies in karyology (chromosome number and morphology), amino acid sequence analysis for certain proteins, and serological tests. The genetic relationships between different primates, as determined by immunologic tests of serum albumin samples, are shown in figure 8.1.

The actual stages through which man's earlier ancestors evolved cannot be traced with accuracy. Nevertheless, a few intriguing fossils that belong to advanced anthropoid apes close to the human line of descent are known from the Miocene of Africa, Europe, and Asia. They are usually grouped in a subfamily Dryopithecinae of the ape family Pongidae. In the late Miocene and early Pliocene (approximately ten million years ago), we find in southern Asia and East Africa *Ramapithecus*, which may well belong to the distinctly hominid lineage leading to *Homo sapiens* after *Ramapithecus* branched out from the ancestors of the living apes (Simpson 1972). The oldest known fossils that, by consensus, belong to the family Hominidae are the australopithecines, who lived over two million years ago for a period probably more than three million years. There were at least two groups in this family: one now referred to as *Australopithecus* (or *Paranthropus*) *robustus*, heavier and less manlike; the other, *A. africanus*, more gracile, and possibly the direct ancestor of *Homo sapiens*.

Another group, known as *Homo habilis*, can be described as an intermediate of an earlier *A. africanus*-like form, and a later one similar to *Homo erectus*. Opinions differ as to whether *Homo habilis* should be referred to as *A. africanus*, a phyletic intermediate between *A. africanus* and *Homo erectus*, or whether it should be considered as being of human ancestral lineage, and thus distinct from, and contemporaneous with, *Australopithecus*. In any event, it is generally agreed that *Homo* evolved from some ancestors very much like *A. africanus*, and the older specimens of *Homo habilis* furnish additional evidence for this thesis (Simpson 1972).

The next stage in human ancestry, as established through definite fossil identification, is represented by the famous creature, *Pithecanthropus*, now technically termed *Homo erectus*. They flourished approximately 500,000 years ago in Asia and Africa and probably also in Europe. Their presence in this time period reflects an incompletely known but long process of transition out of *Australopithecus*, and a shorter transition to *Homo sapiens* thereafter. How old is *Homo sapiens*? Most investigators would choose to mark the period of origin somewhere between 20,000 to 500,000 years ago. Taking thirty years as an estimated average generation turnover time for modern man, we arrive at a range of approximately 700 to 17,000 generations of *Homo sapiens* survived. There is, of course, a tendency toward lengthening generation turnover time following the evolutionary advance; the above-estimated range represents a lower limit.

The Evolution of Bipedalism

Man's biological distinction from other mammalian species results from two major characteristics: bipedalism and a larger brain size. With regard to these, Sir Julian Huxley stated that "man stands alone." Actually, as far as bipedalism is concerned,

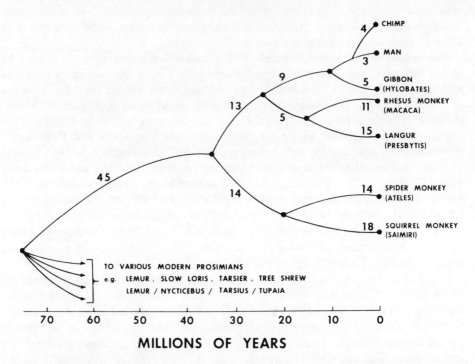

Fig. 8.1. Genetic relationships between primates based on immunological distances. The number adjacent to each lineage segment represents the amount of albumin evolution in immunological distance units. It is assumed that immunological distance is proportional to the number of amino acid replacement for groups since the time of separation from a common ancestor. The construction of the cladogram is calculated by a method (see Sarich 1971; Farris 1972) yielding the minimum total distance for all species included in the analysis. The resulting divergence dates, based on immunological distances, differ from those generally agreed on by other investigators. It should be kept in mind that immunological distances are based on a single protein, and that there are variations between species with respect to selection pressure for a specific mutation and random drift, which is related to population size. For the genetic analysis of closely related species, however, the amino acid sequence analysis discussed in chapter 2 is a useful tool in evolutionary studies. More reliable information is expected when a larger number of proteins have been studied. (Reprinted from Lovejoy and Meindl 1972 by permission of the American Association of Physical Anthropologists.)

birds walking on their hind limbs and kangaroos and other mammals frequently hopping about on their hind legs exhibit other instances. Man is not the only bipedal animal, but he alone stands erect without conscious effort. The evolution of upright posture in humans has resulted in many changes in their skeletal and muscular structure and function. For example, the shape of the human foot evolved from that of a walking and grasping organ to one entirely designed for walking and body support; individual bone sizes and articulation have altered considerably from those of other primates, such as monkeys and apes, which human ancestors undoubtedly resembled. Man's leg comprises about half of his total stature (longer than in apes), providing for a lower center of gravity. The neck and head of the femur, which articulates the pelvic bone, extends obliquely upward to facilitate the transmission of weight into the legs from the pelvic arch. The human pelvis has been shortened and broadened for the transmission of the body weight and for facilitation of childbirth with relatively large head size.

The vertebral column of man has undergone less noticeable changes, but important modifications have been made to accommodate erect posture. From the neck to the tailbones the vertebrae form a sigmoid arch for stabilizing the attachment of internal viscera and for maintaining balance in standing and walking. The accompanied modifications lie in the shoulders being pulled back and the rib cage flattened. Compared to corresponding features in monkeys, both man's lumbar and tail vertebrae are reduced in number.

In newborn human infants and adult apes, the hind legs measure little more than one-third of the total body length. Human legs grow in length and girth during childhood much more rapidly than the trunk. There are also other modifications in the front limbs and neck more fully described in texts on comparative anatomy and human biology. Such evolutionary changes in *Homo sapiens* constitute essentially modifications in (1) size of individual bones, (2) number (for homologous bones such as the vertebrae and ribs), and (3) angle of articulation. These changes can also be described as an alteration of the relative growth rate of one organ with respect to another, termed "allometric growth."

Variations in the size of individual skeletons can be effected through a number of genes of different forms normally present in a population; these genes are referred to as "polygenes" and have been discussed in chapters 5 and 6. A substantial amount of evidence has demonstrated their presence and effects on different biological systems. For example, Lerner and Dempster (Lerner 1954) selectively bred chickens for large shank length and achieved an increase of 15% over the control population in twelve years of selection. This change involves modification of the growth rate of bones and muscles in the chest and lumbar region with respect to the other body parts.

The number of lumbar vertebrae commonly observed is five in man, three in apes, and seven in monkeys. It is generally agreed that the shortening of the lumbar region results in, or is associated with, the evolution of bipedalism or brachiation (that is, the Asiatic apes are known to be the most versatile arboreal acrobats and the best brachiators) (Schultz 1969). Chai (1969) has reported the number of lumbar vertebrae to range from six to eight in four inbred strains of rabbits. Green (1951) found variations in the number of presacral vertebrae variations among inbred strains of mice. All these variations are partly environmental and partly genetic, resulting from differences in a small number of genes. As far as fitness is concerned, these variations in laboratory species seem to have no noticeable effects. It can thus be assumed that natural selection for shift in the number of lumbar vertebrae in either direction for primate species can be achieved with ease.

The role of polygenes in quantitative variations in natural populations is not well appreciated, partly because individual effects are small. But these effects, in turn, become additive and accumulative and, for this reason, polygenes are assumed to be the genetic units responsible for microevolution. The concept of polygenes has been discussed in chapter 6. Their presence in natural populations is assumed to occur in varying frequencies in the form of genes for polymorphisms. The percentage of the total number of genetic loci segregating in a population was not known until it was discovered that an average heterozygosity of 40% for twenty-seven isozyme loci in *Drosophila* (Prakash, Lewontin, and Hubby 1969) and 16% for twenty arbitrary isozyme loci in human populations (Hopkinson and Harris 1971) were segregating. It has also been assumed that each biological system in an individual is affected by many genetic loci. Thus, at this point it becomes possible to visualize the large genetic variations available for natural selection to operate on, and the importance of polygenes in biological evolution.

Accepting the function of polygenic variation and microevolution should not be taken as excluding the possibility of genes with quantal effect being involved in bipedalism evolution. However, on the polygenic concept we can explain that the partial bipedalism of apes constitutes a genetically intermediate stage between tetrapods (for example, monkeys) and bipedalism as it occurs in man. Polygenic characters studied in laboratory animals usually result in intermediate types when a crossbreeding is made between two stocks having a large difference in the character. But genetic hypotheses on bipedalism cannot be tested because crosses between species cannot be achieved; nevertheless, the polygenic explanation for bipedalism evolution seems to be most plausible and compatible with Darwinism.

THE EVOLUTION OF THE BRAIN

Man occupies a unique position among the primates and among all species of animals in general. He relies on weapons for defense, uses tools, and develops elaborate languages and culture. He remembers his past and plans for his future. He thinks in abstractions and perceives and possesses a sense of good and evil. These abilities are fundamentally dependent on his mental faculty, which, in turn, results from the evolution of his brain size and structure; with respect to body size, he has the largest brain among all vertebrates.

The estimated cranial capacity of hominids as a measure of brain size is given in table 8.1 (from *Australopithecus* to *Homo sapiens*). According to Campbell (1972), the fossils found at Chok'outien, China, have a cranial capacity that varies between 915 cu cm and 1,225 cu cm, which is close to twice that of *Australopithecus*. Those found at Vertesszollos, Hungary, at least half a million years old, have a cranial capacity of about 1,550 cu cm, approximately three times the mean of *Australopithecus*. It is claimed that the people of Chok'outien, with greater stature and well-developed bipedalism, were more advanced anatomically than *Australopithecus*. On the basis of nerve cell counts in a variety of primates, and following certain assumptions, Jerison (1963) estimated the number of cortical nerve cells in various hominid taxa. In making his estimate, he separated the brain into two components. One is related to the body size and is larger in individuals of greater body size. The other component varies independently of this factor. It comprises the "extra" nerve cells over those required for satisfaction of the environment through a wider range of brain-behavior and adjustment mechanisms. He developed equations for the calculation of the number of these nerve cells and computed the number of extra neurons in each stage from ape to modern man (fig. 8.2). This comparison shows a remarkable stepwise progression from *Australopithecus* to *Homo sapiens*. The method of estimation involves many assumptions and does not take into consideration differences between different parts of the brain. As Tobias (1971) pointed out, however, it provides an approximate gauge of cortical development and underlines the rank order, if not the extent, of differences in functional capacity of the brain.

TABLE 8.1 Cranial Capacity of Hominids

Taxa	No. of Individuals	Mean and Range (cu cm)
*Australopithecus africanus**	6	494 (435-540)
Homo habilis†	4	664 (590-727)
	3	656 (633-684)
Homo erectus erectus†	4	890 (815-943)
	6	859 (750-975)
*Homo erectus pekinesis**	4	1,043 (915-1,225)
*Homo sapiens**	1,000	1,350 (1,000-2,000)

*Tobias 1971.
†Ralph L. Holloway 1973: personal communication.

Holloway (1969) has stressed the fact that the internal structure of the brain must have undergone evolution to account for behavioral differences in mammalian species. He points out that evidence of structural changes correlated with complex behavior of different extant primates is important. For instance, one gross morphological feature, Broca's area (lying in the third frontal gyrus of the cerebral cortex), is perhaps present solely in man (Rensch 1960). It is an area, connected by a bundle of nerve fibers with Wernicke's area (lying adjacent to the cortical region that receives auditory stimuli), for language articulation, passing auditory stimuli on to the motor area that controls the movement of the muscles for speech (Geschwind 1972). The evolution of speech functions must be associated with the development of Broca's area.

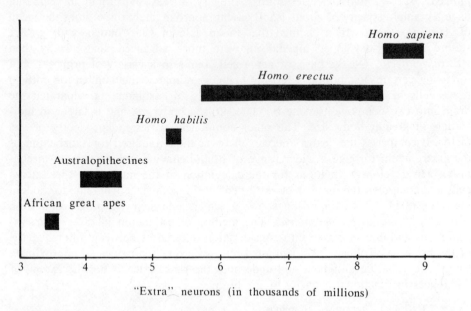

"Extra" neurons (in thousands of millions)

Fig. 8.2. Estimates of extra neurons in thousands of millions in primates. (Reprinted, by permission, from Tobias, 1971.)

Additional structural differences in human brains are the expanded frontal lobes and the deeply convoluted surfaces. The surface of the frontal lobe comprises 47% of the cerebral cortex in humans, in comparison with about 33% in large apes and 23% in lemurs. No other creatures have such an extensive convolution of the cortex as do humans; apes and porpoises have complex patterns of convolutions, but these are less complicated than those of *Homo sapiens* (Hulse 1965). The expansion of man's parietal lobe is also very impressive (Von Bonin 1963). The expansion is mainly by intercalation, and the main addition is in the center between the somesthetic, acoustic, and visual fields. The area, called the parietal association field, is said to be quite small in monkeys, bigger in anthropoids, and very large in man (Von Bonin 1963).

Before formulating genetic explanations for the brain evolution in hominids, let us first examine brain volume variations in mice (*M. musculus*). Roderick et al. (1973) measured the brain weight (closely correlated with volume) of mice in a large number of inbred strains of the subspecies *Mus mus musculus*, and mice of a stock of the subspecies *Mus m. molossinus*. The genetic relationships for some of the strains and the stock of *Mus m. molossinus*, together with the average brain weight for each, are shown in figure 8.3. It should be noticed that the brain weight is correlated with genetic relationships, and that the variation is greater between subspecies than within subspecies. The heritability (variance between strains/ [variance between strains + variance within strains]) is estimated at 0.65, which is high compared to most quantitative genetic traits. Roderick et al. concluded that a few genes are involved in the variation between strains.

One can assume that in the course of hominid evolution there were genetic variations in brain size at each stage, as in the mouse species. There was continual

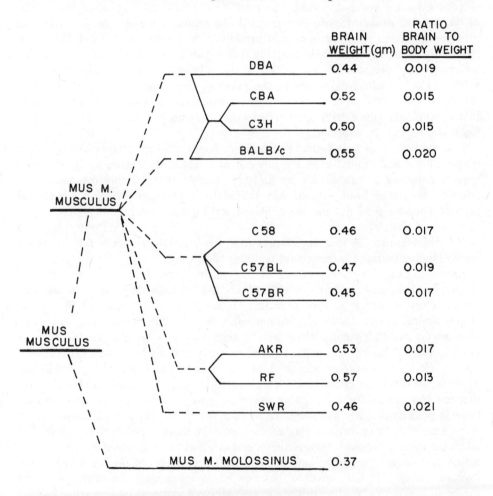

Fig. 8.3. Mean brain weights and ratios of brain to body weight for different inbred strains of *Mus m. musculus* and a stock of *Mus m. molossinus*. Note that the differences and similiarities are associated with their origin of derivation. Data from Roderick et al. 1973.

selection for large brain volume concurrent with continual mutation to supply new genes. One can expect that through a large number of generations, a brain volume such as that of the present *Homo sapiens* could be reached. Two factors are probably related to the intensity of natural selection: environmental temperature changes and bipedalism. It is known that the early ancestors of man survived cold climates during the glacial and interglacial periods about one million years ago (there seems to be no definite information on the length of these). Assuming a total of 600,000 years, this corresponds to about 20,000 generations in man. From skeletal evidence at Chok'outien, Campbell (1972) asserts that more than 50% of the population died before they reached full reproductive age. Such a selection pressure can even be found in certain present tribal populations (Chai 1967). In laboratory experiments of selection for differences in certain quantitative traits, with a selection pressure of 80% (20% of the individuals are saved for breeding), a twofold to threefold difference in the resulting lines can often be obtained in twenty or thirty generations. In comparison with 20,000 generations of selection in the hominids (even with smaller selection pressure) a substantial change in brain volume can be expected.

It is possible that pelvis size is a limiting factor on further increase in infant brain volume and cranial capacity, since man is generally known to be the only species requiring assistance for birth. It is possible that for this reason natural selection for large brain volume has shifted from prenatal growth to postnatal growth. Thus the prolongation of childhood development emerges as another characteristic of man.

If brain evolution in man is different from his bipedal evolution, the difference would involve increase in structural complexity.

Examining the life history of one of the existing great apes, and comparing its development with that of man, one finds that the infantile stages of man and ape are far more similar than their mature phases. At birth, the brain of the gorilla closely approximates that of the human infant in size. The human child and newborn gorilla are much more alike facially than they will be in adult life. This is because the sutures of the gorilla skull close early and his brain will expand very little, while the brain of the human child will grow considerably during childhood. There is, however, no concrete evidence suggesting that man has additional genes that are responsible for long-lasting postnatal growth. Evolution of genes occurs through gene duplication and subsequent mutation by nucleotide base substitutions. New genes that arise through this process possess slightly different functions but affect the same biological systems: hemoglobin genes are originated by gene duplication and some of them become active at different stages in the life process, for instance (chap. 2). It is possible that new genes influencing the development of the brain in man evolved to this state, that is, were not present in earlier human stages.

It should be pointed out that the genetics of development in higher organisms remains insufficiently understood. Recent studies on gene regulation and activation have uncovered the fact that many genes in the cells of an organism are inactive; perhaps some were functional at certain stages of development. It is also possible that some were active in the earlier ancestors of the same phyletic line. Besides determining the selection of genes for organ growth and differentiation, the evolution of a genetic system may involve selection on the activation or depression of individual genes.

It is known that the evolution of bipedalism began earlier than that of brain expansion in the hominids. But through transmission of brain impulses, bipedalism has stimulated the development of the brain. Thus, through the hands of man, the mental faculty was transformed into cultured expression. As Wallace pointed out, man has transferred to his machines and tools many of the alterations of parts that in animals take place through body alterations. In other words, the genes that influence man's mental ability achieve a final expression in his instruments and culture.

THE EVOLUTION OF CONSCIOUSNESS

Man's consciousness forms the basis for his cultural achievements and unique characteristics. It is this essential self-awareness that specifically distinguishes him from all other species. Numerous figures in the history of thought have singled out this peculiar human ability as constituting, to a large extent, a definiens for man, beginning perhaps with Socrates; the range and magnitude of the power of human consciousness have often been extolled. But rarely has the subject even been approached in terms of a scientific study of evolution. The difficulty of providing foundations for such an approach is obvious: first, consciousness as an empirical phenomenon is poorly understood and hard to perform experimental tests on; second, the evolution of functional traits is always more difficult to trace than that of gross morphological traits.

As a paleontologist and theologian, Teilhard de Chardin was perhaps one of the few to examine the question of self-consciousness systematically, and from a scientific/evolutionary viewpoint. He regards consciousness as the central phenomenon of reflection and states: "Reflection is, as the word indicates, the power acquired by a consciousness to turn in upon itself, to take possession of itself as of an object endowed with its own particular consistence and value: no longer merely to know, but to know oneself; no longer merely to know, but to know that one knows. . . . Admittedly, the animal knows, but it cannot know that it knows."

Teilhard de Chardin regarded psychical evolution as constituting a separate system in itself. Like the evolution of gross morphological traits or the nervous system, that of "psychism" (a term he uses to mean, roughly, the sum of "inborn" human behavioral patterns) had followed the "branching pattern" of different taxa for more than 500 million years by the end of the Tertiary. With the primates, so much more developed than earlier groups, a final leap forward took place on the evolutionary line: great transformation of the organs resulted in the faculty of consciousness. Teilhard de Chardin further pointed out that access to thought represents a threshold that had to be crossed in a single stride, one by which man was transported onto an entirely new biological plane. He considered the threshold of reflection a critical transformation, a mutation from zero to infinity.

From a genetic standpoint, one would probably object to characterizing the evolution of consciousness as a zero-infinity mutation. The concept of a "threshold of reflection," however, bears a certain similarity to so-called threshold characters frequently observed in experimentation on various species. The idea of inheritance of such threshold characters is usually conceived as the result of the combined effect of both genes and environment. Both are necessary; without a specifically favorable environment, the trait may not be phenotypically exhibited, even though the proper gene is present in a population.

The "threshold" hypothesis remains poor (on methodological grounds) because there is no way in which it can be proved either true or false experimentally. Nevertheless, it does serve to emphasize that environment must be considered as both genetic and cultural, no matter how primitive or simplified. We can, I think, rightly compare the evolutionary progress across the "threshold of consciousness" with that of brain size and structure: in both, there is a gradual (as opposed, partly, to Teilhard de Chardin's notion of a single momentous transformation) accumulation of new genes that all contribute to the total effect. Of course, present living populations have certainly advanced far beyond any "threshold" of reflection or consciousness.

I would like to explain in somewhat more detail the idea of a gradual progression toward self-awareness. From a developmental point of view, we know that newborn babies possess no such ability, but rather develop it gradually over a period of time. We can conceive of the functions of the central nervous system of infants (shortly after birth) as closely resembling that of apes. It is important to note that the development of consciousness occurs *after* birth, and parallels physical growth of the brain. We have supposed that the extended period of this growth in humans (as compared to that in animals) results from genetic evolution. In addition, it would appear that the progression toward self-awareness might stem from the same source. Thus, this theory I have just sketched constitutes an attempt at a genetic explanation of the evolution of human consciousness.

INTERPLAY BETWEEN CULTURAL AND BIOLOGICAL EVOLUTION

Culture is defined as the "act of developing by education, discipline, training, etc." In addition, we could assert that the cultivation or rearing of products or crops, and a particular stage of advancement of civilization, are also referred to as culture: "arbitrary standards, symbols, and ethical judgments are some of the essential characteristics of culture" (Hulse 1965).

Culture is shared and transmitted from generation to generation. The early ancestors of man began to modify their environment by new inventions, such as cooking, the construction of shelter, and clothes making; their offspring learned these techniques and in the meantime adapted to the environment. Man's ancestors have continuously succeeded in modifying the surrounding circumstances; the ability to do so has proved advantageous in all hominid groups since early times. The division of labor, tools, fire, language, and many other aspects of culture began to form a protective screen so that, while some components of the selection pressure from the physical and biotic environment are reduced, man attempts to adapt to the environment of his own creation. Natural selection for mental ability in mankind is unidirectional for all groups, at least in earlier evolutionary periods. And by far the greater part of human evolution took place in earlier stages, although not necessarily at the same rate in different populations. During the last 300 generations, technological revolution and the development of ethical and religious thought have taken place in different fashions and with different rates among societies. The protective screen of culture has thus become more complex in some populations than in others.

Ever since man developed culture, his genetic constitution has been involved in interaction with cultural forces: as a product of biological evolution, culture (specifically, for example, written and verbal language) feeds back into the genetic system to cause further biological changes. This system of interaction is continuous and irreversible. By analogy, we could say that human evolution has been rolling on cross-linked double tracks:

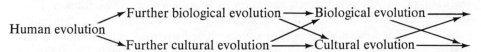

As Pilbeam and Simons (1965) pointed out, "brains expanded as the cultural environment became more and more complex, and larger brains enable more complex culture to develop"—an interplay of two partially dependent variables, as Teilhard de Chardin puts it. This general pattern of human evolution also holds in modern society, although there has been no indication of recent increases in brain size.

EFFECTS OF INDUSTRIALIZATION ON MAN AND HIS ENVIRONMENT

The early ancestors of man participated in a more or less stable natural ecosystem. They adapted to the particular conditions of each biome. There was no sign of overkill of other animal species in the regions where man appeared. The ecosystem can be characterized as a dynamic equilibrium following climatic and other geological changes. In the past 5,000 years man has altered the ecosystem in many parts of the world and destroyed the natural balance (Campbell 1972).

The impact of cultural influence on man parallels cultural advancement. Negligible at first, it gradually acquired momentum through agricultural development and the industrial revolution, reaching a maximum for modern societies in the present atomic age. For technologically advanced states, the amount of energy currently drawn by an average person from his environment is approximately several thousand times greater than in pre-Neolithic times. In the United States, technological growth has been generally exponential, doubling every twenty years (Starr 1969).

As we have just realized, one consequence of industrial advancement is environmental degradation. White (1967) proposed the following model:

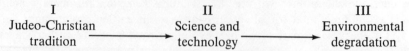

I	II	III
Judeo-Christian tradition	Science and technology	Environmental degradation

Moncrief (1970) suggested that capitalism and democratization, urbanization, increased wealth, increased population, and ownership of individual resources are intermediate steps linking the Judeo-Christian tradition with environmental degradation. The direct relationship between technological advance and environmental degradation is obvious, in addition to the effect of technological advance on population structure.

Man has been able to adapt to a broader range of physical environments than other animals. Such a capacity stems partly from cultural creativity. His response to environmental stress generally includes biological adjustments and modification of behavior patterns, as well as artificial devices—temperature and altitude acclimatization are well-documented examples of these (Weiner 1964; Lasker 1969). In certain circumstances, nevertheless, adjustment approaches a limit and eventually genetic modification must occur. For a population, environmental stress is a selection pressure that tends to favor genotypes with better coordinated adaptive biological modifications. New genes that arise by mutation and facilitate such adaptive modification will easily be incorporated into the population (Waddington 1959, 1961). This is, in essence, the cause of evolution.

The effect of the cultural environment on evolution is different from that of the physical in both nature and magnitude. Man's response involves mental adjustments of various kinds, some of which concern a morality for human behavior, which complements the desire for a proper balance of his inner and outer self. With reference particularly to the effect of technology in general, Leo Marx (1964) claimed that new industrial order is to be feared as dangerously threatening the necessary balance. Thomas Carlyle found men's behavior increasingly determined by social forces alien to their inward impulses. Hegel had called this condition "self-estrangement," signifying a conflict of interests between the social and the natural self. Machine technology is, in fact, creating what Karl Marx called "alienated labor," in which man's work becomes meaningless, bearing little or no relation to him (cf. L. Marx, *The Machine in the Garden*). Nowadays, men vacation to hunt and fish, occupations of earlier societies.

Effects of industrial development are difficult to describe, involving in certain cases not only mental adjustment but also—directly or indirectly—other biological systems. For instance, when a new industry is established, pollution of air and water soon follows, along with altered occupational hazards for workers, and emotional upsets arising from unfamiliar working conditions and other social changes (Mesthene 1968) or the disruption of previous habits and practices. These disturbances provide much of the etiology of chronic ailments such as mental illness, vascular diseases, and perhaps certain types of cancers. In a broad sense, we may conclude that such diseases are consequences of the genetic/environmental interaction.

There is anthropological evidence that most diseases of modern men were also present in prehistoric man, but the relative prevalence of certain types has changed historically and continues to do so. As biological indicators reflect socioeconomic changes in the course of man's evolution, the time trends for a few major diseases are illustrated in figures 8.4, 8.5, and 8.6.

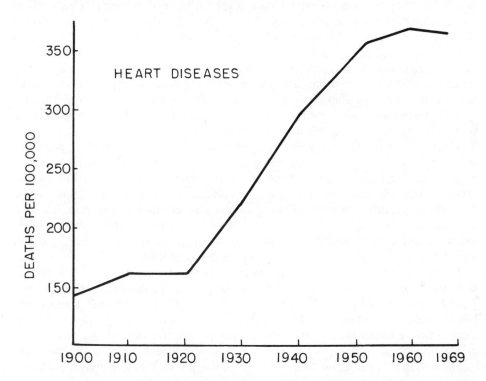

Fig. 8.4. Death rate due to heart diseases per 100,000 population in the United States. Unpublished data, courtesy of Dr. E. J. Glass.

Man possesses a large brain and a well-equipped endocrine system with which to meet stress quickly and effectively. Nevertheless, he consciously or unconsciously creates an environment that challenges his own survival. To meet such a challenge, Dubos (1967) states, "most human beings will not behave as passive receptors of the influences exerted on them by the modern world; they will, wisely or dangerously for the future, select some environments and ways of life and shun others on the basis of innate attributes and historically acquired tendencies." But in many instances, the change occurs on such a large scale as to preclude escape, and in a manner contradictory to man's innate and historical tendencies. Gilula and Daniels (1969) argued, "We now are seeing an unprecedented acceleration of various man-made changes which call for accompanying changes in man, changes which we are having difficulty in making. While biological change is extremely slow, cultural change . . . occurs at least every generation, although some aspects of culture (such as technology) change faster than others (for example, beliefs and customs)."

In evolutionary terms, man generates his own selection pressure and criteria, which differ somewhat from those of previous generations. The situation is comparable to that in an artificial selection experiment. For instance, a cattle herd continually bred on a selective basis for meat production now must be selected for milk production, or for work such as drafting and plowing. Most of the cattle in the herd would not be qualified and would have to be eliminated. If the owner selects for both milk production and working, and if they are negatively correlated, he will be able to produce cattle qualified for only one type of performance and not the other. In human societies, this corresponds to a situation in which a type of person who performed well in past generations or years manages poorly today for physical, mental, or other reasons.

The replacement of one gene by another, except in the case of a dominant lethal, cannot be achieved within one generation. Haldane (1957) drew attention to the fact that in the course of evolution the substitution of one allele for another, at any selection intensity, always involves a substantial loss of individuals from the population; he referred to it as the cost of natural selection. Thus to increase the rate of the genetic process to anywhere near the rate of great environmental changes there would have to be a rise in the selection coefficient, which would lead to a larger loss of individuals in the population.

The genetic basis of behavior and disease resistance is very complex. The above-cited bases constitute only a few examples. In general, in the event of death after reproductive age, natural selection will have no effect on the gene frequencies. Natural selection reduces the frequency of "unfit" genes through deaths of the carriers before termination of the reproductive period; this process, however, takes a large number of generations to reduce the gene frequency to a negligible level.

The above discussion on the effects of industrial evolution merely points out some of the negative elements and appears to take a somewhat dim view of the whole. In the long course of man's evolution, natural selection of slight deviation from the mean always occurs under certain circumstances and in particular societies. These can be viewed as temporary and peripheral. We must admit, on the other hand, that industrial evolution has definitely advanced our general living conditions and created a broad spectrum of culture; of this there is no need for discussion. Whatever social imperfections exist as a result, it is hoped that our foresight will be capable of making the necessary adjustments.

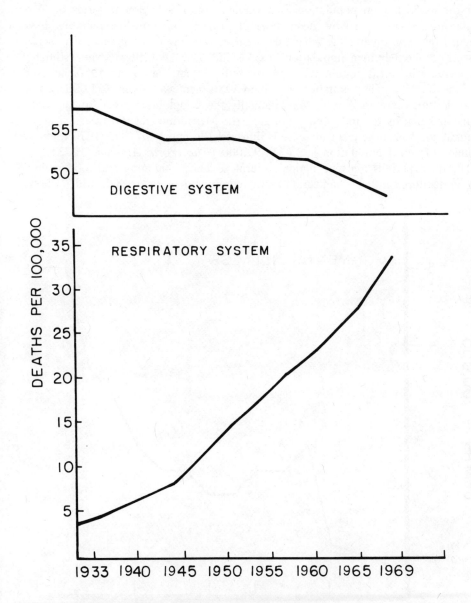

Fig. 8.5. Death rate due to cancer of the digestive system (*top*) and of the respiratory system (*bottom*) per 100,000 population in the United States. Unpublished data, courtesy of Dr. E. J. Glass; Potter 1947.

EFFECT OF CULTURE ON POPULATION GROWTH AND STRUCTURE

For the world human population as a whole, natural selection in terms of Darwinian fitness (measured by the number of progeny per breeding pair that have reached the reproductive age) has been relaxed. On the contrary, modern society is worried about human population "explosion." Deevey (1960), who estimated the increase in world population from one million years ago up to 1950, believed that the total human population a million years ago was about 125,000 people located somewhere in Africa; that gradually the population spread to all continents; and that by ten thousand years ago (the Mesolithic stage) it reached about five million. The subsequent increase is illustrated in figure 8.7. The 1970 midyear estimate of world population is 3,631,797,000 (*The World Almanac* 1973).

The three periods marking major cultural advances—hunting and food gathering, agriculture, and mechanization—are represented in figure 8.7, which shows

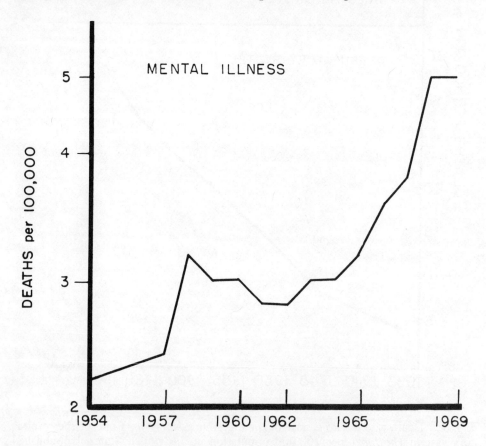

Fig. 8.6. Death rate due to mental diseases per 100,000 population in the United States. Unpublished data, courtesy of Dr. E. J. Glass.

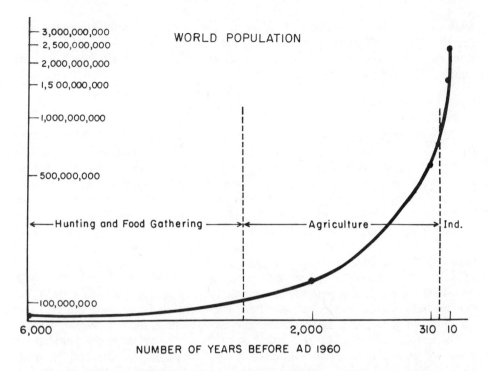

Fig. 8.7. World population from 1950 back 6,000 years. Data from Deevey 1960.

the relationship of cultural advances to population growth. The lengths of these periods are roughly estimated for convenience in examining the trend of population growth. As we know, the transition from one period to another occurred gradually, and the technical advances involved varied in degree and extent between different populations. In what we now consider the period of industrialization, there are still tribes and indigenous groups present in the world who survive mainly by hunting and food gathering. The curve is plotted on the basis of statistics from a few selected years separated by long-term intervals, and it is not fitted mathematically. Its purpose is simply to show the general trend of population growth.

Pearl and Reed (1920) attempted to find a mathematical formulation for the growth of the U.S. population in order to predict further population size. They conceived of the rate of population growth as resembling that of an individual plant or animal, or an autocatalysis in a chemical system. Growth starts at a slow rate, then gradually accumulates momentum until it reaches a maximum; thereafter, the rate declines somewhat and gradually achieves a plateau. Pearl and Reed treated this type of growth phenomenon as a geometric progression and described it by a logarithmic function,

$$y = \frac{b}{e^{-ax} + c},$$

where y is the population size; $a, b,$ and c are constants that can be computed from the statistics for different years; and x is the time in years. For the U.S. population, they based their calculations on the census from 1790 to 1910 and obtained values for constants as follows:

$$y = \frac{2{,}931}{e^{-0.0313x} + 0.0148}.$$

Using the above equation, they computed theoretical values and plotted a curve from 1790 up to the year 1910. The curve fits the actual census very well (fig. 8.8). For this equation, they computed the point of inflection (where the population reaches its maximum growth rate and after which it declines) at 98,637,000 and the maximum population size (the asymptote of the curve) at about 197,274,000. Following their equation, I plotted their curve up to the year 1970.

We can see from the census for recent years that their predictions are incorrect. They did caution that if some factors came into play that were not known at that time and had never operated during the past history of the country, their mathematical extrapolation for future statistics might not hold. This has, in fact, happened. There are factors underlying organic laws of U.S. population growth, not present before 1920 and occurring after. These are evidently very complex, but they may be generalized as socioeconomic changes since the two World Wars.

The purpose of this discussion has been to show that the rate of a population's growth does not necessarily follow the growth curve of an individual organism or a simple autocatalytic reaction. A population is strictly limited in physical space, but not to such an extent in its socioeconomic development. Population growth relates to both factors.

Demographic prospects in any country are connected not only to socioeconomic conditions, but also to political issues and the direction of official policies. The evolution of those aspects of a society affects vital rates in a complex interaction. In mainland China, the birth control programs have been somewhat cyclic and are reflected in variations in birth rates (table 8.2). In spite of a general trend toward birth reduction, increase in population size has only been slightly affected. This is due to the simultaneous decrease in mortality resulting from improvements in medical services, which counterbalances the decreasing birth rate. The correlation may exist in many countries that are undergoing transformation from a basically agricultural society to an industrial one.

We will now examine the effect of socioeconomic differences on population structures. The population pyramids in figure 8.9 show a sharp contrast between three groups. The socioeconomically backward Taiwan aborigines' pyramid has the widest base and tapers off rapidly with age, indicating a high birth rate accompanied by high infant mortality, which differs greatly from U.S. figures and, to some extent, from those of the Chinese population of Taiwan. Since individuals who die before reaching reproductive age do not contribute genetically to the population of the next generation, natural selection operates strongly in the Taiwan

TABLE 8.2 Estimates of the Population and Vital Rates in Mainland China, 1953–1970

Year	Population	Birth Rate	Death Rate	Natural Increase Rate
1953	582,603	45.0	22.5	22.5
1955	610,881	44.0	19.5	24.5
1960	688,811	39.9	20.1	19.8
1965	750,532	37.2	16.5	20.7
1970	836,036	37.3	15.0	22.3

SOURCE: Data from Aird 1972.
NOTE: Population figures in thousands as of July 1; vital rates per 1,000 population.

aborigines, less so in the Chinese, and least so in the U.S. population. These population pyramids demonstrate the connection of socioeconomic levels with population structures.

The two parameters of natural selection, fertility and mortality, can be estimated by the following equations (Neel and Schull 1972):

$$I_m = P_d/P_s \qquad\qquad \textbf{8.1}$$

$$I_f = \frac{(V_f/x_s^2)}{P_s} \qquad\qquad \textbf{8.2}$$

where I_m represents selection intensity due to mortality prior to reproductive age and I_f represents that due to fertility differences among women. In equation 8.1, P_d is the proportion of all individuals dying before reproductive age, and P_s the proportion of all individuals surviving. As Neel and Schull pointed out, P_d should ideally include all zygote losses between conception and reproduction, but data on losses early in pregnancy are extremely inadequate, and no investigator has attempted to include them in the calculation. In equation 8.2, x_s is the mean number of surviving individuals reaching reproductive age, and V_f represents the variance in the number of such individuals. This equation is a modification of the index for potential selection intensity (I_t) in Crow (1958):

$$I_t = \frac{V}{\overline{W}^2}, \qquad\qquad \textbf{8.3}$$

where V is the variance of number of children per female, and $\overline{W}$ is the mean number. This index is composed of both I_m and I_f.

Estimates for these different selection indexes were made for several populations by Neel and Schull (1972). These populations are classified as early, agricultural, and industrialized societies, with their estimates and some birth statistics given in table 8.3.

From early to industrialized societies, the trend of change in fertility selection is not as consistent as that for mortality. This occurs because in industrialized societies the change in fertility index does not necessarily reflect natural selection.

Rather, traditions and social factors often interplay to bring about this change. For instance, a woman with a small number of children may have the genetic potential of those who produce many but choose not to have a large family; similarly, the existence of a childless couple does not mean that either one of them is sterile. There are difficulties in collecting true fecundity data in human populations, as pointed out by Neel and Schull (1972).

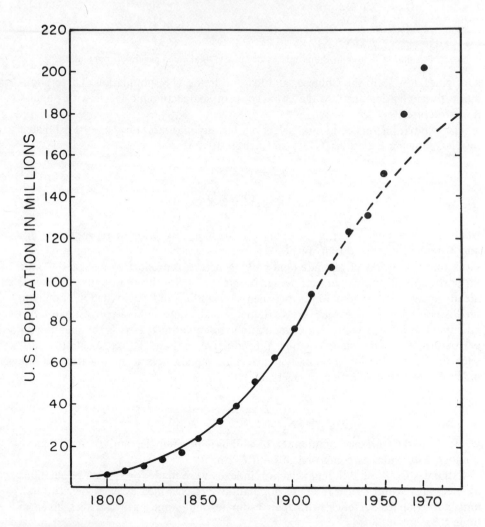

Fig. 8.8. The growth of the U.S. population from 1800 to 1970. The growth curve is fitted according to the equation of Pearl and Reed (1920) derived from the census of 1800 to 1910. Note the unpredicated greater rate of growth since 1940. Redrawn with modifications from Pearl and Reed 1920.

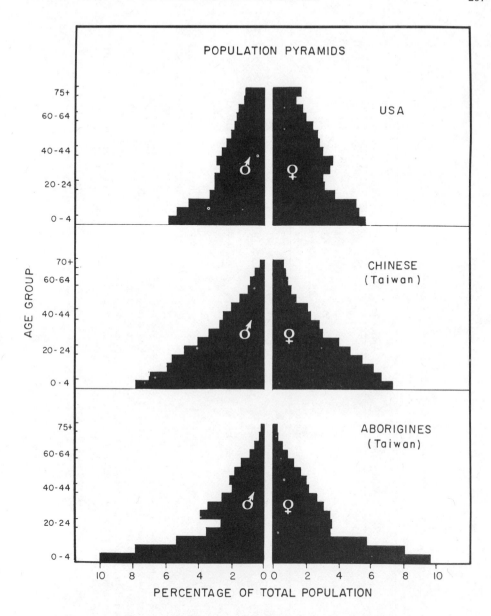

Fig. 8.9. Population pyramids. Note the wide base and sharp top of Taiwan aborigines pyramid contrasted to that of the United States, which has a narrow base and broad top, and of the Chinese which is intermediate. This indicates the high birth rate and mortality for young and old in the aborigines, characteristic for socioeconomical backward societies in comparison with more advanced societies, such as the Chinese and the United States. Redrawn from Chai 1967.

TABLE 8.3 A Comparison of the Reproductive Performance and Frequency of Death Prior to Age 15 and Crow's Index of Potential Selection

	Early Man		Agricultural Man			Man of Industrialized Countries		
			Uganda, Basoga Bantu (Roscoe 1950)	Nygasaland, Yao (Mitchell 1949)	Gold Coast, Ashanti (Meyers 1954)	Contemporary United States (U.S. Dept. HEW 1966)		Rural Japan (Schull et al. 1968)
	Xavantes	Yanomama				White	Nonwhite	
Mean live births per female reproduction complete ($M \pm \sigma^2$)	4.7 ± 6.1*	2.63 ± 3.7*	7.0 ± 7.2†	5.4 ± 9.7†	6.5 ± 10.0†	2.2 ± 4.3‡	2.7 ± 9.8‡	5.1 ± 9.5§
Estimated percentage dying prior to age 15	ca. 0.33	ca. 0.18	ca. 0.64	ca. 0.24	ca. 0.44	ca. 0.03	ca. 0.05	0.22§
Index of potential selection								
I_m	0.49	0.22	1.78	0.32	0.79	0.03	0.05	0.28
I_f	0.41	0.66	0.42	0.43	0.43	0.92	1.41	0.76
I_t	0.90	0.88	2.20	0.75	1.22	0.95	1.46	1.04

SOURCE: Reprinted, by permission, from J. Neel and W. J. Schull, *Evolutionary Biol.* 6(1972):363–78.

*Based on an effort to reconstruct a cohort of women reaching the age of reproduction.

†Based on women living through the reproductive period, that is, to age 40 or 45.

‡Based on women completing their forty-fourth year, who, because of low mortality from ages 20 to 44, approximate a cohort.

§Based on all marriages contracted in 1920–1939 represented by a living spouse residing on Hirado in 1964.

But the selection index for mortality does represent true differences between different populations. What are the consequences of such a difference in intensity of selection due to mortality?

Obviously, major factors that contribute to mortality are famine, epidemics, and war. Although famine and epidemics still exist in economically backward populations, improved medical services have reduced mortality in the world population as a whole, and continue to do so. In the United States, deaths from common infections were approximately 70 per 100,000 people in 1900, as opposed to only 10 in 1970; deaths caused by pneumonia dropped from about 200 people per 100,000 in 1900 to approximately 40 in 1945 even before the widespread use of antibiotics. Many contagious diseases have not appeared for a long time. This trend seems to exist in most industrialized countries. Although there are a large number of viral agents now causing health problems in western countries, the major bacterial pestilences have been kept in check.

Generally speaking, the genetics of human fertility has undergone no major changes since the early history of *Homo sapiens*, despite some slight advance in age of sexual maturation that is apparently a result of better nutrition in certain populations. It can safely be concluded that the reduction of mortality remains the major factor contributing to the world population problem. The reduction of mortality also causes the relaxation of selection intensity for fertility.

What are the effects of reduced selection pressure due to mortality on individual genetic constitution? Let us illustrate this with an arbitrary example. Assume that a single autosomal recessive gene confers physiological fitness on individuals to a lesser extent than its normal allele. Its presence in the population follows the mutation rate and selection coefficient u/s (chap. 6). Let us assume further that in the Taiwan aborigines 10% of the homozygous recessives survive over reproductive age, 50% in the Chinese, and 90% in the United States. If the gene mutation rate is 0.0001, the gene frequency will be 0.001, 0.005, and 0.009 for the Taiwan aborigines, the Chinese, and the U.S. populations, respectively. Thus it appears that technological advances in society foster survival of the mutant gene. This applies to all genes contributing mutational loads in one way or another, and indicates that technologically advanced societies constantly carry more mutant genes than do the less advanced societies. Maintenance of some of these is a social burden to the society.

It should be clear from the above discussion that reduced selection due to mortality in youth is a major factor for the rapid increase in population density. Reduced selection has also caused a decline in genetic resistance to many environmental factors, especially bacterial pestilences, and has increased the mutational load.

Another obvious consequence of cultural evolution, especially industrialization, is urbanization. Previously, self-supporting social units of approximately 50 to 100 individuals, organized as tribes or agricultural villages, began merging into larger social groups. Now much of the world's population is contained in cities of many millions of people, who live in interdependency within an artificially controlled environment. Some barriers and individual biological characteristics remain in this large social collectivity, but the general tendency is toward mixing in both culture and gene pool. Thus the city acts essentially as a ground for a kind of accelerated migration (with respect to its effects on the genetic constitution of populations). It results in increasing genetic variations within a group, counteracting random genetic fixation. It should be remarked that these are characteristics of *Homo sapiens,* and do not so often occur in other species.

EUGENIC MEASURES

The wisdom of applying genetic knowledge gained from experiments with animals to human eugenics has long been debated, especially since the recent advance of knowledge concerning genes and mutation, and the possibility of using eugenic measures ranging from artificial selection to gene insertion. Among scientists, one faction strongly favors the application of such measures, while the other strongly opposes all or some of them. The ultimate resolution of the issue may dramatically effect the future of man.

H. J. Muller was perhaps the leading advocate of the use of artificial methods in human eugenics. He believed that man could direct and thereby accelerate his own evolution through conscious, rather than natural, selection; that is, that by applying the principles of heredity and variation to keep the unfit from being born, and simultaneously working to extend the production of genetically superior beings, the genetic quality of man could be brought to higher levels. Muller reflected on these ideas while still a college student. In later years he wrote (1935*b*) and spoke on the subject with regard to its social, genetic, and evolutionary aspects. In *Out of the Night* he stated: "But man is the first being yet evolved on earth who has the power to note this changefulness, and if he will, to turn it to his own advantage, to work out genetic methods, eugenic ideals, yes, to invent new characteristics, organs, and biological systems that will work out to further the interests, the happiness, the glory of the god-like beings whose meagre foreshadowings we present ailing creatures are."

Recently, science fiction writers and news reporters have become greatly interested in the questions of the feasibility of new technical and eugenic procedures. Current ideas concerning eugenic measures may be summarized as follows:

1. Gene insertion or genetic surgery would essentially involve isolation of human genes, followed by a synthesis in vitro in large numbers, or even a modification of such genes for better quality that would be incorporated into human cells. This might be done much the same way as introducing the viral genome into the bacterial genome (chap. 3). The procedure could be applied to somatic cells for curing certain genetic defects, and to germ cells for the treatment of genetic disease, as well as for improving the quality of man. Attempts to produce desirable alterations in DNA by mutagenesis are also being made (Davis 1970). Thus far, known mutagenic agents have caused random mutation in virtually all cases, most of them proving to be harmful rather than beneficial.

2. By asexual reproduction or cloning the regulatory mechanism of cell differentiation would be reversed so that somatic cells could be used to start the embryo. Individuals could be developed in a foster mother or test tube, with the somatic cells being obtained from individuals of superior genetic endowment. An alternative cloning procedure suggests using the undifferentiated cells of an embryo. Some experimental success in mice has resulted from the use of this technique (Mintz 1964).

3. Selective reproduction, as practiced in animal husbandry, would increase the spread of genes from those with superior genetic endowment while reducing those containing hereditary defects. Mass production could be accomplished through the frozen preservation of sperm and by artificial insemination. These were the early ideas of Muller.

From the scientific point of view, two basic problems deserve consideration here: the first concerns a number of technical questions; the second, evolutionary consequences. Previously I mentioned that genetic variations may be classified into two types, discrete and continuous. Some hereditary defects due either to single recessive or dominant genes can produce discrete variations; such genes have been located in the chromosomes of many experimental animals and some have been identified in man. Their precise location and chemical variation have been discovered in many structural genes; so it may be possible that duplicate copies of desirable genes can be produced for curing certain genetic defects. For the improvement of the behavioral or intellectual quality of man, however, we must deal with polygenes, a great number of which are distributed on different chromosomes, each having numerous allelic forms. As yet we know little of their chemical or biological properties and, thus far, polygenic inheritance has received little public attention. But the importance of such investigation must be emphasized if the public is to be able to distinguish between realistic programs and wild projections for human betterment. One could argue that the behavior of an organism rests basically with physiology and anatomy, and that inheritance of behavior signifies the inheritance of certain bodily factors. But the phylogeny and ontogeny of behavior are complicated in their manifestations, and the measurement of intelligence has always been subject to criticism. It is interesting to speak of inserting genetically superior genes for the improvement of the intelligence of human stock, but, in practice, it is premature.

Asexual reproduction is normally irreversible in higher animal cells, although differentiation can be reversed in plant cells, as well as in transfer cuttings. The transfer of nuclei by microsurgery into the cytoplasm of different cells has already been accomplished. The stability and activity of nuclear genes depends on the interaction of the nucleus with its surrounding cytoplasmic medium in which genes may also be found. Parthenogenesis has been effected in mice[1]; but none of the eggs have yet developed to a stage involving distinct organs. We should remember that sexual evolution and the viviparous development in Eutharia, with

1. Leroy C. Stevens 1972: personal communication.

the delicate hormonal balance required for maintaining normal development of the young, has taken millions of years. The mechanisms of development and differentiation remain only sparsely understood. To attempt test tube production of man through copying in vivo systems is an idea involving difficulties analogous to fishing out polygenes for gene insertion. Feasible techniques for either genetic surgery or cloning remain to be demonstrated, although artificial insemination of humans has often been performed without encountering any great difficulties.

As far as genetic consequences are concerned, all eugenic measures are essentially related to the application of selective breeding as carried out in domestic plants and animals. Dobzhansky (1962) called Muller's eugenic measures "Muller's bravest new world," inquiring:

> Are we hastily made-over apes, ready to agree what the ideal man ought
> to be like? Granted that mankind would profit immeasurably from
> the birth of more persons with the mental stamina of Einstein, Pasteur,
> and even Lenin, do we really want to live in a world with millions of
> Einsteins, Pasteurs, and Lenins? Muller's implied assumption that
> there is, or can be, the ideal human genotype which it would be desirable
> to bestow upon everybody is not only unappealing but almost certainly
> wrong—it is human diversity that acted as a leaven of creative effort
> in the past and will so act in the future.

Moreover, assuming that the intelligence level could be raised, the question remains as to whether or not superior intellect implies greater emotional stability or happiness.

Selective breeding in man could be applied either positively or negatively. Positively, it could lead to the increased production of individuals with certain desirable characteristics; negatively, it could result in decreased production of undesirables. The result of one application may not necessarily concur with that of another. (Actually, selective social breeding has been subtly practiced in man since neolithic times, polygamy and caste systems being but two examples.) However, the consequences of selective breeding on a genetic level would be to increase genetic uniformity.

Selective breeding for one or two specific traits in laboratory animals frequently results in reduced fitness. Animals with better physiological and reproductive fitness do not occur in the inbred, nor in the selectively bred stocks, but rather in those randomly bred. In natural populations, including man, it is always desirable to have a minimum level of homozygosity. The advantage of this feature for maintaining normal physiological homeostasis in development, as well as for genetic shifting of the population in cases of environmental change, has become an established fact (Lerner 1954).

Genes in natural populations generally occur in frequencies conferring maximum fitness within that specific time and environment. Disruption of such a balance is bound to produce a greater percentage of genetically weak individuals.

However, one may not reject the practice of selective breeding measures on the small scale, which would not result in the production of monotonous phenotypes or in the disturbance of the coadapted gene pool.

Perhaps the use of negative eugenic measures for the elimination of genes known to contribute to the mutational load need not be questioned. Deleterious genes pollute the gene pool, contributing nothing. Where do these genes originate, and how are they maintained in the population? Gene mutation arises either spontaneously or through induction by physical and chemical agents. Spontaneous mutation, resulting from random errors in DNA is, by our present knowledge, unavoidable. Induced mutations, caused by radiation exposure, chemicals, or drugs, can be reduced.

In technically advanced countries, mutation rates increase; yet elimination of harmful mutations remains negligible. Because of our improved living conditions and medical services, plus the social practice of caring for the sick in our population regardless of the type of their illness, many hereditary disorders and weaknesses have been perpetuated. Defective individuals produce children who will become carriers of the genes in the next generation, so genetic loads continue to rise.

China's Yellow River has been described in the past as running on top of a wall. Each year the people raised the height of the dikes to prevent flooding, thus further raising the level of the riverbed where soil was continuously deposited. Eventually, when the river overflowed, the damage was even greater than it would have been were it not for the extra soil that elevated the riverbed by continually building the dikes. The situation is analogous to that of any society attempting to maintain genetically defective individuals, and the latter has a relatively simple remedy: the carriers of mutant genes should not have progeny. Although law should not be used as the means of restriction, mutant gene carriers should be made aware of their condition and its consequences, and educated to act in the interest of the health of the population as a whole and for the future of man. Since pollution of the gene pool is ultimately worse than pollution of the environment, negative eugenic measures are in this case unquestionably necessary, and can effectively be carried out if we so desire.

Another important eugenic measure is the control of population density. It has been observed that in small mammals reproduction is progressively inhibited as population density increases. Animals under crowded conditions are found to have larger adrenal glands and smaller male sex organs. These observations lead to the conclusion that the endocrine effects of crowding act as an automatic mechanism for regulating population density (Christian 1955). Recent studies (Bruce 1970) reveal that in mice, pregnancy can be blocked by a foreign male introduced into the cage a few days after the female has been mated. This fact has been interpreted as a pheromonal stimulation of an endocrine imbalance. The exact physiological mechanisms remain to be studied. The fact that there is an automatic mechanism for population density regulation in species of small mammals is of great interest.

Another population density regulator mechanism in *M. musculus* was found in that the *t* alleles, causing tail abnormality, are present in all natural populations. As mentioned in previous chapters, the mutant genes are practically lethal when homozygous, but when heterozygous there are generally some heterotic effects in physiological fitness. This is perhaps the most efficient genetic mechanism in control of population density, that is, reduction of litter size on the one hand, and increasing vigor of surviving individuals on the other. The litter size of polytocous (giving multiple birth) species corresponds to the clutch size in birds. That intermediate clutch size is favored by natural selection is an established fact.

Let us return to our own species. Have we such physiological and genetic regulation mechanisms as are found in rodents? We do not know. As has been pointed out, we do carry many deleterious genes, and we protect them from natural selection. An average woman has the potential of producing twenty or more children. Such potential is desirable for the survival of the species and was highly favored by natural selection during early stages of evolution. Cultural and medical advances have enabled us to keep at least 90% of newborn babies, and our usual desire is to keep all if we can. Under such circumstances, the only way to control population density is to limit the number of births.

The ultimate size of the world's population is difficult to predict, but unquestionably its increase must be slowed down within the near future, either through family planning or by consequence of catastrophe. Based on our present life expectancy in the Western hemisphere, the population size can be kept in check in the long run if the number of children does not exceed 2.3 per married couple (Dubos 1967). Such a limitation will create human problems genetically, physiologically, and emotionally. As Dubos pointed out, "Many couples will continue to have more children than desirable and . . . others, hopefully, will choose or be compelled to have only one child or to remain childless. Many genetic and social determinants will play a role in this choice, or compulsion, concerning the number of children." But in any case, whatever social and genetic measures should be taken, a less crowded population is still better than a congested population, the consequences of which would be much worse.

Man's Fate

The final destiny of all species in the past has been identical—extinction (Romer 1949; Wright 1970). This fact, translated into the language of prophecy, means that *Homo sapiens* cannot escape this end. Human evolution, however, has, and perhaps will continue to have, a different history from that of any other species because of man's social heritage and genetic endowment. Is it possible to realize a different end for our particular and unique development?

Most theoreticians recognize that interspecific competition is a major factor in species extinction. Man, however, has not only conquered all other species on earth, he has also utilized them to his advantage. Human history can claim the development of agriculture and aquaculture, and even the employment of microorganisms in fermentation and in the production of antibiotics and vaccines. To an equal extent, man has utilized other species for shelter and clothing. He has conquered his physical environment with the development of technology for protection against meteorological changes and the establishment of energy sources. By these means, the human species has almost freed itself from various components of natural selection, and should survive with much greater facility than its non-human ancestors if it should have to undergo another glacial age, for instance.

Intraspecific deterioration remains a second important factor contributing to extinction. In genetic terms, random drift and inbreeding in natural populations are the evolutionary processes that metaphorically drive a genetic system to the end of a one-way street. In small, isolated populations, only a slight chance of producing favorable mutations exists; the level of homozygosity rises continuously. Such genetic deterioration in sexual organisms, a factor especially harmful in small populations, affects all species.

As a conscious being, man seeks to outbreed when the danger of extinction threatens; immigrations have occurred throughout human history. In general, human populations remain relatively large, and inbreeding is prohibited.

History has shown us that isolated human groups tend to disappear with small populations or merge into larger collectives. These include genetic, social, and even political groups. The merging of groups naturally involves the merging of cultures, living customs, religious beliefs, and so on. Genetic assimilation goes hand in hand with cultural assimilation. The trend for *Homo sapiens* to become one large population, having an integrated gene pool with free exchange of genes between individuals within a universal culture, is constantly increasing. A member of this population will not find others biologically or socially foreign to himself. In such a situation, there would be less likelihood for genetic drift and gene fixation. *Homo sapiens*—as a species—will eventually become one massive interbreeding population with a large amount of genetic variation on which natural selection can operate with more freedom.

One can assert, with a certain assurance, that the merging of genetic systems is a prerequisite for the establishment of a stable and enduring social structure. If the utopian concept can be realized in human history, it will not be contained within the confines of a select social group, but instead will embrace the entire human population.

The constituents of humanity's intraspecific dilemmas do not lie within man's genetic system. We know for certain that mankind now carries a greater mutational burden than in the past, yet human populations as a whole continue to flourish. Civilization can and will endure, absorbing the slowly accumulating genetic imperfections. These are peripheral issues.

Instead, the threat to the survival of the human species lies within its intrasocial stress: the rioting, war, and mental instability that have become the plight of a population insufficiently adjusted to rapid technological advances. This touches on the basic issue of morality and ethics in man.

Human morality and ethics provide the basis on which man can claim superiority to all other animal species. This ethical consciousness is now generally understood to be a social acquisition; man is not born with morality and does not have genes for the automatic expression of such behavior. He simply possesses the mental capacity to acquire the sense for it. Thus the human heritage is, to a large extent, cultural.

However, as a guiding principle for human conduct, morality and ethics are in sharp conflict with an innate behavioral pattern in man: aggression. Freud (1930) so stated: "In all that follows I take up the standpoint that the tendency to aggression is an innate, independent, instinctual disposition of man, and I come back now to the statement that it constitutes the most powerful obstacle to culture." If we face the problem squarely, we need to realize that man has, as a result of evolution, conquered every animal species through intelligence and aggressive behavior. This disposition is evolutionary and genetic. It is questionable whether in modern societies aggressive individuals actually possess greater Darwinian fitness than the less aggressive, but the contemporary social structure in many ways still encourages such behavior. The word *competition,* as used in every profession, consists of a partially transformed mode of aggression. The innate disposition plus social stimulation results in a more positive expression.

It is clear that in order to reduce violence between individuals, groups, or political systems, morality and ethics must be strengthened and reinforced to eventually overcome some of the innate nature of man. Initial solutions to these problems will occur, if at all, in the social and political realm; as such they are beyond the scope of this discussion.

Both the origin and the end point of human evolution remain uncertain and/or unknown. Nevertheless, we might justly regard man's development of erect posture and expansion of brain volume as the crucial events in the unfinished progress. It is precisely this physiological-anatomical alteration that has gradually evolved into a fundamental distinction between man and the animal realms, a distinction based on the remarkable phenomeon of human self-consciousness. The implications of consciousness, in turn, have brought about a transformation of the role of mankind, placing it outside the natural order.

In Kant's celebrated *Critique of Pure Reason* (1929, p. 379) he writes: "If we judged according to *analogy with the nature* of living beings in this world, in dealing with which reason must necessarily accept the principle that no organ, no faculty, no impulse, indeed nothing whatsoever is either superfluous or disproportioned to its use, and that therefore nothing is purposeless, but everything exactly conformed to its destiny in life—if we judged by such an analogy we should have to regard man, who alone can contain in himself the final end of all this order, as the only creature that is expected from it."

Indeed the transference of cultural heritage has finally begun to outweigh the effects of the biological. But, to reemphasize a point made earlier, if human evolutionary progress is to continue, man must recognize the implications of his self-consciousness—specifically, the development of a morality and ethics as well as their attendant cultural basis—the "roads to [his] freedom." I would like to conclude with the following passage from Kant (1929, p. 380):

> Man's natural endowments—not merely his talents and the impulses
> to enjoy them, but above all else the moral law within him—go so far
> beyond all the utility and advantage which he may derive from
> them in this present life, that he learns thereby to prize the mere
> consciousness of a righteous will as being, apart from all advantageous
> consequences, apart even from the shadowy reward of posthumous
> fame, supreme over all other values; and so feels an inner call to fit
> himself, by his conduct in this world, and by the sacrifice of many
> of its advantages, for citizenship in a better world upon which he lays
> hold in idea. This powerful and incontrovertible proof is reinforced
> by our ever-increasing knowledge of purposiveness in all that we see
> around us, and by contemplation of the immensity of creation, and
> therefore also by the consciousness of a certain illimitableness in
> the possible extension of our knowledge, and of a striving commensurate
> therewith.

bibliography

Abelson, P. H. 1966. Chemical events on the primitive earth. *Proc. Nat. Acad. Sci. U.S.* 55:1365–72.

Adamson, J. W., and Stamatoyannopoulos, G. 1974. The A → C switch in sheep hemoglobin in vitro studies with erythropoietin. In *Comparative molecular biology models for the study of diseases*, eds. H. Kitchen and S. Boyer. New York: New York Academy of Sciences.

Aird, J. S. 1972. Population policy and demographic prospects in the People's Republic of China. In *People's Republic of China: An economic assessment*. A compendium of papers submitted to the Joint Economic Committee, Congress of the United States, 1972.

Allard, R. W., and Wehrhahn, C. 1964. A theory which predicts stable equilibrium for inversion polymorphisms in the grasshopper *Moraba scurra*. *Evolution* 18:129–30.

Allen, S. L. 1965. Genetic control of enzymes in *Tetrahymena*. In *Genetic control of differentiation*, pp. 427–46. New York: Academic Press Inc.

Allison, A. C. 1955. Aspects of polymorphism in man. *Cold Spring Harbor Symp. Quant. Biol.* 20:239–52.

Ames, B. N., and Hartman, P. E. 1963. The histidine operon. *Cold Spring Harbor Symp. Quant. Biol.* 28:349–61.

Anderson, P. K. 1964. Lethal alleles in *Mus musculus*: Local distribution and evidence for isolation of demes. *Science* 145:177–78.

Arey, L. B. 1946. *Developmental anatomy*. Philadelphia: W. B. Saunders Co.

Astraurov, B. L. 1969. Experimental polyploidy in animals. *Ann. Rev. Genet.* 3:99–126.

Avery, O. T.; MacLeod, C. M.; and McCarty, M. 1944. Studies on the chemical nature of the substance inducing transformation of pneumococcal types. *J. Exp. Med.* 79:137–58.

Babcock, E. B., and Stebbins, G. L. 1938. The American species of *Crepis*. Publication 504. Washington, D. C.: Carnegie Institution of Washington.

Bada, J. L., and Miller, S. L. 1968. Ammonium ion concentration in the primitive ocean. *Science* 159:423–25.

Baglioni, C. 1962. The fusion of two peptide chains in hemoglobin Lepore and its interpretation as a genetic deletion. *Proc. Nat. Acad. Sci. U.S.* 48:1880–86.

————. 1967. Molecular evolution in man. In *Proceedings of the Third International Congress of Human Genetics*, eds. J. F. Crow and J. V. Neel. Baltimore: Johns Hopkins University Press.

Barghoorn, E. S., and Schopf, J. W. 1965. Microorganisms from the late pre-Cambrian of central Australia. *Science* 150:337–39.
————. 1966. Microorganisms three billion years old from the pre-Cambrian of South Africa. *Science* 152:758–63.
Barr, M., and Bertram, L. F. A. 1949. A morphological distinction between neurons of the male and female and the behavior of the nucleolar satellite during accelerated nucleoprotein synthesis. *Nature* 163:676–77.
Barrington, D. J. W. 1965. *The biology of Hemichordata and Protochordata*. London: Oliver & Boyd.
Barrington, D. J. W., and Franchi, L. L. 1956. Some cytological characteristics of thyroidal function in the endostyle of the *Ammocoete* larva. *Quart. J Microscop. Sci.* 97:393–409.
Barrington, D. J. W., and Thorpe, A. 1963. Comparative observations on iodine binding by *Saccoglossus horsti* (Brambell and Goodhart), and by the tunic of *Ciona intestinalis* (L). *Gen. Comp. Endocrinol.* 3:166–75.
Bastock, M. 1956. A gene mutation which changes a behavior pattern. *Evolution* 10:421–39.
Beadle, G. W., and Coonradt, V. L. 1944. Heterokaryosis in *Neurospora crassa*. *Genetics* 29:291–308.
Beckman, L., and Johnson, F. M. 1964. Variations in larval alkaline phosphatase controlled by *Aph* alleles in *Drosophila melanogaster*. *Genetics* 49:829–35.
Becquerel, P. 1924. La vie terrestre provientelle d'un autre monde? *Soc. Astron. Fr. Bull.* 38:393–417.
Beermann, W. 1967. Gene action at the level of the chromosome. In *Heritage from Mendel*, ed. R. Brink. Madison: University of Wisconsin Press.
Benirschke, K. 1969. *Comparative mammalian cytogenetics*. New York: Springer-Verlag New York, Inc.
Benjamin, T. L. 1966. Virus-specific RNA in cells productively infected or transformed in polyoma virus. *J. Mol. Biol.* 16:359–73.
Bennett, D. 1961. Miniature: A new gene for small size in the mouse. *J. Heredity* 52:95–98.
Bennett, D.; Boyse, E. A.; and Old, L. J. 1972. Cell surface immunogenetics in the study of morphogenesis. In *Cell interaction*, ed. L. G. Silvestri. New York: American Elsevier Publishing Co.
Benzer, S. 1960. On the topography of the genetic fine structure. *Proc. Nat. Acad. Sci. U.S.* 47:403–16.
Bergner, A. D. 1943. Chromosomal interchange among six species of *Datura* in nature. *Am. J. Botany* 30:431–40.
Bern, H. A. 1967. Hormones and endocrine glands of fishes. *Science* 158:455–62.
Bern, H. A., and Nicoll, C. S. 1968. The comparative endocrinology of prolactin. *Recent Progr. Hormone Res.* 24:681–720.
Bernal, J. D. 1951. *The physical basis of life*. London: Routledge & Kegan Paul, Ltd.
Bernstein, N., and Goldschmidt, E. 1961. Chromosome breakage in structural heterozygotes. *Am. Naturalist* 95:53–57.

Beutler, E.; Yeh, M.; and Fairbanks, V. F. 1962. The normal human female as a mosaic of X-chromosome activity: Studies using the gene for G-6-PD deficiency as a marker. *Proc. Nat. Acad. Sci. U.S.* 48:9–16.

Bewley, T. A., and Li, C. H. 1970. Primary structures of human pituitary growth hormone and sheep pituitary lactogenic hormone compared. *Science* 168: 1361–62.

Blout, E. R.; Doty, P.; and Yang, J. T. 1957. Polypeptide: XII. The optical rotation and configurational stability of α-helices. *J. Am. Chem. Soc.* 79:749–50.

Blout, E. R., and Idelson, M. 1956. Polypeptide: IX. The kinetics of strong-base initiated polymerizations of amino acid N-carboxyanhydrides. *J. Am. Chem. Soc.* 78:3857–60.

Bodmer, W. F., and Parsons, P. A. 1962. Linkage and recombination in evolution. *Advan. Genet.* 11:2–87.

Bonin von, G. 1963. *The evolution of the human brain*, pp. 50–53. Chicago: University of Chicago Press.

Bourne, G. H., and Tewari, H. B. 1964. In *Cytology and cell physiology*, ed. G. H. Bourne. 3d ed. New York: Academic Press Inc.

Boyse, E. A. 1971. Cell membrane receptors. In *Immune surveillance*, eds. R. T. Smith and M. Landy. New York: Academic Press Inc.

Brace, C. L. 1963. Structural reduction in evolution. *Am. Naturalist* 97:39–49.

Bradley, T. B.; Wohl, R. C.; and Rieder, R. F. 1967. Hemoglobin Gun Hill: Deletion of five amino acid residues and impaired heme-globin binding. *Science* 157:1581–83.

Brat, S. B. 1966. Genetic systems in *Allium*: II. Sex Differences in meiosis. In *Chromosomes today*, eds. C. D. Darlington and K. R. Lewis. New York: Plenum Publishing Corp.

Braunitzer, G. 1966. Phylogenetic variation in the primary structure of hemoglobin. *J. Cellular Physiol.* 67 (suppl. 1):1–19.

Bridges, C. B. 1936. Salivary chromosome maps. *J. Heredity* 26:60–64.

―――. 1936. The bar "gene": A duplication. *Science* 83:210–11.

Britten, R. J., and Kohne, D. E. 1968. Repeated sequences in DNA. *Science* 161: 529–40.

Brown, S. W. 1966. Heterochromatin. *Science* 151:417–25.

Bruce, H. M. 1970. Pheromones. *Brit. Med. Bull.* 26:10–13.

Brues, A. M. 1969. Genetic load and its varieties. *Science* 164:1130–36.

Bungenberg de Jong, H. G. 1932. Sammelreferat die Koazervation bedeutung fur die biologie. *Protoplasma* 15:110–76.

Cain, A. J., and Sheppard, P. M. 1954. Natural selection in *Cepaea*. *Genetics* 39: 89–116.

Calvin, M. 1969. *Chemical evolution*. Oxford: Oxford University Press.

Campbell, A. M. 1962. Episomes. *Advan. Genet.* 11:101–45.

Campbell, B. 1972. Man for all seasons. In *Sexual selection and the descent of man*, ed. B. Campbell. Chicago: Aldine-Atherton, Inc.

Carothers, W. H. 1936. Polymers and polyfunctionality. *Faraday Soc. Trans.* 32: 39–53.

Carson, H. L. 1959. Genetic conditions which promote or retard the formation of species. *Cold Spring Harbor Symp. Quant. Biol.* 24:87–105.

————. 1970. Chromosome tracers of the origin of species. *Science* 168:1414–18.

Carter, T. C.; Lyon, M. F.; and Phillips, R. J. S. 1958. Genetic hazard of ionizing radiations. *Nature* 182:409.

Cavalli-Sforza, L. L., and Bodmer, W. F. 1971. *Genetics of human populations*, pp. 118–19. San Francisco: W. H. Freeman & Co., Publishers.

Chadwick, C. S. 1941. Further observations on the water drive in *Triturus viridescens*: II. Induction of the water drive with the lactogenic hormone. *J. Exp. Zool.* 86:175–87.

Chai, C. K. 1956. Analysis of quantitative inheritance of body size in mice: II. Gene action and segregation. *Genetics* 41:165–78.

————. 1961. Analysis of quantitative inheritance of body size in mice: IV. An attempt to isolate polygenes. *Genet. Res.* 2:25–32.

————. 1967. *Taiwan aborigines: A genetic study of tribal variations*. Cambridge: Harvard University Press.

————. 1969. Effects of inbreeding in rabbits: Inbred lines, discrete characters, breeding performance and mortality. *J. Heredity* 60:64–70.

————. 1970. Effects of inbreeding in rabbits: Skeletal variations and malformations. *J. Heredity* 61:3–8.

————. 1971. Analysis of palm dermatoglyphics in Taiwan indigenous populations. *Am. J. Phys. Anthropol.* 34:369–76.

————. 1972. Biological distances between indigenous populations of Taiwan. In *The assessment of population affinities in man*, eds. J. S. Weiner and J. Huizinger. Oxford: Clarendon Press.

Chai, C. K., and Crary, D. D. 1971. Conjoined twinning in rabbits. *Teratology* 4:433–44.

Chai, C. K., and Stevens, L. C. 1972. Unpublished data.

Chatterjee, I. B. 1973. Evolution and the biosynthesis of ascorbic acid. *Science* 182:1271–72.

Chevremont, M. 1963. Cytoplasmic deoxyribonucleic acids: Their mitochondrial localization and synthesis in somatic cells under experimental conditions and during the normal cell cycle in relation to the preparation for mitosis. *Symp. Intern. Soc. Cell. Biol.* 2:323–33.

Chevremont, M.; Chevremont-Comhaire, S.; and Baeckeland, E. 1959. Action de desoxyribonucleases neutre et acide sur des cellules somatiques vivantes cultiveas in vitro. *Arch. Biol.* 70:811–29.

Christian, J. J. 1955. Effect of population size on adrenal glands and reproductive organs of male mice in populations of fixed size. *Am. J. Physiol.* 182:292–301.

Clarke, B. 1970. Darwinian evolution of proteins. *Science* 168:1009–11.

Clarke, C. A., and Sheppard, P. M. 1960. The evolution of mimicry in the butterfly *Papilio dardanus*. *Heredity* 14:163–73.

Cloud, P. E., Jr. 1968. Atmospheric and hydrospheric evolution on the primitive earth. *Science* 160:729–36.

————. 1974. Evolution of ecosystems, *Am. Scientist* 62:54–66.

Cohen, S. S. 1970. Are/were mitochondria and chloroplasts microorganisms? *Am. Scientist* 58:281–89.

Comings, D. E. 1967. Histories of genetically active and inactive chromatin. *J. Cellular Biol.* 35:699–708.

Copeland, H. F. 1956. *The classification of lower organisms.* Palo Alto, Calif.: Pacific Books, Publishers.

Corbin, K. W., and Uzzell, T. 1970. Natural selection and mutation rates in mammals. *Am. Naturalist* 104:37–53.

Corey, H. I. 1938. Heteropycnotic elements of orthopteran chromosomes. *Arch. Biol.* 49:159–72.

Court-Brown, W. M., and Smith, P. G. 1969. Human population cytogenetics. *Brit. Med. Bull.* 25:74–80.

Cox, E. C., and Yanofsky, C. 1967. Altered base ratios in the DNA of an *Escherichia coli* mutator strain. *Proc. Nat. Acad. Sci. U.S.* 58:1895–1902.

Crew, F. A. E., and Koller, P. C. 1932. The sex incidence of chiasma frequency and genetical crossing-over in the mouse. *J. Genet.* 26:359–83.

Crick, F. H. C. 1968. The origin of the genetic code. *J. Mol. Biol.* 38:367–79.

Crow, J. F. 1958. Some possibilities for measuring selection intensities in man. *Human Biol.* 30:1–30.

————. 1963. Genetic load: Three views: II. The concept of genetic load: A reply. *Am. J. Human Genet.* 15:310–14.

Crow, J. F., and Kimura, M. 1970. *An introduction to population genetics theory.* New York: Harper & Row, Publishers.

Cumins, H., and Midlo, C. 1961. *Fingerprints, palm and soles: An introduction to dermatoglyphics.* New York: Dover Publications.

Darlington, C. D. 1935. The internal mechanism of the chromosome: I, II, and III. *Proc. Roy. Soc. (London) Ser. B.* 118:33–96.

————. 1939. *The evolution of genetic system.* Cambridge: Cambridge University Press.

————. 1959. The evolution of chromosome systems. *Nat. Acad. Lincei* 47:227–36.

————. 1966. The chromosomes as *we* see them. In *Chromosomes today,* ed. C. D. Darlington and K. R. Lewis. New York: Plenum Publishing Corp.

Davis, B. D. 1970. Prospects for genetic intervention in man. *Science* 170:1279–83.

Davis, B. D.; Dulbecco, R.; Eisen, H.; Ginsberg, H.; and Wood, W. B., Jr. 1967. *Microbiology,* p. 1100. New York: Harper & Row, Publishers.

Dayhoff, M. O. 1969. *Atlas of protein sequence and structure.* Washington, D.C.: National Biomedical Research Foundation.

Dayhoff, M. O., and Eck, R. V. 1968. *Atlas of protein sequence and structure: 1967–1968.* Washington, D.C.: National Biomedical Research Foundation.

DeBusk, A. G. 1968. *Molecular genetics.* New York: Macmillan Co.

Deevey, E. S. 1960. The human population. *Sci. Am.* 203:195–204.

DeHaan, R. L., and Ebert, J. D. 1964. Morphogenesis. *Ann. Rev. Physiol.* 26:15–46.

DeMars, R., and Nance, W. E. 1964. Electrophoretic variants of glucose-6-phos-
phate dehydrogenase and the single-action X in cultivated human cells. *Wistar
Inst. Symp. Monogr.* 1:35–48.

Demerec, M. 1937. Frequency of spontaneous mutations in certain stocks of
Drosophila melanogaster. Genetics 22:467–83.

Demerec, M., and Hartman, P. E. 1959. Complex loci in microorganisms. *Ann.
Rev. Microbiol.* 13:377–406.

Demerec, M., and Hoover, M. E. 1936. Three related X-chromosome deficiencies
in *Drosophila. J. Heredity* 27:207–12.

Dickerson, R. E. 1971. The structure of cytochrome *c* and the rates of molecular
evolution. *J. Mol. Evol.* 1:26–45.

Dobzhansky, T. 1950a. Genetics of natural populations: XIX. Origin of heterosis
through natural selection in populations in *Drosophila pseudoobscura. Genetics*
35:288–302.

————. 1950b. Mendelian populations and their evolution. *Am. Naturalist* 84:
401–18.

————. 1955a. *Evolution, genes, and man*, p. 168. New York: John Wiley &
Sons, Inc.

————. 1955b. A review of some fundamental concepts and problems of population genetics. *Cold Spring Harbor Symp. Quant. Biol.* 20:1–15.

————. 1957. Genetic loads in natural populations. *Science* 126:191–94.

————. 1962. *Mankind evolving*. New Haven: Yale University Press.

————. 1970. *Genetics of the evolutionary process*. New York: Columbia University Press.

Dodge, B. O. 1942. Heterokaryotic vigor in *Neurospora. Bull. Torrey Botan. Club*
69:75–91.

Dodson, E. O. 1952. *Evolution*. Philadelphia: W. B. Saunders Co.

Dounce, A. L. 1971. Nuclear gels and chromosomal structure. *Am. Scientist* 59:
74–83.

Droop, M. 1963. Algae and invertebrae in symbiosis. *Symp. Soc. Gen. Microbiol.*
13:171–99.

Dubos, R. 1967. Nature and nurture. *Progr. Med. Genet.* 5:2–7.

Dunn, L. C. 1956. Complex locus in the mouse. *Cold Spring Harbor Symp. Quant.
Biol.* 21:187–95.

Dunn, L. C.; Beasley, A. B.; and Tinker, H. 1960. Polymorphisms in populations
of wild house mice. *J. Mammalogy* 41:220–29.

Dunn, L. C., and Caspari, E. 1945. A case of neighboring loci with similar effects.
Genetics 30:543–68.

Ebert, J. D. 1965. *Interacting systems in development*. New York: Holt.

Ebert, J. D., and Kaighn, M. E. 1966. The keys to change: Factors regulating
differentiation. In *Major problems in developmental biology*, ed. M. Locke.
New York: Academic Press Inc.

Eck, R. V., and Dayhoff, M. O. 1966. Evolution of the structure of ferredoxin
based on living relics of primitive amino-acid sequences. *Science* 152:263–66.

Edelman, G. M. 1972. Variability, symmetry, and periodicity in the structure of
immunoglobulins. In *Cell interactions*, ed. L. G. Silvestri. Amsterdam: North

Holland Publishing Co.

Eden, M. 1967. Inadequacies of neo-Darwinian evolution as a scientific theory. In *Mathematical challenges to the neo-Darwinian interpretation of evolution*, ed. P. S. Moorhead and M. M. Kaplan. Monograph 5. Philadelphia: Wistar Institute Press.

Egozcue, J. 1969. Primates. In *Comparative mammalian cytogenetics*, ed. K. Benirschke. New York: Springer-Verlag New York, Inc.

Emmer, M.; deCrombrugghe, B.; Pastan, I.; and Perlman, R. 1970. Cyclic AMP receptor protein of *E. coli*: Its role in the synthesis of inducible enzymes. *Proc. Nat. Acad. Sci. U.S.* 66:480–87.

Epstein, R. H.; Bolle, A.; Steinberg, C. M.; Kellenberger, E.; de la Tour, E. B.; Chevalley, R.; Edgar, R. S.; Sussman, M.; Denhardt, G. H.; and Lielausis, A. 1963. Physiological studies of conditional lethal mutants of bacteriophage T4D. *Cold Spring Harbor Symp. Quant. Biol.* 28:375–94.

Ewens, W. J. 1967. Random sampling and the rate of gene replacement. *Evolution* 21:657–63.

Falconer, D. S. 1960. *Introduction to quantitative genetics*. New York: Ronald Press Co.

Farb, P. 1962. *The insects*. New York: Life Nature Library.

Farris, J. S. 1972. Estimating phylogenetic trees from distance matrices. *Am. Naturalist* 106:645–68.

Feldmann, M., and Nossall, G. J. V. 1972. Cellular basis of antibody production. *Quart. Rev. Biol.* 47:269–302.

Fincham, J. R. S. 1966. *Genetic complementation*. Menlo Park, Calif.: W. A. Benjamin, Inc.

Fink, B. 1935. *The lichen flora of the United States*. Ann Arbor: University of Michigan Press.

Fisher, R. A. 1930. *The genetic theory of natural selection*. Oxford: Clarendon Press.

Fitch, W. M. 1966. Improved method for testing for evolutionary homology. *J. Mol. Biol.* 16:9–16.

Fitch, W. M., and Margoliash, E. 1970. The usefulness of amino acid and nucleotide sequences in evolutionary studies. *Evolutionary Biol.* 4:67–109.

Fitch, W. M., and Neel, J. V. 1969. The phylogenetic relationships of some Indian tribes of Central and South America. *Am. J. Human Genet.* 21:384–97.

Fogwill, M. 1958. Differences in crossing-over and chromosome size in sex cells of *Lilium* and *Fritillaria*. *Chromosoma* 9:493–504.

Ford, E. B. 1953. The genetics of polymorphism in the Lepidoptera. *Advan. Genet.* 5:43–87.

———. 1964. *Ecological genetics*, pp. 33–39. London: Methuen & Co., Ltd.

Fox, S. W. 1967. *Radiation and the first biopolymers: Radiation research, 1966*, pp. 714–29. Amsterdam: North Holland Publishing Co.

Fox, S. W., and Dose, K. 1972. *Molecular evolution and the origin of life*. San Francisco: W. H. Freeman & Co., Publishers.

Fox, S. W., and Krampitz, G. 1964. Catalytic decomposition of glucose in aqueous solution by thermal proteinoids. *Nature* 203:1362–64.

Fox, S. W., and Wood, A. 1968. *Origin of life*, pp. 67–76. New York: McGraw-Hill Book Co.

Fox, S. W., and Yuyama, S. 1963. Abiotic production of primitive protein and formed microparticles. *Ann. N.Y. Acad. Sci.* 108:487–94.

Francis, W. 1925. A contribution to the theory of the relationship of iron to the origin of life. *Proc. Roy. Soc. Queensland* 37:98–107.

Freud, S. 1930. Civilization and its discontents. International Psychoanalytical Library no. 17. London: Hogarth Press.

Fuller, J. L., and Thompson, W. R. 1960. *Behavior genetics.* New York: John Wiley & Sons, Inc.

Gabel, N. W., and Ponnamperuma, C. 1967. Model for origin of monosaccharides. *Nature* 216:453–55.

Gaffron H. 1960. The origin of life. In *The evolution of life*, ed. S. Tax. Chicago: University of Chicago Press.

Gall, J. G., and Callan, H. G. 1962. H^3 uridine incorporation in lampbrush chromosomes. *Proc. Nat. Acad. Sci. U.S.* 48:562–70.

Garen, A. 1968. Sense and nonsense in the genetic code. *Science* 160:149–59.

Garren, L. D.; Gill, G. N.; Masui, H.; and Walton, G. M. 1971. On the mechanism of action of ACTH. *Recent Progr. Hormone Res.* 27:433–78.

Garrison, W. M.; Morrison, D. C.; Hamilton, J. G.; Benson, A. A.; and Calvin, M. 1951. Reduction of carbon dioxide in aqueous solution by ionizing radiation. *Science* 114:416–18.

Geitler, L. 1959. Syncyanosen. In *Handbuch der Pflanzenphysiologie, band 11. Heterotrophie,* ed. W. Ruhland. Berlin: Springer.

Geschwind, I. I. 1959. Species variation in protein and polypeptide hormones. In *A textbook of comparative endocrinology,* eds. A. Gorbman and H. A. Bern. New York: John Wiley & Sons, Inc.

————. 1967. Molecular variation and the possible lines of evolution of peptide and protein hormones. *Am. Zool.* 7:89–108.

Geschwind, I. I.; Li, C. H.; and Barnafi, L. 1956. Isolation and structure of melanocyte-stimulating hormone from porcine pituitary glands. *J. Am. Chem. Soc.* 78:4494–95.

————. 1957. The structure of beta-melanocyte-stimulating hormone. *J. Am. Chem. Soc.* 79:620–25.

Geschwind, N. 1972. Language and the brain. *Sci. Am.* 226:76–83.

Gibson, T. C.; Scheppe, M. L.; and Cox, E. C. 1970. Fitness of an *Escherichia coli* mutator gene. *Science* 169:686–88.

Gilula, M. F., and Daniels, D. N. 1969. Violence and man's struggle to adapt. *Science* 164:396–405.

Glass, E. J. 1973. Vital statistics, U. S. Department of Health, Education and Welfare, unpublished data.

Goldschmidt, R. 1940. *The material basis of evolution.* New Haven: Yale University Press.

Gorbman, A., and Bern, H. A. 1959. *A textbook of comparative endocrinology.* New York: John Wiley & Sons, Inc.

Gorovsky, M. A., and Woodard, J. 1967. Histone content of chromosomal loci active in RNA synthesis. *J. Cell Biol.* 33:723–26.

Grant, V. 1963. *The origin of adaptations*. New York: Columbia University Press.

Grant, W. C., Jr., and Pickford, G. C. 1959. Presence of the red eft water-drive factor prolactin in the pituitaries of teleosts. *Biol. Bull.* 116:429–35.

Gray, R. W.; Dreyer, W. J.; and Hood, L. 1967. Evolution of immunoglobulins: Structural homology of kappa and lambda Bence Jones proteins. *Science* 155: 828–35.

Green, E. L. 1951. The genetics of a difference in skeletal type between two inbred strains of mice (BALB C and C57 blk). *Genetics* 36:391–409.

————. 1962. Quantitative genetics of skeletal variations in the mouse: II. Crosses between four inbred strains (C3H, DBA, C57BL, BALB/c). *Genetics* 47:1085–96.

Green, E. L., and Green, M. C. 1942. The development of three manifestations of the short-ear gene in the mouse. *J. Morphol.* 70:1–19.

Green, M. 1970. Oncogenic viruses. *Ann. Rev. Biochem.* 39:701–56.

Gropp, A.; Winking, H.; Zech, L.; and Muller, H. 1972. Robertsonian chromosomal variation and identification of metacentric chromosomes in feral mice. *Chromosoma* 39:265–88.

Gruenberg, B. C. 1929. *The story of evolution*. New York: Garden City Publishing Co.

Grüneberg, H. 1963. *The pathology of development*. New York: John Wiley & Sons, Inc.

Gulick, A. 1955. Phosphorus as a factor in origin of life. *Am. Sci.* 43:479–89.

Haldane, J. B. S. 1922. Sex ratio and unisexual sterility in hybrid animals. *J. Genet.* 12:105–9.

————. 1927. A mathematical theory of natural and artificial selection: Part IV. *Proc. Cambridge Phil. Soc.* 23:235–43.

————. 1932. *The cause of evolution*. London: Harper and Brothers.

————. 1937. The effect of variation on fitness. *Am. Naturalist* 71:337–49.

————. 1949. The rate of mutation of human genes. *Hereditas* 35 (suppl.): 267–73.

————. 1957. The cost of natural selection. *J. Genet.* 55:511–24.

————. 1964. A defense of "bean bag" genetics. *Perspectives Biol. Med.* 7: 343–53.

Hall, W., and Lehman, I. R. 1968. An in vitro transversion by a mutationally altered T4-induced DNA polymerase. *J. Mol. Biol.* 36:321–33.

Handschuh, G. J., and Orgel, L. E. 1973. Struvite and prebiotic phosphorylation. *Science* 179:483–84.

Hansen, H. N., and Smith, R. E. 1932. The mechanism of variation in imperfect fungi: *Botrytis cineria*. *Phytopathology* 22:953–64.

Harada, K., and Fox, S. W. 1960. The thermal copolymerization of aspartic acid and glutamic acid. *Arch. Biochem. Biophys.* 86:274–80.

Harris, H. 1970. *The principles of human biochemical genetics*. New York: American Elsevier Publishing Co.

Harris, I. 1966. The chemistry of intermediate lobe hormones. In *The pituitary gland*, eds. G. H. Harris and B. T. Donovan. Berkeley and Los Angeles: University of California Press.

Harris, J. I. 1959. The structure of alpha-melanocyte-stimulating hormone from pig pituitary glands. *Biochem. J.* 71:451–59.

Harris, J. I., and Roos, P. 1959. The structure of beta-melanocyte-stimulating hormone from pig pituitary glands. *Biochem. J.* 71:434–45.

Harrison, G. A.; Weiner, J. S.; Tanner, J. M.; and Barnicot, W. A. 1964. *Human biology*. Oxford: Oxford University Press.

Hawley, E. S., and Wagner, R. P. 1967. Synchronous mitochondrial division. *J. Cell. Biol.* 35:489–99.

Helevig, E. R. 1942. Unusual integration of the chromatin in *Machaerocera* and other genera of the Acrididae (Orthoptera). *J. Morphol.* 71:1–33.

Hershey, A. D., and Chase, M. 1952. Independent functions of viral protein and nucleic acid in growth of bacteriophage. *J. Gen. Physiol.* 36:39–56.

Hill, R. L.; Delaney, R.; Fellows, R. E., Jr.; and Lebovitz, H. E. 1968. The evolutionary origins of the immunoglobins. *Proc. Nat. Acad. Sci. U.S.* 56:1762–69.

Hof van't, J., and Sparrow, A. H. 1963. A relationship between DNA content, nuclear volume, and minimum mitotic cycle time. *Proc. Nat. Acad. Sci. U.S.* 49:897–902.

Holloway, R. L. 1969. Some questions on parameters of neural evolution in primates. *Ann. N.Y. Acad. Sci.* 167:332–40.

Holloway, R. L., Jr. 1967. Structural reduction through the "probable mutation effect": A critique with questions regarding human evolution. *Am. J. Phys. Anthropol.* 25:7–12.

Holtfreter, J. 1947. Changes of structure, and the kinetics of differentiating embryonic cells. *J. Morphol.* 80:57–91.

Hopkinson, D. A., and Harris, H. 1971. Recent work on isozymes in man. *Ann. Rev. Genet.* 5:5–32.

Horowitz, N. H. 1965. The evolution of biochemical synthesis: Retrospect and prospect. In *Evolving genes and proteins*, eds. V. Bryson and H. J. Vogel. New York: Academic Press Inc.

Hoyer, B. H.; McCarthy, B. J.; and Bolton, E. T. 1965. A molecular approach in the systematics of higher organisms. *Science* 144:959–67.

Huehns, E. R.; Hecht, F.; Keil, J. V.; and Motulsky, A. G. 1964. Developmental hemoglobin anomalies in a chromosomal triplication: D_1 trisomy syndrome. *Proc. Nat. Acad. Sci. U.S.* 51:89–97.

Hull, D. E. 1960. Thermodynamics and kinetics of spontaneous generations. *Nature* 186:693–94.

Hulse, F. S. 1965. *The human species*. New York: Random House, Inc.

Hutt, F. B. 1949. *Genetics of the fowl*. New York: McGraw-Hill Book Co.

Hyman, L. H. 1951. *The invertebrates: Platyhelminthes and Rhynchocoela*. New York: McGraw-Hill Book Co.

Ibanez, J. D.; Kimball, A. P.; and Oro, J. 1971. Possible prebiotic condensation of mononucleotides by cyanamide. *Science* 173:444–45.

Ingalls, A. M.; Dickie, M. M.; and Snell, G. D. 1950. Obese: A new mutant in the house mouse. *J. Heredity* 41:317–18.

Ingram, V. M. 1963. *The hemoglobins in genetics and evolution.* New York: Columbia University Press.

Irwin, M. R., and Cumley, R. W. 1945. Suggestive evidence for duplicate genes in a species hybrid in doves. *Genetics* 30:363–75.

Itikawa, N. 1955. An example of the silkworm moth which has the third wing incompletely developed. *Acta Sericol (Tokyo)* 12:13–15 (in Japanese).

Ives, P. T. 1950. The importance of mutation rate genes in evolution. *Evolution* 4:236–52.

Jackson, R. C. 1964. Preferential segregation of chromosomes from a trivalent in *Haplopappus gracilis. Science* 145:511–13.

Jacob, F., and Monod, J. 1961. On the regulation of gene activity. *Cold Spring Harbor Symp. Quant. Biol.* 26:193–211.

Jardine, N., and McKenzie, D. 1972. Continental drift and the dispersal and evolution of organisms. *Nature* 235:20–24.

Jerison, H. J. 1963. Interpretation of the evolution of the brain. *Human Biol.* 35:263–91.

Johnson, F. M., and Denniston, C. C. 1964. Genetic variation of alcohol dehydrogenase in *Drosophila melanogaster. Nature* 204:906–7.

Jones, R. T.; Brimbhall, B.; Huisman, T. H. J.; Kleihauer, E.; and Betke, K. 1966. Hemoglobin Freiburg: Abnormal hemoglobin due to the deletion of a single amino-acid residue. *Science* 154:1025–27.

Kacian, D. L.; Mill, D. R.; Kramer, F. R.; and Spiegelman, S. 1972. A replicating RNA molecule suitable for a detailed analysis of extracellular evolution and replication. *Proc. Nat. Acad. Sci. U.S.* 69:3038–42.

Kandutsch, A. A., and Coleman, D. L. 1967. Inherited metabolic variations. In *The biology of the laboratory mouse,* ed. E. L. Green. New York: McGraw-Hill Book Co.

Kant, I. 1929. *Critique of pure reason,* trans. Norman Kemp Smith. New York: Macmillan Co.

Kastritsis, C. D. 1966. A comparative chromosome study in the incipient species of the *Drosophila paulistorum* complex. *Chromosoma* 19:208–22.

————. 1967. A comparative study of the chromosomal polymorphs in the incipient species of the *Drosophila paulistorum* complex. *Chromosoma* 23:180–202.

Keosian, J. 1964. *The origin of life.* New York: Reinhold Book Corp.

Keyl, H. G. 1965. Duplikationen von center einherten der chromosomalen DNS während der evolution von chiromomus thumime. *Chromosoma* 17:139–80.

Kimura, M. 1956a. A model of genetic system which leads to closer linkage by natural selection. *Evolution* 10:278–87.

————. 1956b. Random genetic drift in a tri-allelic locus: Exact solution with a continuous model. *Biometrics* 12:57–66.

————. 1968. Evolutionary rate at the molecular level. *Nature* 217:624–26.

————. 1970. The length of time required for a selectively neutral mutant to reach fixation through random frequency drift in a finite population. *Genet. Res.* 15:131–33.

Kimura, M., and Ohta, T. 1969. The average number of generations until fixation of a mutant gene in a finite population. *Genetics* 61:763–71.

Kimura, M., and Weiss, G. H. 1964. The stepping-stone model of population structure and the decrease of genetic correlation with distance. *Genetics* 49:561–76.

King, J. L., and Jukes, T. H. 1969. Non-Darwinian evolution. *Science* 164:788–98.

Kirk, R. L. 1968. *The haptoglobin groups in man.* Basel: S. Karger.

Kit, S.; Dubbs, D. R.; and Somers, K. 1971. Strategy of Simian virus 40. In *Strategy
of the viral genome*, eds. G. E. W. Wolstenholme and M. O'Connor. Edin-
burgh and London: Churchill Livingstone.

Kliss, R. M., and Matthews, C. N. 1962. Hydrogen cyanide dimer and chemical
evolution. *Proc. Nat. Acad. Sci. U.S.* 48:1300–5.

Kojima, K. 1965. The evolutionary dynamics of two gene systems. In *Computers
in biomedical research*, eds. R. W. Stacy and B. Waxman. New York: Academic Press Inc.

Kubai, D. F., and Ris, H. 1969. Division in the dinoflagellate *Gyrodinium cohnii*
(Schiller). *J. Cell Biol.* 40:508-28.

Kuo, T. H. 1967. Human genetic study on Yami tribe in Orchid Island. *J. For-
mosan Med. Assoc.* 66:1–20.

Lamoreux, W. F. 1941. The autosexing ancobar. *J. Heredity* 32:221–26.

LaMotte, M. 1959. Polymorphism in natural populations of *Cepaea nemoralis*.
Cold Spring Harbor Symp. Quant. Biol. 24:65–86.

Landauer, W. 1956. Rudimentation and duplication of the radius in the duplicate
mutant form of fowl. *J. Genet.* 54:199–218.

Lane, P. W., and Dickie, M. M. 1968. Three recessive mutations producing disproportionate dwarfing in mice: Achondroplasia, brachymorphic, and stubby.
J. Heredity 59:300–8.

Lanham, U. N. 1968. The Blochmann bodies: Hereditary intracellular symbionts
of insects. *Biol. Rev.* 43:269–86.

Lascelles, J. 1965. Comparative aspects of structure associated with electron transport. *Symp. Soc. Gen. Microbiol.* 15:32–56.

Lasker, G. W. 1969. Human biological adaptability. *Science* 166:1480–86.

Laughnan, J. R. 1949. The action of allelic forms of the gene *A* in maize: II. The
relationship of crossing-over to mutation of A^b. *Proc. Nat. Acad. Sci. U.S.*
35:167–78.

Law, C. N. 1967. The location of genetic factors affecting a quantitative character
in wheat. *Genetics* 56:445–61.

Leadbeater, B., and Dodge, J. D. 1966. The fine structure of *Wolosznskia micra*
sp. nov.: A new marine dinoflagellate. *Brit. Phycol. Bull.* 3:1–17.

———. 1967. An electron microscope study of nuclear and cell division in a
dinoflagellate. *Arch. Mikrobiol.* 57:239–54.

Lederberg, J. 1952. Cell genetics and hereditary symbiosis. *Physiol. Rev.* 32:403–
30.

Lerner, I. M. 1954. *Genetic homeostasis.* New York: John Wiley & Sons, Inc.

Lett, J. T.; Klucis, E. S.; and Sun, C. 1970. On the size of the DNA in the mammalian chromosome: Structural subunits. *Biophys. J.* 10:277–92.

Lewis, E. B. 1964. Genetic control and regulation of developmental pathways. In
The role of chromosomes in development, ed. M. Locke. New York: Academic
Press Inc.

————. 1967. Genes and gene complexes. In *Heritage from Mendel*, ed. R. A. Brink. Madison: University of Wisconsin Press.

Lewis, H. 1953. The mechanism of evolution in the genus *Clarkia*. *Evolution* 16:257–71.

Lewontin, R.; Kirk, D.; and Crow, J. 1968. Selective mating, assortative mating, and inbreeding: Definitions and implications. *Eugen. Quart.* 15:141–43.

Lewontin, R. C., and Kojima, K. 1960. The evolutionary dynamics of complex polymorphisms. *Evolution* 14:458–72.

Lewontin, R. C., and White, M. J. D. 1960. Interaction between inversion polymorphisms of two chromosome pairs in the grasshopper *Moraba scurra*. *Evolution* 14:116–29.

Li, C. C. 1955. *Population genetics*. Chicago: University of Chicago Press.

Li, C. H. 1959. The relationship of chemical structure to biologic activity of pituitary hormones. *Lab. Invest.* 8:574–87.

————. 1963. The way the load ratio works. *Am. J. Human Genet.* 15:316–21.

Lindegren, C. C. 1934. The genetics of *Neurospora*: V. Self-sterile bisexual heterokaryons. *J. Genet.* 28:425–35.

Livingstone, D. A. 1971. Speculation on the climatic history of mankind. *Am. Scientist* 59:332–37.

Livingstone, F. B. 1960. Natural selection, disease, and ongoing human evolution as illustrated by the ABO blood groups. *Human Biol.* 32:17–27.

Lohrmann, R., and Orgel, L. E. 1971. Urea-inorganic phosphate mixtures as prebiotic phosphorylating agents. *Science* 171:490–94.

Lovejoy, C. O., and Meindl, R. S. 1972. Eukaryote mutation and the protein clock. *Yearbook Phys. Anthropol.* 16:18–30.

Lubs, H. A., and Ruddle, F. H. 1970. Chromosomal abnormalities in the human population: Estimation of rates based on New Haven newborn study. *Science* 169:495–70.

Lush, J. L. 1945. *Animal breeding plans*. 3d ed. Ames: Iowa State University Press.

Lyon, M. F. 1961. Gene action in the X-chromosome of the mouse (*Mus musculus*). *Nature* 190:372–73.

Magni, G. E., and Sora, S. 1969. Relationships between recombination and mutation. In *Mutation as cellular process*, eds. G. E. W. Wolstenholme and Maeve O'Connor. London: J. & A. Churchill.

Makino, S. 1936. The spiral structure of chromosomes in the meiotic division of Podisma (Orthoptera). *J. Fac. Sci. Hokkaido Univ. Ser. VI* 5:29–40.

————. 1951. *An atlas of the chromosome numbers in animals*. Ames: Iowa State University Press.

Malde, H. E. 1964. Environment and man in arid America. *Science* 145:123–29.

Malecot, G. 1948. *Les mathématiques de l'hérédité*. Paris: Masson et Cie.

Margolin, P. 1963. Genetic fine structure of the leucine operon in *Salmonella*. *Genetics* 48:441–57.

Margulis, L. 1968. Evolutionary criteria in thallophytes: A radical alternative. *Science* 161:1020–22.

————. 1974. Five-kingdom classification and the origin and evolution of cells. *Evol. Biol.* 7:46–78.

Markert, C. L. 1964. Developmental genetics. *Harvey Lectures* 59:187–218.

Markert, C. L., and Faulhauber, I. 1965. Lactate dehydrogenase isozyme patterns of fish. *J. Exp. Zool.* 159:319–32.

Marmur, J., and Lane, D. 1960. Strand separation and specific recombination in deoxyribonucleic acids: Biological studies. *Proc. Nat. Acad. Sci. U.S.* 46:453–61.

Marr, A. G. 1960. Enzyme localization in bacteria. *Ann. Rev. Microbiol.* 14:241–60.

Marx, L. 1964. *The machine in the garden.* Oxford: Oxford University Press.

Mather, K. 1943. Polygenic inheritance and natural selection. *Biol. Rev.* 18:32–64.

————. 1964. *Human diversity*, pp. 84–91. New York: Free Press.

Mayr, E. 1942. *Systemics and the origin of species.* New York: Columbia University Press.

————. 1959. Where are we? *Cold Spring Harbor Symp. Quant. Biol.* 24:1–14.

————. 1963. *Animal species and evolution.* Cambridge: Harvard University Press.

McClintock, B. 1933. The association of non-homologous parts of chromosomes in the mid-prophase of meiosis in *Zea mays. Z. Zellforsch. Mikrosk. Anat.* 19:191–237.

————. 1941. The stability of broken ends of chromosomes in *Zea mays. Genetics* 26:234–82.

————. 1951. Chromosome organization and expression. *Cold Spring Harbor Symp. Quant. Biol.* 16:13–63.

————. 1967. Genetic systems regulating gene expression during development. In *Control mechanisms in developmental process*, ed. M. Locke, pp. 84–111. New York: Academic Press Inc.

McKusick, V. A. 1960. *Heritable disorders of connective tissue.* St. Louis: C. V. Mosby Co.

Menkes, B.; Deleanu, M.; and Ilies, A. 1965. Comparative study of some areas of physiological necrosis in the embryo of man, some laboratory mammals and fowl. *Rev. Roumaine Embryol. Cytol.* 1:69–77.

Mesthene, E. G. 1968. How technology will shape the future. *Science* 165:135–43.

Mettler, L. E., and Gregg, T. G. 1969. *Population genetics and evolution*, p. 33. Englewood Cliffs, N.J.: Prentice-Hall, Inc.

Metzenberg, R. L. 1972. Genetic regulatory systems in *Neurospora. Ann. Rev. Genet.* 6:111–32.

Milaire, J. 1963. Étude morphologique et cytochimique de développment des membres chez la souris et chez la taupe. *Arch. Biol. (Liege)* 74:131–317.

Milkman, R. D. 1967. Heterosis as a major cause of heterozygosity in nature. *Genetics* 55:493–95.

Miller, O. L., Jr., and Beatty, B. R. 1969. Portrait of a gene. *J. Cellular Physiol.* 74(suppl.):225–32.

Miller, S. L. 1953. A production of amino acids under possible primitive earth conditions. *Science* 117:528–29.

Miller, S. L., and Parris, M. 1964. Synthesis of pyrophosphate under primitive earth conditions. *Nature* 204:1248–50.

Mills, D. R.; Kramer, F. R.; and Spiegelman, S. 1973. Complete nucleotide sequence of a replicating RNA molecule. *Science* 180:916–27.

Mintz, B. 1964. Formation of genetically mosaic mouse embryos and early development of lethal (t^{12}/t^{12}) normal mosaics. *J. Exp. Zool.* 157:273–91.

————. 1967. Gene control of mammalian pigmentary differentiation: I. Clonal origin of melanocytes. *Proc. Nat. Acad. Sci. U.S.* 58:341–51.

Mishra, N. C. 1971. Heterokaryosis in *Neurospora sitophila*. *Genetics* 67:55–59.

Mitchell, H. K., and Mitchell, M. B. 1952. A case of "maternal" inheritance in *Neurospora crassa*. *Proc. Nat. Acad. Sci. U.S.* 38:442–49.

Moncrief, L. W. 1970. The cultural basis for our environmental crisis. *Science* 170:508–12.

Morgan, T. H. 1917. The theory of the gene. *Am. Naturalist* 51:513–44.

Morton, J. K. 1966. The role of polyploidy in the evolution of a tropical flora. In *Chromosomes today*, eds. C. D. Darlington and K. R. Lewis. New York: Plenum Publishing Corp.

Morton, N. E.; Crow, J. F.; and Muller, H. J. 1956. An estimate of the mutational damage in man from data on consanguinous marriages. *Proc. Nat. Acad. Sci. U.S.* 42:855–63.

Moses, M. J.; Counce, S. J.; and Paulson, D. F. 1975. Synaptonemal complex complement of man in spreads of spermatocytes with details of the sex chromosome pair. *Science* 187:363–65.

Mourant, A. E.; Kopec, A. C.; and Domaniewska-Sobczak, K. 1976. *The distribution of the human blood groups*. London: Oxford University Press.

Mudal, S. 1951. The chromosomes of the earthworms: I. The evolution of polyploidy. *Heredity* 6:55–76.

Mueller, G. C.; Voderhaar, B.; Kim, U. H.; and Le Mahieu, M. 1972. Estrogen action: An inroad to cell biology. *Recent Progr. Hormone Res.* 28:1–45.

Mukai, T. 1964. The genetic structure of natural populations of *Drosophila melanogaster*: I. Spontaneous mutation rate of polygenes controlling viability. *Genetics* 50:1–19.

Mulder, M. P.; van Duijn, P.; and Gloor, H. J. 1968. The replicative organization of DNA in polytene chromosomes of *Drosophila hydei*. *Genetics* 39:385–428.

Muller, H. J. 1925. Why polyploidy is rarer in animals than in plants. *Am. Naturalist* 59:346–53.

————. 1932. Further studies on the nature and cause of gene mutations. In *Proceedings of the Sixth International Congress on Genetics*, vol. 1, pp. 213–55. New York: Brooklyn Botanical Gardens.

————. 1935a. A viable two-gene deficiency. *J. Heredity* 26:469–77.

————. 1935b. *Out of the night: A biologist's view of the future*. New York: Vanguard Press Inc.

————. 1950. Our load of mutations. *Am. J. Human Genet.* 2:111–76.

————. 1962. *Studies in genetics*. Bloomington: Indiana University Press.

Nason, J. 1965. *Textbook of modern biology*. New York: John Wiley & Sons, Inc.

Nass, M. M. K. 1969. Mitochondrial DNA: Advances, problems, and goals. *Science* 165:25–35.

Nass, S. 1969. The significance of the structural and functional similarities of bacteria, and mitochondria. *Intern. Rev. Cytol.* 25:55–129.

Navyvary, J., and Nagpal, K. L. 1972. Oligothymidylates: Formation by thermal condensation of O^2, 5'-cyclothymidine 3'-phosphate. *Science* 177:272–74.

Neel, J., and Schull, W. J. 1972. Differential fertility and human evolution. *Evolutionary Biol.* 6:363–78.

Nicoll, C. S.; Pfeiffer, E. S.; and Revold, H. R. 1967. Prolactin and nesting behavior in phalaropes. *Gen. Comp. Endocrinol.* 8:61–65.

Novitski, E. 1961. Chromosome breakage in inversion heterozygotes. *Am. Naturalist* 95:250–52.

Noyes, W. A., and Leighton, P. A. 1967. *The photochemistry of gases.* New York: Dover Publications.

Ohno, S. 1969. Evolution of sex chromosomes in mammals. *Ann. Rev. Genet.* 3:495–524.

————. 1972. Gene duplication, mutation, and mammalian genetic regulatory systems. *J. Med. Genet.* 9:254–62.

Ohno, S.; Muramoto, J.; Christian, L.; and Atkin, N. B. 1967. Diploid-tetraploid relationship among Old-World members of the fish family Cyprinidae. *Chromosoma* 23:1–9.

Ohno, S.; Wolf, U.; and Atkin, N. B. 1968. Evolution from fish to mammals by gene duplication. *Hereditas* 59:169–87.

Olson, E. C. 1965. *The evolution of life.* New York: New American Library Inc.

Oparin, A. I. 1957. *The origin of life on the earth.* New York: Academic Press Inc.

Orgel, L. E. 1968. Evolution of the genetic apparatus. *J. Mol. Biol.* 38:381–93.

Oro, J. 1963. Synthesis of organic compounds by electric discharge. *Nature* 197: 862–67.

————. 1965. Stages and mechanisms of prebiological organic synthesis. In *Origins of prebiological systems and of their molecular matrices*, ed. S. W. Fox. New York: Academic Press Inc.

Oro, J., and Cox, A. C. 1962. No enzyme synthesis of 2-deoxyribose. *Fed. Proc.* 21:80.

Oro, J., and Kimball, A. P. 1962. Synthesis of purines under possible primitive earth conditions: II. Purine intermediates from hydrogen cyanide. *Arch. Biochem. Biophys.* 96:293–313.

Osborn, H. F. 1918. *The origin and evolution of life.* New York: Charles Scribner's & Sons.

Paigen, K. 1970. The genetics of enzyme realization. In *Enzyme synthesis and degradation in mammalian systems*, ed. M. Rechcigl. Basel: S. Karger.

Pastor, J. B., and Callan, H. G. 1952. Chiasma formation in spermatocytes of turbellarian *Dendrocoeleum lacteum. J. Genet.* 50:449–54.

Patterson, J. T., and Stone, W. S. 1952. *Evolution in the genus* Drosophila. New York: Macmillan Co.

Paul, J., and Gilmour, R. S. 1968. Organ-specific restriction of transcription in mammalian chromatin. *J. Mol. Biol.* 34:305–16.

Pauling, L., and Corey, R. B. 1956. Specific hydrogen-bond formation between pyrimidines and purines in deoxyribonucleic acids. *Arch. Biochem. Biophys.* 65:164–81.

Pearl, R., and Reed, L. J. 1920. On the rate of growth of the population of the United States since 1790 and its mathematical representation. *Proc. Nat. Acad. Sci. U.S.* 6:275–88.

Pelling, C. 1966. A replicative and synthetic chromosomal unit: The modern concept of the chromomere. *Proc. Roy. Soc. (London) Ser. B* 164:279–89.

Penrose, L. S. 1963. Finger-prints, palms, and chromosomes. *Nature* 197:933–38.

Perutz, M. F.; Kendrew, J. C.; and Watson, H. C. 1965. Structure and function of haemoglobin: II. Some relations between polypeptide chain configuration and amino-acid sequence. *J. Mol. Biol.* 13:669–78.

Pfeil, E., and Ruckert, H. 1961. Formation of sugars from formaldehyde by the action of alkalies. *Ann. Chem.* 641:121–31.

Pierce, J. G. 1971. Eli Lilly Lecture: The subunits of pituitary thyrotropin: Their relationship to other glycoprotein hormones. *Endocrinology* 89:1331–44.

Pilbeam, D. R., and Simons, E. L. 1965. Some problems of hominid classification. *Am. Scientist* 53:237–59.

Polanyi, M. 1968. Life's irreducible structure. *Science* 160:1308–12.

Pongs, O., and Ts'o, P. O. 1971. Polymerization of unprotected 2'-deoxyribonucleoside 5'-phosphates at elevated temperature. *J. Am. Chem. Soc.* 93:5241–50.

Ponnamperuma, C. 1965. *A biological synthesis of some nucleic acid constituents*, ed. S. W. Fox, pp. 221–36. New York: Academic Press Inc.

Ponnamperuma, C., and Mariner, R. 1963. The formation of ribose and deoxyribose by ultraviolet irradiation of formaldehyde in water. *Radiation Res.* 19:183.

Ponnamperuma, C.; Sagan, C.; and Mariner, R. 1963. Synthesis of adenosine triphosphate under possible primitive earth conditions. *Nature* 199:222–26.

Pontecorvo, G. 1946. Genetic systems based on heterokaryosis. *Cold Spring Harbor Symp. Quant. Biol.* 11:193–201.

Potier, P. 1918. *Les symbiotes*. Paris: Masson et Cie.

Potter, E. A. 1947. The changing cancer death rate. *Cancer Res.* 7:351–55.

Prakash, S.; Lewontin, R. C.; and Hubby, J. L. 1969. A molecular approach to the study of genic heterozygosity in natural populations: Part IV. *Genetics* 61:841–58.

Rabinowitz, J.; Chang, S.; and Ponnamperuma, C. 1968. Phosphorylation on the primitive earth. *Nature* 218:442–43.

Rajalakshmi, R.; Deodhar, A. D., and Ramakrishnan, C. V. 1965. Vitamin C secretion during lactation. *Acta Paediat. Scand.* 54:375–82.

Rao, C. R. 1955. *Advanced statistical methods in biometric research*. New York: John Wiley & Sons, Inc.

Ratcliffe, S.; Stewart, A.; Melville, M.; Jacobs, P.; and Keay, A. J. 1970. Chromosome studies on 3,500 newborn male infants. *Lancet* 1:121–22.

Raven, P. H. 1970. A multiple origin for plastids and mitochondria. *Science* 169:641–46.

Rendel, J. M. 1951. Mating of ebony vestigial and wild type *Drosophila melanogaster* in light and dark. *Evolution* 5:226–30.

Rensch, B. 1959. Trends toward progress of brains and sense organs. *Cold Spring Harbor Symp. Quant. Biol.* 24:291–304.

———. 1960. *Evolution above the species level*. New York: Columbia University Press.

Reznikoff, W. S. 1972. The operon revisited. *Ann. Rev. Genet.* 6:133–56.

Ridley, S. M., and Leech, R. M. 1970. Division of chloroplasts in an artificial environment. *Nature* 227:463–65.

Rieder, R. F., and Bradley, T. B. 1968. Hemoglobin Gun Hill: An unstable protein associated with chronic hemolysis. *Blood* 32:355–69.

Rieger, R.; Michaelis, A.; and Green, M. M. 1968. *A glossary of genetics and cytogenetics*. New York: Springer-Verlag New York, Inc.

Riggs, A. D.; Reiness, G.; and Zubay, G. 1971. Purification and DNA-binding properties of the catabolite gene activator protein. *Proc. Nat. Acad. Sci. U.S.* 68:1222–25.

Ris, H. 1961. Ultrastructure and molecular organization of genetic systems. *Can. J. Genet. Cytol.* 3:95–120.

Ris, H., and Kubai, D. F. 1970. Chromosome structure. *Am. Rev. Genet.* 4:263–94.

Roberts, J. A. Fraser. 1961. Some association between blood groups and diseases. *Br. Med. Bull.* 15:129–33.

Robertson, S. G. 1967. The nature of quantitative genetic variation. In *Heritage from Mendel*, ed. R. A. Brink, pp. 265–80. Madison: University of Wisconsin Press.

Robinson, R. 1958. Genetic studies of the rabbit. *Bibliog. Genet.* 17:229–558.

Roderick, T. H., and Hawes, N. L. 1970. Two radiation-induced chromosomal inversions in mice (*Mus musculus*). *Proc. Nat. Acad. Sci. U.S.* 67:961–67.

Roderick, T. H.; Wimer, R. E.; Wimer, C. C.; and Schwartzkroin, P. A. 1973. Genetic and phenotypic variation in weight of brain and spinal cord between inbred strains of mice. *Brain Res.* 64:245–53.

Romer, A. S. 1949. Time series and trends in animal evolution. In *Genetics, paleontology, and evolution*, eds. G. L. Jepsen, E. Mayr, and G. G. Simpson, pp. 103–20. Princeton: Princeton University Press.

———. 1968. *The procession of life*. New York: World Publishing Co.

Rowe, W. 1973. Genetic transmission of murine leukemia viruses. Paper read at the Thirteenth International Genetic Congress, Berkeley, Calif.

Rubey, W. W. 1951. Geographic history of sea water. *Bull. Geol. Soc. Am.* 62:1111–48.

———. 1955. Development of the hydrosphere and atmosphere, with special reference to the probable composition of the early atmosphere. *Geol. Soc. Am. Spec. Papers* 62:631–50.

Rudkin, G. T., and Schultz, J. 1961. Disproportionate synthesis of DNA in polytene chromosome regions in *Drosophila melanogaster*. *Genetics* 46:893–94.

Russell, E. S. 1948. A quantitative histological study of the pigment found in the coat-color of the house mouse: I. Variable attributes of the pigment granules. *Genetics* 31:327–46.

Russell, R. L., and Coleman, D. L. 1963. Genetic control of hepatic δ-aminolevuli-nate dehydratase in mice. *Genetics* 48:1033–39.

Russell, W. L. 1963. The effect of radiation dose rate and fractionation on muta-tion in mice. In *Repair from genetic radiation damage*, ed. F. H. Sobels, pp. 205–17. London: Pergamon Press, Ltd.

Salthe, S. N.; Chilson, O. P.; and Kaplan, N. O. 1965. Hybridization of lactic de-hydrogenase in vivo and in vitro. *Nature* 207:723–26.

Sambrook, H.; Westphal, H.; Sprinivasan, P. R.; and Dulbecco, R. 1968. The inte-grated state of viral DNA in SV40-transformed cells. *Proc. Nat. Acad. Sci. U.S.* 60:1288–95.

Sanchez, R. A.; Ferris, J. P.; and Orgel, L. E. 1966. Cyanoacetyline in prebiotic synthesis. *Science* 154:784–85.

Sanghvi, L. D. 1963. The concept of genetic load: A critique. *Am. J. Human Genet.* 15:298–309.

Sarich, V. M. 1971. A molecular approach to the question of human origins. In *Background of man*, eds. V. M. Sarich and P. Dolhinow. Boston: Little, Brown & Co.

Sarich, V. M., and Wilson, A. C. 1967. Immunological time scale for hominoid evolution. *Science* 158:1200–3.

Saunders, J. W., Jr., and Fallon, J. F. 1966. Cell death in morphogenesis. In *Major problems in developmental biology*, ed. M. Locke, pp. 289–312. New York: Academic Press Inc.

Sawyer, W. H. 1966. Biological assays for neurohypophysial principles in tissues and in blood. In *The pituitary gland*, eds. G. H. Harris and B. T. Donovan. Berkeley and Los Angeles: University of California Press.

Schlager, G., and Dickie, M. M. 1967. Spontaneous mutations and mutation rates in the mouse. *Genetics* 57:319–30.

Schmitt, F. O. 1951. Structural and chemical studies on collagen. *J. Am. Leather Chemists' Assoc.* 46:538–47.

Schull, W. J., and Neel, J. V. 1965. *The effects of inbreeding on Japanese children.* New York: Harper & Row, Publishers.

Schultz, A. H. 1969. *The life of primates*, p. 246. New York: Universe Books.

Schultz, J. 1965. Genes, differentiation, and animal development. In *Genetic con-trol of differentiation*, pp. 116–47. Upton, N.Y.: Brookhaven National Labor-atory.

Schwartz, A., and Ponnamperuma, C. 1968. Phosphorylation of adenosine with linear polyphosphate salts in aqueous solution. *Nature* 218:443.

Schwartz, D., and Laughner, W. J. 1969. A molecular basis for heterosis. *Science* 166:626–27.

Scott, J. J. 1956. Synthesis of crystallizable porphobilinogen. *Biochem. J.* 62:6–7.

Searle, A. G. 1968. *Comparative genetics of coat colour in mammals.* London: Logos Press.

Sheppard, P. M. 1952. A note on non-random mating in the moth *Panaxia domin-ula* (L). *Heredity* 6:239–41.

―――――. 1959. The evolution of mimicry. *Cold Spring Harbor Symp. Quant. Biol.* 24:131–40.

―――――. 1969. Evolutionary genetics of animal populations. In *Proceedings of the Twelfth International Congress of Genetics*, ed. C. Oshima, vol. 3, pp. 261–80. Tokyo: Science Council of Japan.

Shreffler, D. C. 1972. Genetic organization of *H-2* and its relationship to the Ir and MLC genes. In *Genetic control of immune responsiveness*, eds. H. O. Mc-Devitt and M. Landy. New York: Academic Press Inc.

Sillen, L. G. 1965. Oxidation state of earth's ocean and atmosphere: A model calculation on earlier states and myths of the "prebiotic soup." *Arkiv. Kemi.* 24:431–56.

Simpson, G. G. 1944. *Tempo and mode in evolution.* New York: Columbia University Press.

———. 1950. History of the fauna of Latin America. *Am. Scientist* 38:361–89.

———. 1951. *Horses: The story of the horse family in the modern world and through sixty million years of evolution.* Oxford: Oxford University Press.

———. 1953. *The major features of evolution.* New York: Columbia University Press.

———. 1961. *Principles of animal taxonomy.* New York: Columbia University Press.

———. 1972. The evolutionary concept of man. In *Sexual selection and the descent of man*, ed. B. Campbell. Chicago: Aldine-Atherton, Inc.

Singer, E. R. 1968. Lysogeny: The integration problem. *Ann. Rev. Microbiol.* 22:452–84.

Singer, S. F. 1966. *The origin and dynamical evolution of the moon.* Tarzana, Calif.: American Astronautical Society.

Singer, S. J., and Doolittle, R. F. 1966. Antibody active sites and immunoglobulin molecules. *Science* 153:13–25.

Slizynski, B. M. 1960. Sexual dimorphism in mouse gametogenesis. *Genet. Res.* 1:477–86.

Sloper, J. C. 1966. The experimental and cytopathological investigation of neurosecretion in the hypothalamus and pituitary. In *The pituitary gland*, eds. G. H. Harris and B. T. Donovan. Berkeley and Los Angeles: University of California Press.

Smith, D.; Muscatine, L.; and Lewis, D. 1969. Carbohydrate movement from autotrophs to heterotrophs in parasitic and mutualistic symbiosis. *Biol. Rev.* 44:17–90.

Smith, E. L. 1968. The evolution of proteins. *Harvey Lectures* 62:231–56.

Smith, E. L., and Margoliash, E. 1964. Evolution of cytochrome *c. Fed. Proc.* 23:1243–47.

Smith, K.; Church, R. B.; and McCarthy, B. J. 1969. Template specificity of isolated chromatin. *Biochemistry* 8:4271–77.

Smithies, O. 1964. Chromosomal rearrangements and protein structure. *Cold Spring Harbor Symp. Quant. Biol.* 29:309–19.

———. 1967. Antibody variability. *Science* 157:267–73.

Smithies, O.; Connell, G. E.; and Dixon, G. H. 1962. Chromosomal rearrangements and the evolution of haptoglobin genes. *Nature* 196:232–36.

Snell, G. D., and Stimpfling, J. H. 1966. Genetics of tissue transplantation. In *Biology of the laboratory mouse*, ed. E. L. Green. 2d ed. New York: McGraw-Hill Book Co.

Snyder, L. H., and David, P. R. 1957. *The principles of heredity*. Boston: D. B. Heath Co.

Somerville, R. L., and Yanofsky, C. J. 1964. On the translation of the *A* gene region of tryptophan messenger RNA. *Mol. Biol.* 8:616–19.

Sonneborn, T. M. 1948. The determination of hereditary antigenic differences in genetically identical paramedium cells. *Proc. Nat. Acad. Sci. U.S.* 34:413–18.

Spickett, S. J. 1963. Genetic and developmental studies of a quantitative character. *Nature* 199:870–73.

Spiegelman, S. 1973. Molecular genetics of human cancer. Paper read at the Thirteenth International Genetics Congress. Berkeley, California.

Srb, A. M., and Owen, R. D. 1952. *General genetics*. San Francisco: W. H. Freeman & Co., Publishers.

Stadler, L. J. 1942. Some observations on gene variability and spontaneous mutation. In *Spragg Memorial Lectures*. 3d ser. East Lansing: Michigan State College.

Stalker, H. D. 1966. The phylogenetic relationships of the species in the *Drosophila melanica* group. *Genetics* 53:327–42.

Stanier, R. Y. 1954. *Cellular metabolism and infections*, ed. E. Racker. New York: Academic Press Inc.

Starr, C. 1969. Social benefit versus technological risk. *Science* 165:1232–38.

Stebbins, G. L. 1966. Chromosomal variation and evolution. *Science* 152:1463–69.

———. 1967. Gene action, mitotic frequency, and morphogenesis in higher plants. *Develop. Biol.* 1(suppl):113–35.

Steinberg, A. G.; Bleibtreu, H. K.; Kurszynski, T. W.; Martin, A. O.; and Kurczynski, E. M. 1967. Genetic studies on an inbred human isolate. In *Proceedings of the Third International Congress of Human Genetics*, eds. J. F. Crow and J. V. Neel, pp. 267–89. Baltimore: Johns Hopkins University Press.

Steinman, B.; Lemmon, R. M.; and Calvin, M. 1964. Cyanamide: A possible key compound in chemical evolution. *Proc. Nat. Acad. Sci. U.S.* 52:27–38.

Stent, G. S. 1971. *Molecular genetics*. San Francisco: W. H. Freeman & Co., Publishers.

Stephens, S. G. 1951. Possible significance of duplication in evolution. *Advan. Genet.* 4:247–65.

Stephen-Sherwood, E.; Oro, J.; and Kimball, A. P. 1971. Thymine: A possible prebiotic synthesis. *Science* 173:446–47.

Stern, C. 1960. *Principles of human genetics*. San Francisco: W. H. Freeman & Co., Publishers.

Stern, H. 1963. Intracellular regulatory mechanisms in chromosome replication and segregation. *Fed. Proc.* 22:1097–1102.

Stewart, A. D. 1971. Genetic variation in the neurohypophysial hormones of the mouse, *Mus musculus*. *J. Endocrinol.* 51:191–201.

Stimpfling. J. H., and Richardson, A. 1965. Recombination within the histocompatibility-2 locus of the mouse. *Genetics* 51:831–46.

Stone, W. S. 1962. The dominance of natural selection and the reality of super species (species groups) in the evolution of *Drosophila*. In *Studies in genetics*, ed. M. R. Wheeler. Publication 6205, pp. 507–36. Austin: University of Texas Press.

Stone, W. S.; Guest, W. C.; and Wilson, F. D. 1960. The evolutionary implications of the cytological polymorphism and phylogeny of the virilis group of *Drosophila*. *Proc. Nat. Acad. Sci. U.S.* 46:350–61.

Streamer, F. A., and Pimental, D. 1961. Effects of immigration on the evolution of populations. *Am. Naturalist* 95:201–10.

Sturtevant, A. H. 1925. The effects of unequal crossing-over at the bar locus in *Drosophila*. *Genetics* 10:117-47.

———. 1929. Contributions to the genetics of *Drosophila stimulans* and *Drosophila melanogaster*: I. The genetics of *Drosophila simulans*. Publication 399, pp. 1–62. Washington, D. C.: Carnegie Institution of Washington.

Sturtevant, A. H., and Dobzhansky, T. 1936. Inversions in the third chromosome of wild races of *Drosophila pseudoobscura* and their use in the study of the history of these species. *Proc. Nat. Acad. Sci. U.S.* 22:448–50.

Sturtevant, A. H., and Morgan, T. H. 1923. Reverse mutation of the bar gene correlated with crossing-over. *Science* 57:746–47.

Sutherland, D. W. 1972. Studies on the mechanism of hormone action. *Science* 177:401–8.

Sutherland, E. D.; Øye, I.; and Butcher, R. W. 1965. The action of epinephrine and the role of the adenyl cyclase system in hormone action. *Recent Progr. Hormone Res.* 21:623–42.

Sutton, H. E. 1960. Human heredity and its cytologic bases. In *The metabolic basis of inherited disease*, eds. J. B. Standbury; J. B. Wyngaarden; and D. S. Fredrickson. New York: McGraw-Hill Book Co.

Sved, J. A.; Reed, T. E.; and Bodmer, W. F. 1967. The number of balanced polymorphisms that can be maintained in a natural population. *Genetics* 55:469–81.

Szutka, A. 1966. Formation of pyrrolic compounds by ultraviolet irradiation of delta-aminolevulinic acid. *Nature* 212:401–2.

Szutka, A.; Hazel, J. F.; and McNabb, W. 1959. Synthesis of porphine-like substances induced by gamma-irradiation. *Radiation Res.* 10:597–603.

Tan, C. C. 1946. Mosiac dominance in the inheritance of color patterns in the lady-bird beetle, *Harmonia axyridis*. *Genetics* 13: 195–210.

Teilhard de Chardin, P. 1959. *The phenomenon of man*. London: William Collins Sons & Co., Ltd.

Thoday, J. M. 1961. The location of polygenes. *Nature* 191:368–70.

Thoday, J. M., and Boam, T. B. 1961. Regular responses to selection: I. Description of responses. *Genet. Res.* 2:161–76.

Thomas, C. A., Jr. 1966. Recombination of DNA molecules. *Progr. Nucl. Acid. Res. Mol. Biol.* 5:315–43.

Thomas, C. A., Jr.; Hamkalo, B. A.; Misra, D. N.; and Lee, C. S. 1970. Cyclization of eucaryotic deoxyribonucleic acid fragments. *J. Mol. Biol.* 51:621–32.

Tice, S. C. 1914. A new sex-linked character in *Drosophila*. *Biol. Bull.* 26:221–30.

Tobias, P. V. 1971. *The brain in hominid evolution*. New York: Columbia University Press.

Trautner, T. A.; Swartz, M. N.; and Kornberg, A. 1962. Enzymatic synthesis of deoxyribonucleic acid: X. Influence of bromouracil substitutions on replication. *Proc. Nat. Acad. Sci. U.S.* 48:449–55.

Urey, H. C. 1952. On the early chemical history of the earth and the origin of life. *Proc. Nat. Acad. Sci. U.S.* 38:351–63.

————. 1966. Some general problems relative to the origin of life on earth or elsewhere. *Am. Naturalist* 100:285–88.

Ursprung, H. 1966. The formation of patterns in development. In *Major problems in developmental biology*, ed. M. Locke. New York: Academic Press Inc.

Wacker, A., and Chandra, P. 1969. Aspects of modification of nucleic acids in mutational processes. In *Mutation as cellular process*, eds. G. E. W. Wolstenholme and Maeve O'Connor. London: J. & A. Churchill.

Waddington, C. H. 1957. *The strategy of the genes.* New York: Macmillan Co.

————. 1959. Evolutionary adaptation. *Perspectives Biol. Med.* 2:379–401.

————. 1960. Evolutionary adaptation. In *Evolution of life*, ed. S. Tax, pp. 381–402. Chicago: University of Chicago Press.

————. 1961. *The nature of life.* London: George Allen and Unwin.

————. 1966. Mendel and the study of development. *Proc. Roy. Soc. (London) Ser. B* 164:219–29

Waehneldt, T. V., and Fox, S. W. 1967. Condensation of cytidylic acid in the presence of polyphosphoric acid. *Biochim. Biophys. Acta* 134:9–16.

Wagner, R. P., and Mitchell, H. K. 1955. *Genetics and metabolism.* New York: John Wiley & Sons, Inc.

Wald, G. 1954. The origin of life. *Sci. Am.* 191:44–53.

————. 1957. The origin of optical activity. *Ann. N.Y. Acad. Sci.* 69:352–68.

Wallace, B. 1963. The elimination of an autosomal lethal from an experimental population of *Drosophila melanogaster. Am. Naturalist* 97:65–66.

————. 1968. *Topics in population genetics.* New York: W. W. Norton & Co., Inc.

Wallace, B., and Vetukhiv, M. 1955. Adaptive organization of the gene pools of *Drosophila* populations. *Cold Spring Harbor Symp. Quant. Biol.* 20:303–10.

Wallin, I. E. 1927. *Symbioticism and the origin of species.* Baltimore: Williams & Wilkins Co.

Wang, A. C., and Fudenberg, H. H. 1974. IgA and evolution of immunoglobulins. *J. Immunogenet.* 1:3–31.

Wasserman, M. 1960. Cytological and phylogenetic relationships in the repleta group of the genus *Drosophila. Proc. Nat. Acad. Sci. U.S.* 46:842–59.

————. 1963. Cytology and phylogeny of *Drosophila. Am. Naturalist* 97:333–52.

Wasserman, M. V. 1962 Cytological studies of the repleta group of the genus *Drosophila*: III. The mercatorum subgroup. IV. The hydei subgroup. V. The mulleri subgroup. VI. The fasciola subgroup. Publication 6205, pp. 63–134. Austin: University of Texas Press.

Watson, J. D. 1970. *Molecular biology of the gene.* 2d ed. New York: W. A. Benjamin, Inc.

Watson, J. D., and Crick, F. H. C. 1953. A structure for deoxyribonucleic acid. *Nature* 171:737–38.

Weaver, K. F. 1973. Have we solved the mysteries of the moon? *Nat. Geograph.* 144:309–25.

Weiner, J. S. 1964. Climate adaptation. In *Human biology*, eds. G. A. Harrison; J. S. Weiner; J. M. Tanner; and N. A. Barnicot. Oxford: Oxford University Press.

Weiss, M. C., and Ephrussi, B. 1966. Studies of interspecific (rat X mouse) somatic hybrids: I. Isolation, growth, and evolution of the karyotype. *Genetics* 54: 1095–1109.

Welshons, W. J., and Russell, L. B. 1959. The Y-chromosome as the bearer of male determining factors in the mouse. *Proc. Nat. Acad. Sci. U.S.* 45:560–66.

Wettstein von, D. 1961. Plastid. In *Encyclopedia of the biological sciences*, ed. P. Gray. New York: Reinhold Publishing Corp.

White, A.; Handler, P.; Smith, E. M.; and Stetten, D. 1959. *Principles of biochemistry*. New York: McGraw-Hill Book Co.

White, L., Jr. 1967. The historical roots of our ecologic crisis. *Science* 155:1203–7.

White, M. J. D. 1945. *Animal cytology and evolution*. Cambridge: Cambridge University Press.

————. 1964. Principles of karyotype evolution in animals. In *The Proceedings of the Eleventh International Congress of Genetics*, vol. 2, pp. 391–97. London: Pergamon Press, Ltd.

————. 1973. *Animal cytology and evolution*. Cambridge: Cambridge University Press.

Whitfield, H. J., Jr.; Martin, R. G.; Ames, B. N. 1966. Classification of aminotransferase (*c* gene) mutants in the histidine operon. *J. Mol. Biol.* 21:335–55.

Whittaker, R. H. 1969. New concepts of kingdoms of organisms. *Science* 163:150–60.

Willmer, E. N. 1970. *Cytology and evolution*. New York: Academic Press Inc.

Wilson, E. B. 1928. *The cell in development and heredity*. New York: Macmillan Co.

Woese, C. 1967. *The genetic code*. New York: Harper & Row, Publishers.

Wolfe, G., and Coleman, D. L. 1967. Pigmentation. In *The biology of the laboratory mouse*, ed. E. L. Green. New York: McGraw-Hill Book Co.

Wright, S. 1921. Systems of mating: II. The effects of inbreeding on the genetic composition of a population. *Genetics* 6:124–43.

————. 1931. Evolution of Mendelian populations. *Genetics* 16:97–159.

————. 1932a. General, group, and special factors. *Genetics* 17:603–9.

————. 1932b. The roles of mutation, inbreeding, crossbreeding, and selection in evolution. In *Proceedings of the Sixth International Congress of Genetics*, vol. 1, pp. 356–66, ed. D. F. Jones. New York: Brooklyn Botanic Gardens.

————. 1935. The analysis of variance and the correlations between relatives with respect to deviation from an optimum. *J. Genet.* 30:243–56.

————. 1942. Statistical genetics and evolution. *Bull. Am. Math. Soc.* 48:223–46.

————. 1943. Isolation by distance. *Genetics* 28:114–38.

————. 1946. Isolation by distance under diverse systems of mating. *Genetics* 31:39–59.

————. 1951. The genetical structure of populations. *Ann. Eugen.* 15:323–54.

————. 1960. Genetics of vital characteristics of the guinea pig. *J. Cellular Comp. Physiol.* 56(suppl. 1):123–52.

————. 1967. The foundations of population genetics. In *Heritage from Mendel*, ed. R. A. Brink. Madison: University of Wisconsin Press.

———. 1969. *Evolution and the genetics of populations: The theory of gene frequencies,* vol. 2. Chicago: University of Chicago Press.

———. 1970. Random drift and the shifting balance theory of evolution. In *Mathematical topics in population genetics,* ed. K. I. Kojima, vol. 1. New York: Springer-Verlag New York, Inc.

Wurster, D. H., and Benirschke, K. 1970. Indian muntjac, *Muntiacus muntjak*: A deer with a low diploid chromosome number. *Science* 168:1364–66.

Yanofsky, C.; Carlton, B. C.; Guest, J. R.; Helenski, D. R.; and Henning, U. 1963. On the colinearity of gene structure and protein structure. *Proc. Nat. Acad. Sci. U.S.* 51:266–72.

Yasuda, N., and Morton, N. E. 1967. Studies on human population structure. In *Proceedings of the Third International Congress of Human Genetics,* eds. J. F. Crow and J. V. Neel, pp. 249–65. Baltimore: Johns Hopkins University Press.

Zinder, N. D., and Lederberg, J. 1952. Genetic exchange of *Salmonella. J. Bacteriol.* 64:679–99.

author index

subject index